Nanoengineering: Advances and Applications

Nanoengineering: Advances and Applications

Edited by **Peggy Rusk**

CWILLFORD PRESS

New York

Published by Willford Press,
118-35 Queens Blvd., Suite 400,
Forest Hills, NY 11375, USA
www.willfordpress.com

Nanoengineering: Advances and Applications
Edited by Peggy Rusk

International Standard Book Number: 978-1-68285-110-4 (Hardback)

Printed in the United States of America.

Contents

Permissions

List of Contributors

Preface

In my initial years as a student, I used to run to the library at every possible instance to grab a book and learn something new. Books were my primary source of knowledge and I would not have come such a long way without all that I learnt from them. Thus, when I was approached to edit this book; I became understandably nostalgic. It was an absolute honor to be considered worthy of guiding the current generation as well as those to come. I put all my knowledge and hard work into making this book most beneficial for its readers.

Nanoengineering is a rapidly advancing field of engineering which has enabled mankind to harness processes that occur at nanoscale. This book elucidates new techniques in the field of nanotechnology and their applications in a multidisciplinary approach. While understanding the long-term perspectives of topics like nanomaterials, molecular assembly, molecular nanotechnology, etc., the book makes an effort in highlighting their impact as a modern tool for the growth of nanotechnology as a discipline. Written and edited by internationally renowned scholars, this book will be ideal for the students of engineering and nanoscience.

I wish to thank my publisher for supporting me at every step. I would also like to thank all the authors who have contributed their researches in this book. I hope this book will be a valuable contribution to the progress of the field.

Editor

Nanostructuring effect of multi-walled carbon nanotubes on electrochemical properties of carbon foam as constructive electrode for lead acid battery

Rajeev Kumar · Saroj Kumari ·
Rakesh B. Mathur · Sanjay R. Dhakate

Abstract   In the present study, nanostructuring effect of multi-walled carbon nanotubes (MWCNTs) on electrochemical properties of coal tar pitch (CTP) based carbon foam (CFoam) was investigated. The different weight fractions of MWCNTs were mixed with CTP and foam was developed from the mixture of CTP and MWCNTs by sacrificial template technique and heat treated at 1,400 and 2,500 °C in inert atmosphere. These foams were characterized by scanning electron microscopy, X-ray diffraction, and potentiostat PARSTAT for cyclic voltammetry. It was observed that, bulk density of CFoam increases with increasing MWCNTs content and decreases after certain amount. The MWCNTs influence the morphology of CFoam and increase the width of ligaments as well as surface area. During the heat treatment, stresses exerting at MWCNTs/ carbon interface accelerate ordering of the graphene layer which have positive effect on the electrochemical properties of CFoam. The current density increases from 475 to 675 mA/cm^2 of 1,400 °C heat treated and 95 to 210 mA/cm^2 of 2,500 °C heat-treated CFoam with 1 wt% MWCNTs. The specific capacitance was decreases with increasing the scan rate from 100 to 1,000 mV/s. In case of 1 % MWCNTs content CFoam the specific capacitance at the scan rate 100 mV/s was increased from 850 to 1,250 μF/cm^2 and 48 to 340 μF/cm^2 of CFoam heat treated at 1,400 °C and 2,500 °C respectively. Thus, the higher value surface area and current density of MWCNTs-incorporated CFoam heat treated to 1,400 °C can be suitable for lead acid battery electrode with improved charging capability.

R. Kumar · S. Kumari · R. B. Mathur · S. R. Dhakate (✉)
Physics and Engineering of Carbon, Division of Materials
Physics and Engineering, CSIR-National Physical Laboratory,
Dr. K. S. Krishnan Marg, New Delhi 110012, India
e-mail: dhakate@mail.nplindia.org

Keywords   Carbon foam · Multi-walled carbon
nanotubes · Surface area · Electrical conductivity ·
Electrochemical properties

Introduction

Secondary batteries are next generation energy storage alternatives for wide variety of portables personal computers, cellular phones, digital cameras, items of military electronic equipment, etc. Battery technology has emerged over the past few years as one of the most advanced power sources meeting the requirements of portable, high energy density, specific capacity, high working potential, good cycling behavior, power environment friendly, etc. (Broussely 1999; Tamura and Horiba 1999; Tanake et al. 2001; Marsh et al. 2001). The most efficient hybrid vehicles lack power and models that accelerate quickly do so with the assistance of large internal combustion engines. This significantly degrades their fuel efficiency, barely making the extra cost worthwhile in terms of fuel efficiency. While batteries in these vehicles are capable of storing large quantities of energy, they cannot be charged or discharged quickly. This lack of power density requires the battery packs to be oversized, resulting in increased vehicle weight and reduced efficiency. As with poor discharge rate, battery charging is limited by the same kinetics, thus reducing efficiency gains through full regenerative braking. Moreover, the biggest drawback of lead acid batteries is the heavy weight due to the use of lead as a current collector. Lead grid constitutes 30–40 % of the battery weight. To reduce the weight of lead acid batteries, light weight electrically conducting materials are to be used as a potential substitute for it (Czerwinski and Zelazowska 1996; Das and Mondal 2000; Gyenge et al.

2002). Carbon materials have been playing a significant role in the development of alternative clean and sustainable energy technologies. Various carbon materials such as natural graphite, hard carbon from various polymer precursor, petroleum coke, and mesocarbon microbeads (MCMB) and light weight carbon foam (CFoam) have been utilized for improving battery performance (Flandrois and Simon 1999; Wu et al. 1998; Chang et al. 1998; Imanishi et al. 2008; Yang et al. 2005).

The carbon foam (CFoam), a lightweight material, three-dimensional network with open cell structure, large specific surface area, thermally and electrically conductive along with good corrosion resistance, is a promising candidate for current collector for lead acid batteries (Jang et al. 2006; Chen et al. 2008). The electrical conductivity of CFoam derived from different organic and inorganic precursors can be tailored by controlling processing parameters. Initially, CFoam was prepared from thermosetting polymeric material by heat treatment under controlled atmosphere (Cowlard and Lewis 1967). Later on, coal tar and petroleum pitches were used for the development of CFoam (Chen et al. 2006). To make highly crystalline CFoam of high electrical and thermal conductivity, generally mesophase pitch is used as the starting material (Klett et al. 2000, 2004) and it is prepared by high temperature and pressure foaming process. It is an expensive process, therefore in the present study, using the simple sacrificial template (Chen et al. 2007) technique CFoam is developed from the coal tar pitch (CTP) (Yadav et al. 2011). But CTP-based CFoam does not have high electrical conductivity and hence might have poor electrochemical properties.

To overcome the electrochemical properties of pitch-based CFoam, new approach of nanostructuring is adopted to improve the electrochemical properties of CFoam by taking the advantage of outstanding properties micro- and nano-forms of carbon, e.g., CTP and carbon nanotubes (CNTs). Nanostructured materials including metal nanoparticles, nanowires and carbon nanotubes have been demonstrated for high energy/power density with significantly improved durability (Arico et al. 2005; Kumar et al. 2013). Since their first observation by Iijima (1991) CNTs attracted considerable attention because of their excellent electrical conductivity (10^3–10^6 S/cm), ultra-high strength, high surface area and large aspect ratio that made them an ideal reinforcing additive for development of high-performance advanced materials (Moniruzzaman and Winey 2006; Spitalskyy et al. 2010; Ajayan et al. 1994). A combination of extraordinary electrical, thermal and mechanical properties makes CNTs not only attractive materials in nanoelectronic devices, but also excellent building blocks for assembling new hybrid materials for widespread applications, particularly energy storage devices (Snow et al. 2005).

In the present investigation, an effort has been made to develop MWCNTs-incorporated CFoam with improved electrochemical properties, so that it can be a constructive electrode material in lead acid battery with enhanced charging capacity, reduced weight of battery and increased life time. The effect of MWCNTs is ascertained by characterizing CFoam by scanning electron microscopy, surface area by Autosorb 3B, cyclic voltammetry (CV), X-ray diffraction.

Experimental

Development of CFoam

The CFoam was developed by sacrificial template technique from modified CTP. The modified pitch was synthesized from CTP by heat treating at 400 °C for 20 h. MWCNTs were procured from Nanocyl, Belgium. They were dispersed in an organic solvent ethanol by magnetic stirring for 10–15 h and mixed with the modified CTP in different weight fractions (0, 0.5, 1.0 and 2.0 wt%). The water slurry of MWCNTs mixed modified pitch with 3 % polyvinyl alcohol (PVA) was impregnated into a polyurethane foam (density 0.030 g/cc and average pore size 0.45 mm) template under vacuum. MWCNTs mixed modified CTP impregnated polyurethane foam was converted into CFoam by several heat treatments in air as well as in an inert atmosphere up to 2,500 °C (Kumar et al. 2013).

Characterization

The morphology of the CFoam, MWCNTs-incorporated CFoam was observed by scanning electron microscope (SEM model LEO 440). Electrical conductivity of CFoam was measured using the four-probe technique. Kiethley 224 programmable current source was used for providing constant current (I). The voltage drop (V) in between two pinpoints with a span of 1.2 cm was measured by Keithley 197A auto ranging microvolt digital multimeter.

The CFoam was also characterized by X-ray diffractometer (XRD, RIGAKU Tokyo) to understand the structural changes that take place due to the incorporation of MWCNTs.

The (110) reflection is used to calculate average crystallite width (La) by Scherrer equation (Braun and Huttinger 1996). La = $k\lambda/\beta\cos\theta$, where k is Scherrer constant = 1.84

The (002) reflection is used to obtain average crystallite height (Lc), λ wave length of CuK$_\alpha$ radiation is 1.5418 Å, Lc = $k\lambda/\beta\cos\theta$, where $k = 0.9$, is the Scherrer constant, β is the corrected full width at half maxima (FWHM) value in radians.

The electrochemical properties of CFoam were studied with the help of a potentiostat PARSTAT 2263 (Princeton Applied Research). The CV was used to examine the electrochemical behavior of the CFoam with different contents of MWCNTs (0–2 %) using 1 M H_2SO_4 solution as an electrolyte. The CFoam was used as working electrode. Ag/AgCl electrode and platinum plate were used as the reference electrode and counter electrode, respectively.

Result and discussion

Physical and mechanical properties of CFoam

Figure 1 shows the density and porosity of CFoam with increasing content of MWCNTs. Initially, bulk density of CFoam heat treated at 2,500 °C is 0.50 g/cm^3 and on addition of MWCNTs, bulk density of CFoam increases slightly. The increase is related to ordering of graphene layer parallel to MWCNTs axis. It was indicated that suitable addition of MWNTs promoted graphitization degree CFoam due to carbon atoms can orderly grow along MWNTs. During pyrolysis the mechanical stresses exert at MWCNT/carbon interface and accelerate ordering of the graphene layer (Li et al. 2011). The increase in the bulk density of CFoam results in the decrease in porosity as shown in Fig. 1.

The mechanical property in the CFoam, i.e., compressive strength is measured by Universal Instron testing machine. The compressive strength of CFoam depends mainly on two factors namely microstructure and bulk density. The microstructure mainly includes width of the ligaments and quantity of micro cracks. The compressive strength of the 1,400 and 2,500 °C heat-treated CFoam is

found to be 6.0 and 5.3 MPa. On the other hand, MWCNTs-incorporated CFoam heat treated at 1,400 °C, compressive strength is not enhanced significantly. However, after heat treatment at 2,500 °C, MWCNTs-incorporated CFoam strength increases from 5.3 to 6.4 and 8.0 MPa for 0.5 and 1.0 wt% of MWCNTs contents. However, at higher MWCNTs content, even though the bulk density of the CFoam increases but the simultaneous aggregation of MWCNTs may restrain the enhancement of the compressive strength of CFoam (Kumar et al. 2013).

Figure 2 shows the electrical conductivity with increasing MWCNTs content in 1,400 and 2,500 °C heat-treated CFoam. The electrical conductivity of 1,400 °C heat-treated CFoam is 58 S/cm and with increasing MWCNTs content it increases and maximum 84 S/cm at 1 wt% of MWCNTs. While in case of 2,500 °C heat-treated CFoam, without MWCNTs electrical conductivity is 80 S/cm and with increasing nanotube content conductivity increases. On incorporation of MWCNTs (0–2 wt%), electrical conductivity increases with increasing content of MWCNTs up to 1 wt%, i.e., from 80 to 135 S/cm. The increase in conductivity is due to the increase in conduction path of electron which is directly related to the structure of reinforcing material. The higher content of MWCNTs incorporation has the negative effect on the electrical conductivity due to agglomeration of nanotubes and formation MWCNTs–MWCNTs interfaces.

The surface area data of CFoam are measured by the sorption of nitrogen; it is observed that with increasing MWCNTs content, the surface area increases continually from 2.439 m^2/g to 5.25 for 1 wt% of MWCNTs and 7.60 m^2/g with 2 wt% of MWCNTs content in CFoam. This clearly shows the surface area increase by two to three

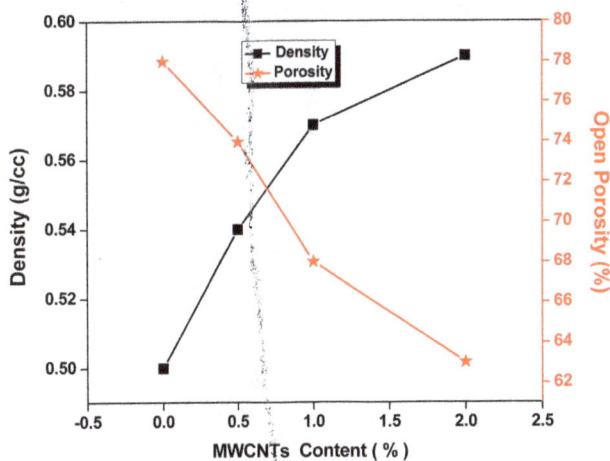

Fig. 1 Variation of bulk density and open porosity of CFoam with increasing MWCNTs content in starting material heat treated at 2,500 °C increasing MWCNTs content

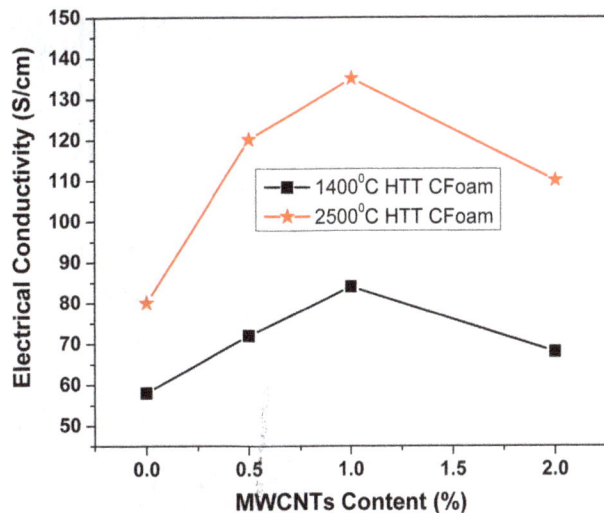

Fig. 2 Electrical conductivity with increasing the MWCNTs content in CFoam of 1,400 and 2,500 °C

times of CFoam on nanostructuring. This may have positive effect on the electrochemical properties of CFoam. Increase in the surface area can enhance utilization level of lead acid chemistry, i.e., liquid diffusion can be increased. Such structure results in much higher power, greater energy delivery and faster recharge process.

Microstructure of CFoam

Figure 3 shows the SEM micrographs of the CFoam with MWCNTs. Figure 2a, show the CFoam without MWCNTs heat treated at 2,500 °C, the pores are uniformly distributed and some pore walls (i.e., ligaments) are broken during machining of samples for SEM characterization because of the brittle nature of material. Figure 3b–d shows the SEM image of CFoam with 0.5, 1.0 and 2.0 wt% of MWCNTs content, respectively. The CFoam encapsulation of MWCNTs in the ligament leads to wider and thicker cell walls as compared to CFoam. Further, the number of cracks got reduced in the Fig. 3b, c due to the presence of MWCNTs that can act as barrier for propagating the cracks. The MWCNTs also act as nucleation site in

modified CTP-derived carbon for the alignment of carbon atoms or graphene layers along the MWCNTs axis (Fig. 3c, MWCNTs-incorporated CFoam with 1 wt%). Figure 3e shows the SEM image of CFoam ligaments which reveal the deposition of MWCNTs over ligaments which influence the ligament thickness. Further, homogeneous and dense distribution of MWCNTs over entire surface of CFoam can be visible at higher magnification in Fig. 3f. It shows that addition of MWCNTs obviously influence the pore structure in terms of cell wall thickness, width of ligament, and open porosity.

Figure 4 shows the XRD spectra of 2,500 °C heat-treated CFoam. The diffraction peaks are observed at $2\theta \sim 26.3°, 43°, 45°, 54°$ and $77°$ correspond to different diffraction planes 002, 100, 101, 004 and 110, respectively. The incorporation of MWCNTs influences the structure of CFoam, and as a consequence peaks registered at diffraction angle and intensity of each peak also change. The interlayer spacing of 0 % MWCNTs CFoam is 0.3387 nm and that of MWCNTs-incorporated foam is 0.3381, 0.3374, 0.3394 nm for MWCNTs loading 0.5, 1.0 and 2.0 wt%, respectively.

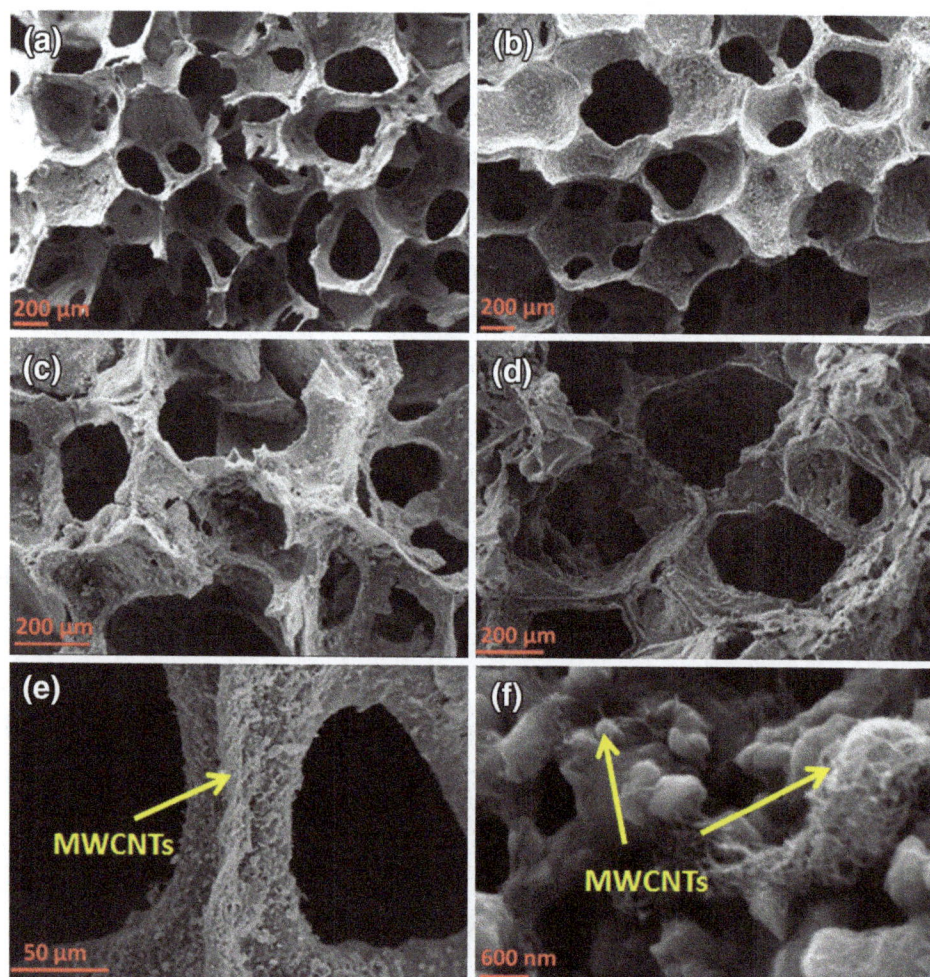

Fig. 3 SEM of CFoam heat treated at 2,500 °C **a** 0 % MWCNTs, **b** 0.5 % MWCNTs, **c** 1.0 % MWCNTs, **d** 2 % MWCNTs, **e** MWCNTs deposited in ligament of CFoam and **f** showing agglomeration of MWCNTs

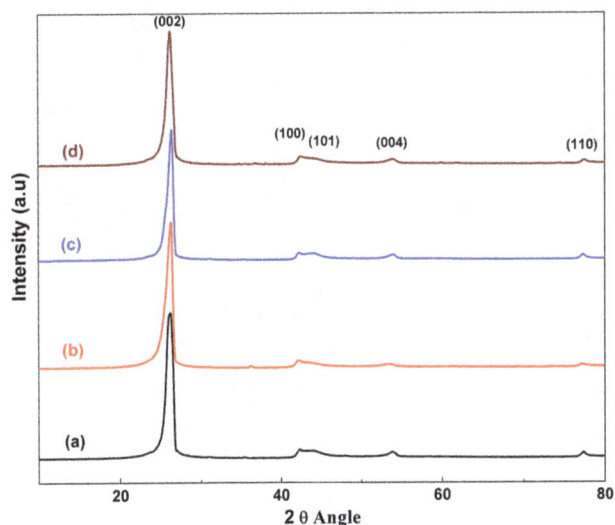

Fig. 4 XRD spectra of CFoam with increasing MWCNTs content, curve (*a*) 0 %, curve (*b*) 0.5 %, curve (*c*) 1.0 % and curve (*d*) 2 % MWCNTs incorporated in starting material and heat treated at 2,500 °C

The crystalline width (La) and height (Lc) is calculated from the diffraction peak of 110 and 002 (Fig. 4), and both increase with increasing content of MWCNTs in CFoam. After certain amount of MWCNTs the La Lc value decreased. The La value of CFoam is 24.1, 26.0, 34.6 and 18.6 nm, while Lc of CFoam is 7.9, 8.8, 11.3 and 8.8 nm for 0, 0.5, 1.0, and 2 wt% of MWCNTs content. This suggests that up to 1 wt% MWCNTs can help in improving the structure of CF, i.e., increases in Lc value. However, higher content of MWCNTs has negative effect on the crystalline parameters. This attributed to the improvement of staking order of graphene layers which can positively influence the electrical conductivity and electrochemical properties of CFoam.

Electrochemical properties of CFoam

The cyclic voltammetry (CV) is used to examine the electrochemical behavior of the CFoam with different contents of MWCNTs (0–2 %) in the voltage range (−1 to +1 V) using 1 M H_2SO_4 solution as an electrolyte. The CFoam is used as working electrode. Ag/AgCl electrode and platinum plate are used as the reference electrode and counter electrode, respectively. Figures 5 and 6 show the CV of MWCNTs containing CFoam electrode in H_2SO_4 electrolyte at different scan rates (from 100 to 1,000 mV/s) heat treated at 1,400 and 2,500 °C. The CV curve attributes to two peaks, i.e., anodic and cathodic peaks. The electrons transferred in a redox (reduction and oxidation) reaction arise from the change of the valance state of materials.

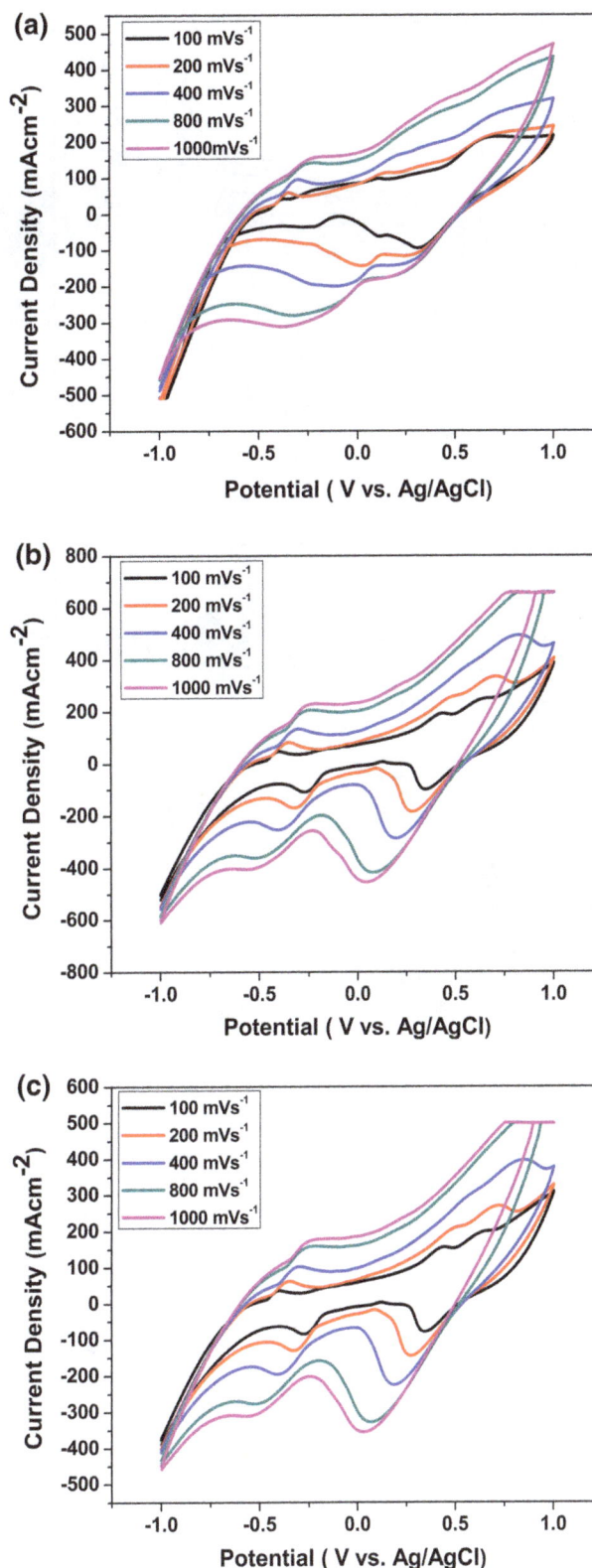

Fig. 5 Cyclic voltammetry behavior in 1 M H_2SO_4 electrolyte solution at different scan rates of 1,400 °C heat-treated CFoams with **a** 0, **b** 1, and **c** 2 % MWCNTs content

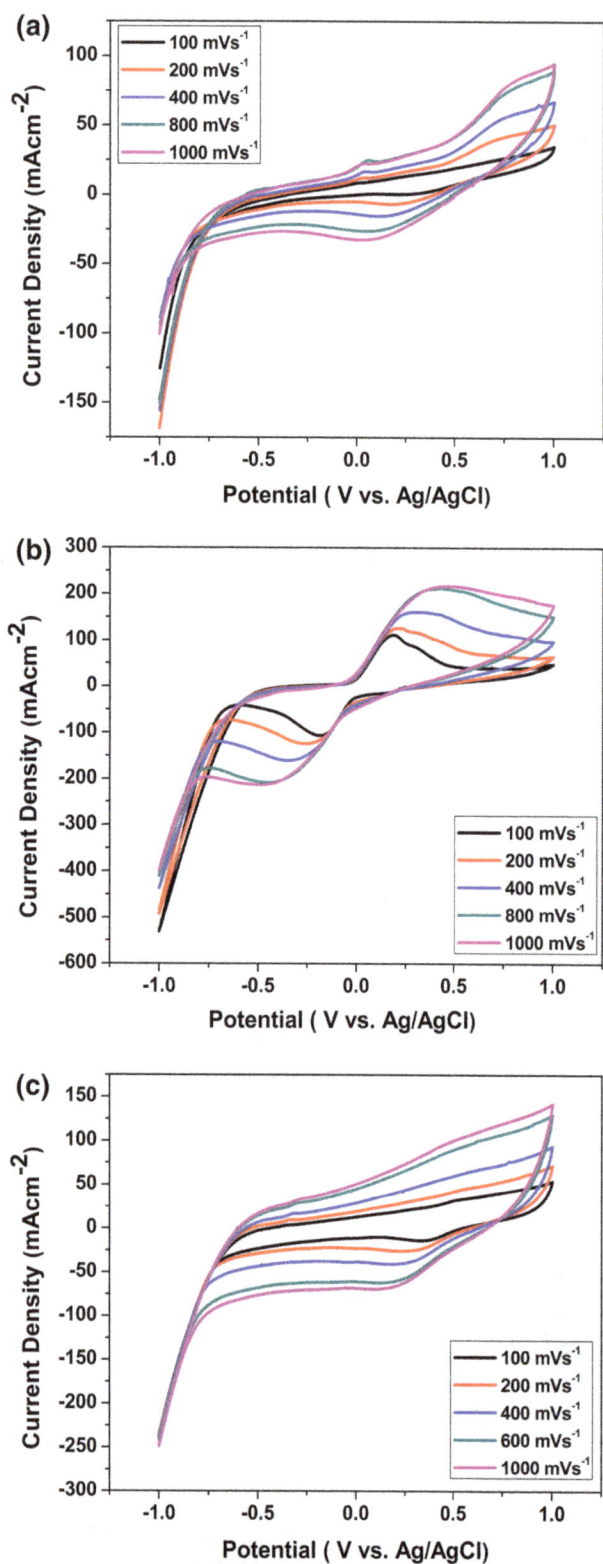

Fig. 6 Cyclic voltammetry behavior in 1 M H$_2$SO$_4$ electrolyte solution at different scan rates of 2,500 °C heat-treated CFoams with **a** 0, **b** 1, and **c** 2 % MWCNTs content

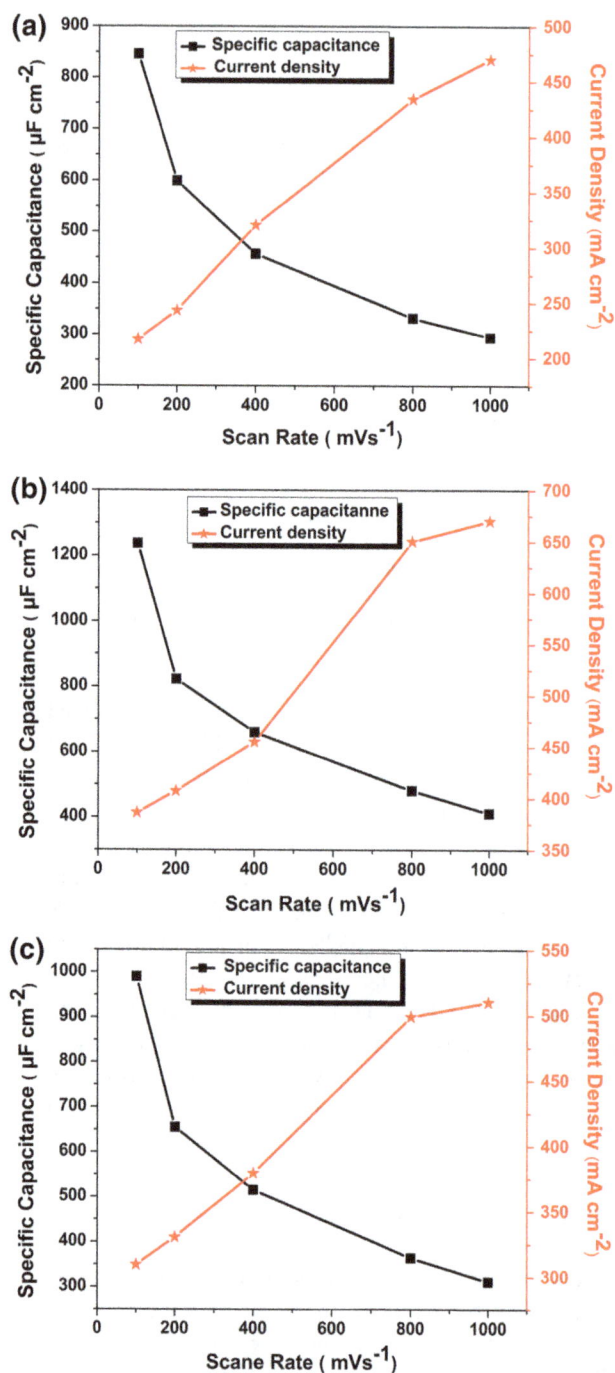

Fig. 7 Specific capacitance and current density of 1,400 °C heat-treated CFoams with **a** 0, **b** 1, and **c** 2 % MWCNTs content at different scan rates

Figure 5, in case of 0 % MWCNTs-incorporated CFoam HTT at 1,400 °C shows the current density 500 mA/cm^2 at the scan rate 1,000 mV/s; on increasing the MWCNTs content (1 %) in the CFoam the current density increases

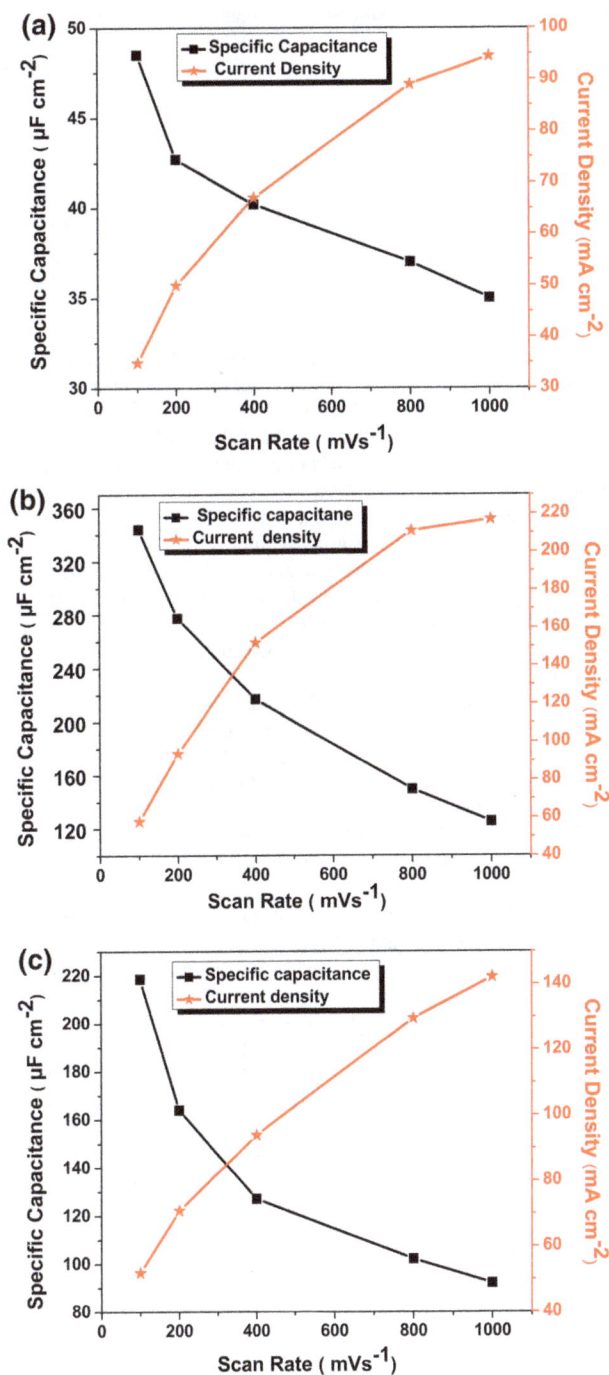

Fig. 8 Specific capacitance and current density of 2,500 °C heat-treated CFoams with **a** 0, **b** 1, and **c** 2 % MWCNTs content at different scan rates

nanotubes that formed MWCNTs–MWCNTs interfaces which inhibit conduction path and as result decreases in electrical conductivity. In case of 1 % MWCNTs-incorporated CFoam HTT at 2,500 °C shows the maximum current density 200 mA/cm^2 at the scan rate 1,000 mV/s. As we compare the current density of 1 % MWCNTs CFoam heat treated at 1,400 and 2,500 °C, it displays that the current density of CFoam heat treated at 1,400 °C is (600 mA/cm), which is higher as the CFoam heat treated at 2,500 °C (200 mA/cm^2). Further, MWCNTs play an important role on the enhanced of electrical properties of CFoam. This is attributed to higher electrical conductivity of the CFoam, this type of CFoam favored for battery electrode with high open circuit. The specific capacitance (SC) of CFoam has small equivalent series resistance, high rate handling capability and excellent electrochemical performance. Moreover, CV curves exhibit nearly mirror-image current response on voltage reversal, indicating a good reversibility (Wu et al. 2010). In this case, anodic and cathodic peak potentials have maintained the proportionality, i.e., reversible electrochemical behavior.

Figures 7 and 8 show the SC and current density of the CFoam with increasing scan rate heat treated at 1,400 and 2,500 °C. The SC is calculated from the CV curve by the following equation (Srinivasan and Weidner 2002).

$$SC = \frac{1}{Av(V_f - V_i)} \int_{V_i}^{V_f} I(V)dV$$

where A is the area of the active electrode material, v is the scan rate, V_f and V_i are the integration potential limits of the voltammetric curve, and $I(V)$ is the voltammetric current. The SC curve of 0 to 2 wt% MWCNTs content CFoam heat treated at 1,400 and 2,500 °C shown in Figs. 7 and 8 respectively. The SC of 1,400 °C HTT CFoam with 0, 1 and 2 wt% of MWCNTs content is 846, 1,237 and 990 µF/cm^2 at the scan rate 100 mV/s and current density at the same scan rate is 217, 386 and 300 mA/cm. The SC is still as high as 1,237 µF/cm^2 even at a high current density 386 mA/cm, in 1 wt% of MWCNTs CFoam due to higher electrical conductivity. The higher content of MWCNTs (2 wt%) has negative effect on electrical properties and also decreases its specific capacitance due to agglomeration effect of MWCNTs in CFoam

The SC of graphitized CFoam with 0, 1 and 2 wt% of MWCNTs content is 48.5, 344 and 218 µF/cm^2 at the scan rate 100 mV/s and current density at same scan rate is 35, 58 and 52 mA/cm^2, respectively. The 1 wt% of MWCNTs content CFoam has maximum SC 344 µF/cm^2. The SC of CFoam increases by 600 % with the addition of 1 % MWCNTs in CFoam as compared to CFoam without MWCNTs. However, CNTs with large surface area have been extensively studied in supercapacitors, with SC of

600 mA/cm, due to increase in its electrical conductivity (Kumar et al. 2013). After a certain amount of MWCNTs (2 wt%) in the CFoam the current density decreases to 447 mA/cm^2. This may be due to the decrease in its electrical conductivity or higher amount of MWCNTs in the matrix of carbon foam or agglomeration effect which is shown in SEM images. Due to the agglomeration of

4–180 F/g in a solution of H_2SO_4 (Chen et al. 2002). Again after certain amount of MWCNTs, the SC starts decreasing due to the poor dispersion of MWCNTs in CFoam. With the increasing scan rate in all the cases, SC decreases and current density increases. As we compare the SC of 1,400 and 2,500 °C HTT CFoam, it displays that 1,400 HTT CFoam with 1 % MWCNTs has a maximum SC 1,237 $\mu F/cm^2$ as compared the 2,500 °C HTT CFoam (344 $\mu F/cm^2$). The decrease in SC is associated with intercalation and de-intercalation of bisulphate ions and sulfuric acid molecules between graphite layers graphitized CFoam, indicating that non-graphitized CFoam is electrochemically stable in the voltage range of lead acid battery. This demonstrates that the MWCNTs-incorporated CFoam heat treated at 1,400 °C is most suitable with as electrode material in lead acid battery.

Conclusions

In this investigation, cost-effective approach for the development of MWCNTs-incorporated CFoam with improved electrochemical properties by simple sacrificial template technique is reported. It is observed that nano-structuring of CFoam by MWCNTs can enhance the surface area and electrical conductivity. The surface area of CFoam increases from 2.439 m^2/g to 5.25 for 1 wt% of MWCNTs and 7.60 m^2/g with 2 wt% of MWCNTs. During the heat treatment, stresses exert at MWCNTs/carbon interface and accelerate ordering of the graphene layer which have positive effect on the electrical conductivity and electrochemical properties of CFoam. The current density increases from 475 to 675 mA/cm^2 of 1,400 °C heat treated and 95 to 210 mA/cm^2 of 2,500 °C heat-treated CFoam with 1 wt% MWCNTs. The specific capacitance was decreases with increasing the scan rate from 100 to 1,000 mV/s. In case of 1 % MWCNTs content CFoam the specific capacitance at the scan rate 100 mV/s was increased from 850 to 1,250 $\mu F/cm^2$ and 48 to 340 $\mu F/cm^2$ of CFoam heat treated at 1,400 °C and 2,500 °C respectively. Thus, the higher value surface area and current density of MWCNTs-incorporated CFoam heat treated to 1,400 °C can be suitable for lead acid battery electrode with higher power, greater energy delivery, faster recharge process and significant reduction in the weight of battery. In the next course of investigation, Pb and PbO_2 will be coated on nanostructured CFoam and it will be characterized by cyclic voltammetry evaluation for charge–discharge cycle in cell.

Acknowledgments Authors are highly grateful to Director, NPL, for his kind permission to publish the results. Also thanks Mr. Jai Tawale for providing SEM characterization facility. One of the authors (Rajeev Kumar) would like to thanks CSIR for SRF fellowship.

References

Ajayan PM, Stephan O, Colliex C, Trauth D (1994) Aligned carbon nanotube arrays formed by cutting a polymer resin–nanotube composite. Science 265:1212

Arico AS, Bruce P, Scrosati B, Tarascon JM, Schalkwijk WV (2005) Nanostructured materials for advanced energy conversion and storage devices. Nat Mater 4:366

Braun M, Huttinger KJ (1996) Sintering of powders of polyaromatic mesophase to high-strength isotropic carbons: III. Powders based on an iron-catalyzed mesophase synthesis. Carbon 34(12):1473

Broussely M (1999) Recent developments on lithium ion batteries at SAFT. J Power Sources 81(82):140

Chang YC, Sohn HJ, Korai Y (1998) Anodic performance of coke from coals. Carbon 36:1653

Chen JH, Li WZ, Wang DJ, Yang SX, Wen GJ, Ren ZF (2002) Electrochemical characterization of carbon nanotubes as electrode in electrochemical double-layer capacitors. Carbon 40:1193–1197

Chen C, Kennel E, Stiller A, Stansberry P, Zondlo J (2006) Carbon foam derived from various precursors. Carbon 44:1535

Chen Y, Chen B, Shi X, Xu H, Hu Y, Yuan Y et al (2007) Preparation of pitch based carbon foam using polyurethane foam. Carbon 45(10):2132

Chen Y, Chen BZ, Shi XC, Xu H, Shang W, Yuan Y, Xiao LP (2008) Preparation and electrochemical properties of pitch-based carbon foam as current collectors for lead acid batteries. Electrochim Acta 53:2245

Cowlard FC, Lewis JC (1967) Vitreous carbon—a new form of carbon. J Mater Sci 2:507

Czerwinski A, Zelazowska M (1996) Electrochemical behavior of lead deposited on reticulated vitreous carbon. J Electroanal Chem 410:55

Das K, Mondal A (2000) Studies on a lead-acid cell with electro-deposited lead and lead dioxide electrodes on carbon. J Power Sources 89:112

Flandrois S, Simon B (1999) Review: carbon materials for lithium-ion re-chargeable batteries. Carbon 37:165

Gyenge E, Jung J, Splinter S, Snaper A (2002) High specific surface area reticulated current collectors for lead-acid batteries. J Appl Electrochem 32:287

Iijima S (1991) Helical microtubules of graphitic carbon. Nature 354:56–58

Imanishi N, Ono Y, Hanai K, Uchiyama R, Liu Y, Hirano A, Takeda Y, Yamamoto O (2008) Surface-modified meso-carbon microbeads anode for dry polymer lithium-ion batteries. J Power Source 178:744

Jang Y, Dudney NJ, Tiegs TN, Klett JW (2006) Evaluation of the electrochemical stability of graphite foams as current collectors for lead acid batteries. J Power Sources 161:1392

Klett J, Hardy R, Romine E, Walls C, Burchell T (2000) High thermal conductivity, mesophase-pitch-derived carbon foams: effect of precursor on structure and properties. Carbon 38:953

Klett JW, McMillan AD, Gallego NG, Burchell TD, Walls CA (2004) The role of structure on thermal properties of graphitic foams. Carbon 42:1849

Kumar R, Dhakate SR, Marhur RB (2013a) The role of ferrocene on the enhancement of the mechanical and electrochemical properties of coal tar pitch-based carbon foams. J Mater Sci 48:7071

Kumar R, Dhakate SR, Gupta T, Saini P, Singh BP, Mathur RB (2013b) Effective improvement of the properties of light weight carbon foam by decoration with multi-wall carbon nanotubes. J Mater Chem A 1:5727–5735

Li WQ, Zhang HB, Xiong X (2011) Properties of multi-walled carbon nanotubes reinforced carbon foam composite. J Mater Sci 46:1143–1146

Marsh RA, Vukson S, Surampudi S, Ratnakumar BV, Smart MC, Manzo M, Dalton PJ (2001) Li ion batteries for aerospace applications. J Power Sources 97(98):25

Moniruzzaman M, Winey KI (2006) Polymer nanocomposites containing carbon nanotubes. Macromolecules 39:5194

Snow ES, Perkins FK, Houser EJ, Badescu SC, Reinecke TL (2005) Chemical detection with a single walled carbon-nanotubes capacitor. Science 30:1942

Spitalskyy Z, Tasis D, Papagelis K, Galiotis C (2010) Carbon nanotube–polymer composites: chemistry, processing, mechanical and electrical properties. Progress in Poly Sci 35:357

Srinivasan V, Weidner JW (2002) Capacitance studies of cobalt oxide films formed via electrochemical precipitation. J Power Sources 108:15–20

Tamura K, Horiba T (1999) Large-scale development of lithium batteries for electric vehicles and electric power storage applications. J Power Sources 81(82):156

Tanake T, Ohta K, Arai N (2001) Year 2000 R&D status of large-scale lithium ion secondary batteries in the national project of Japan. J Power Sources 97(98):2

Wu Y, Fang S, Jiang Y (1998) Carbon anode materials based on melamine resin. J Mater Chem 8:2223

Wu ZS, Ren W, Wang DW, Li F, Liu B, Cheng HM (2010) High-energy MnO_2 nanowire/graphene and graphene asymmetric electrochemical capacitors. ACS Nano 4(10):5835

Yadav A, Kumar R, Bhatia G, Verma GL (2011) Development of mesophase pitch derived high thermal conductivity graphite foam using a template method. Carbon 49:3622–3630

Yang J, Shen ZM, Xue RS, Hao Z (2005) Study of mesophase-pitch-based graphite foam used as anodic materials in lithium ion rechargeable batteries. J Mater Sci 40:1285

2

Extracellular biosynthesis of silver nanoparticle using *Streptomyces* sp. 09 PBT 005 and its antibacterial and cytotoxic properties

P. Saravana Kumar · C. Balachandran ·
V. Duraipandiyan · D. Ramasamy ·
S. Ignacimuthu · Naif Abdullah Al-Dhabi

Abstract The application of microorganisms for the synthesis of nanoparticles as an eco-friendly and promising approach is welcome due to its non-toxicity and simplicity. The aim of this study was to synthesize silver nanoparticle using *Streptomyces* sp. (09 PBT 005). 09 PBT 005 was isolated from the soil sample of the agriculture field in Vengodu, Thiruvannamalai district, Tamil Nadu, India. 09 PBT 005 was subjected to molecular characterization by 16S rRNA sequence analysis. It was found that 09 PBT 005 belonged to *Streptomyces* sp. The isolate *Streptomyces* sp. 09 PBT 005 was inoculated in fermentation medium and incubated at 30 °C for 12 days in different pH conditions. The 0.02 molar concentration showed good antibacterial activity against Gram-positive and Gram-negative bacteria at pH-7. The synthesis of silver nanoparticles was investigated by UV–Vis spectroscopy, scanning electron microscopy and Fourier Transform Infrared analysis. The synthesized AgNPs sizes were found to be in the dimensions ranging between 198 and 595 nm. The cytotoxicity of the synthesized nanoparticles was studied against A549 adenocarcinoma lung cancer cell line. It showed 83.23 % activity at 100 µl with IC 50 value of 50 µl. This method will be useful in the biosynthesis of nanoparticles.

Keywords AgNPs · *Streptomyces* sp. 09 PBT 005 · Biosynthesis · Antibacterial activity · A549 cell line

P. Saravana Kumar · C. Balachandran · S. Ignacimuthu (✉)
Division of Microbiology, Entomology Research Institute,
Loyola College, Nungambakkam, Chennai 600 034,
Tamil Nadu, India
e-mail: entolc@hotmail.com

P. Saravana Kumar
e-mail: savanah.kumar@gmail.com

V. Duraipandiyan · S. Ignacimuthu · N. A. Al-Dhabi
Department of Botany and Microbiology, Addriyah Chair for
Environmental Studies, College of Science, King Saud
University, Po. Box. No. 2455, Riyadh 11451, Saudi Arabia

D. Ramasamy
Regional Research Institute of Unani Medicine, Royapuram,
Chennai 600 013, India

Introduction

Nanotechnology is an emerging field of science which involves synthesis and development of various nanomaterials (Basavaraj et al. 2012). At present, different types of metal nanomaterials are being produced using copper, zinc, titanium, magnesium, gold, alginate and silver. These nanomaterials are used in various fields such as optical devices (Anderson and Moskovits 2006), catalytic (Zhongjie et al. 2005), bactericidal (Rai and Yadav et al. 2009), electronic (Rao and Kulkarni et al. 2000), sensor technology (Vaseashta et al. 2005), biological labelling (Nicewarner-Pena and Freeman et al. 2001) and treatment of some cancers (Sriram and Manikanth et al. 2010). Currently, there is a growing need to develop environmentally benign nanoparticles that do not use toxic chemicals in the synthesis protocol. As a result, researchers in the field of nanoparticles have turned to biological systems for inspiration. Biosynthetic methods have been investigated as alternatives to chemical and physical ones. This is not surprising given that many organisms, both unicellular and multicellular, are known to produce inorganic materials either intra- or extracellularly (Simkiss et al. 1989; Mann 1996).

The metal–microbe interactions have important role in several biotechnological applications including the fields of bioremediation, biomineralization, bioleaching and microbial corrosion. However, it is only recently that microorganisms have been explored as potential biofactory for synthesis of metallic nanoparticles such as cadmium sulphide, gold and silver (Sastry et al. 2003; Ahmad et al. 2003). An important area of research in nanotechnology is the biosynthesis of nanoparticles such as nanosilver of different chemical compositions, sizes and controlled monodispersity. Silver nanoparticles are undoubtedly the most widely used nanomaterials among all. Silver nanoparticles are used as antimicrobial agents, in textile industries, water treatment, sunscreen lotions, etc. (Rai et al. 2009; Sharma et al. 2009). Microorganisms such as bacteria, moulds, yeasts, and viruses in the living environment are often pathogenic and cause severe infections in human beings. There is a pressing need to search for new antimicrobial agents from natural substances (Kim et al. 1998; Cho et al. 2005). Therefore, biological and biomimetic approaches for the synthesis of nanomaterials are being explored. Cell mass or extracellular components from microorganisms such as *Klebsiella pneumoniae, Bacillus licheniformis, Fusarium oxysporum, Aspergillus flavus, Cladosporium cladosporioides, Aspergillus clavatus,* and *Penicillium brevicompactum* have been utilized for the reduction of silver ions to AgNPs (Ahmad et al. 2003; Shahverdi et al. 2007; Kalishwaralal et al. 2008; Balaji et al. 2009; Shaligram et al. 2009; Verma et al. 2010).

Actinomycetes are microorganisms that share important characteristics of fungi and prokaryotes such as bacteria (Okami et al. 1988). Even though they are classified as prokaryotes due to their close affinity with mycobacteria and the coryneforms (and thus amenable to genetic manipulation by modern recombinant DNA techniques), they were originally designated as 'ray fungi' (Strahlenpilze). Focus on actinomycetes has primarily centred on their phenomenal ability to produce secondary metabolites (Sasaki et al. 1988). The present study was aimed at using *Streptomyces* sp. (09 PBT 005) to synthesis silver nanoparticles (AgNPs) and to screen the antibacterial and cytotoxic activities.

Materials and methods

Chemicals

$AgNO_3$ was obtained from Qualigen Mumbai, India. All other chemicals were purchased from Himedia, Mumbai, India. Freshly prepared doubly distilled water was used throughout the experimental work.

Isolation of actinomycetes from soil samples

Soil sample from sugarcane rhizosphere was collected from Vengodu, Thiruvannamalai district, Tamil Nadu, India (Latitude: $12°54'2383''$, North; Longitude: $79°69'9216''$, East elevation ft/m 227.5/65.4). The samples were collected from 5 to 25 cm depth in sterile plastic bags and transported aseptically to the laboratory. The soil samples were air-dried for 1 week at room temperature. Isolation and enumeration of actinomycetes were performed by serial dilution and spread plate technique (Elliah et al. 2004). One gram of soil was suspended in 9 ml of sterile double-distilled water. The dilution was carried out up to 10^{-5} dilutions. Aliquots (0.1 ml) of $10^{-2}, 10^{-3}, 10^{-4}$, and 10^{-5} were spread on the Actinomycetes isolation agar (Himedia, Mumbai). To minimize the fungal and bacterial growth, actidione 20 mg/l and nalidixic acid 100 mg/l were added. The plates were incubated at 30 °C for 10 days. Based on the colony morphology, the actinomycetes cultures were selected and purified on ISP2 (International *Streptomyces* Project 2) medium. In our pilot scale screening, a total of 27 actinomycetes were isolated and designated as 09 PBT 001 to 09 PBT 027. They were used for the screening of AgNPs synthesis; effective synthesizer (09 PBT 005) was further characterized by 16S rRNA sequencing technique.

Morphological, physiological and biochemical observations

Cultural and morphological features of 09 PBT 005 were characterized following the directions (Shirling and Gottlieb 1966). Cultural characteristics of pure isolates in various media (AIA—actinomycetes isolation agar, MHA—Mueller–Hinton agar, SCA—starch casein agar, SDA—Sabouraud dextrose agar, STP—*Streptomyces* agar, YPG—yeast peptone glucose agar, ZMA—Zobell marine agar, ISP—International *Streptomyces* Project) were recorded after incubation at 30 °C for 7–14 days. Morphology of spore bearing hyphae with entire spore chain was observed with a light microscope (Model SE; Nikon) using coverslip method in ISP medium (ISP 3–6). The shape of cell, Gram-stain, colour determination, the presence of spores, and colony morphology were assessed on solid ISP agar medium. Biochemical reactions, different temperatures, NaCl concentration, pH level, pigment production, enzyme reaction and acid or gas production were done following standard methods (Balachandran et al. 2012a; Valanarasu et al. 2009).

Biological synthesis of silver nanoparticles

The *Streptomyces* sp. (09 PBT 005) strain was grown in 500-ml Erlenmeyer flasks containing 150 ml of fermentation medium which was composed of tryptone (7.0 g), peptone (3.0 g), sodium chloride (5.0 g) and potassium hydrogen phosphate (1.25 g); the pH of the medium was adjusted to pH-3.0, 5.0, 7.0, 9.0 and 11 using 1 M HCl and 1 M NaOH. The culture was grown with continuous shaking on a rotary shaker (150 rpm) at 30 °C for 12 days. After the fermentation of the culture, biomass was harvested by centrifugation (5,000 rpm) at 20 °C for 20 min, and then the mycelia were washed thrice with sterile distilled water under aseptic conditions. The biomass was brought into contact with 100 ml sterile double-distilled water for 24 h at 30 °C in an Erlenmeyer flask and agitated at 150 rpm. After incubation the cell filtrate was filtered by Whatman No. 1 filter paper. A carefully weighed quantity of silver nitrate was added to the Erlenmeyer flask containing 100 ml of cell filtrate to yield an overall Ag+ ion concentration of 0.01, 0.02, 0.03, 0.04, 0.05 M and the reaction was carried out under dark conditions. A control experiment containing only 0.01, 0.02, 0.03, 0,04 and 0.05 M of silver nitrate solution was also performed. Formation of AgNPs was characterized using UV–visible spectroscopy, Fourier Transform Infrared (FT-IR) Spectroscopy analysis and scanning electron microscopy (SEM).

Microbial organisms

The following Gram-positive and Gram-negative bacteria were used for the experiment. Gram positive: *Micrococcus luteus* MTCC 106, *Bacillus subtilis* MTCC 441, *Staphylococcus epidermidis* MTTC 3615 and Methicillin resistance *Staphylococcus aureus* (MRSA). Gram negative: *K. pneumoniae* MTCC 109, *Enterobacter aerogenes* MTCC 111, *Salmonella typhimurium* MTCC 1251, *Shigella flexneri* MTCC 1457, *Proteus vulgaris* MTCC 1771 and *Salmonella typhi-B*. The reference cultures were obtained from the Institute of Microbial Technology, Chandigarh, India-160 036. Bacterial inoculums were prepared by growing cells in Mueller–Hinton broth (Himedia) for 24 h at 37 °C.

Antibacterial assay

The antibacterial activity of the silver nanoparticles was assayed using the standard Kirby–Bauer disc diffusion method (Bauer and Kirby et al. 1966). Petri plates were prepared with 20 ml of sterile MHA (Himedia, Mumbai). The test cultures were swabbed on the top of the solidified media and allowed to dry for 10 min. One hundred micro litres of the synthesized AgNPs was filled into the well and kept for incubation overnight at 37 °C and left for 30 min at room temperature for AgNPs diffusion. Streptomycin (10 μg/well) and culture supernatant (100 μl/well) were used as a positive and negative control, respectively. The plates were incubated for 24 h at 37 °C. Diameters of the zones of inhibition were measured using a zone scale from Himedia and expressed in millimetres.

Cell line maintenance and growth conditions

A549 adenocarcinoma lung cancer cell line was obtained from National Institute of Cell Sciences, Pune. A549 adenocarcinoma lung cancer cell line was maintained in complete tissue culture medium (Dulbecco's modified eagle's medium) with 10 % foetal bovine serum and 2 mM L-Glutamine, along with antibiotics (about 100 IU/ml of penicillin, 100 μg/ml of streptomycin) with the pH adjusted to 7.2. The cell lines were maintained at 37 °C at 5 % CO_2 in CO_2 incubator (Hsu et al. 2011). Cultures were viewed using an inverted microscope to assess the degree of confluency and the absence of bacterial and fungal contaminants was confirmed.

Cytotoxic properties

The cytotoxicity was determined according to an available method with some changes (Balachandran et al. 2012b). Cells (5000 cells/well) were seeded in 96 well plates containing medium with different concentrations such as 100, 75, 50, 25, 12.5 and 6.25 μl. The cells were cultivated at 37 °C with 5 % CO_2 and 95 % air in 100 % relative humidity. After various durations of cultivation, the solution in the medium was removed. An aliquot of 100 μl of medium containing 1 mg/ml of 3-(4, 5-dimethylthiazol-2-yl)-2, 5-diphenyl-tetrazolium bromide (MTT) was loaded to the plate. The cells were cultured for 4 h and then the solution in the medium was removed. An aliquot of 100 μl of DMSO was added to the plate, which was shaken until the crystals were dissolved. The cytotoxicity against cancer cells was determined by measuring the absorbance of the converted dye at 570 nm in an ELISA reader. Cytotoxicity of each sample was expressed as IC_{50} value. The IC_{50} value is the concentration of test sample that causes 50 % inhibition of cell growth, averaged from three replicate experiments.

Characterization

UV–vis spectroscopy

The reduction of silver ions was confirmed by measuring the UV–visible spectrum of the reaction medium. Three millilitres of supernatant was withdrawn after 72 h and absorbance was measured using UV–visible spectrophotometer Thermo Fisher, UV–vis double beam, Serial No.3628/0509 and software version 6.89 in the wavelength range from 200 to 600 nm. The absorption in the visible range directly reflects the perceived colour of the chemical involved.

Fourier Transform Infrared (FT-IR) Spectroscopy analysis

The sample was subjected to FT-IR Spectroscopy analysis on a PerkinElmer grating spectrophotometric instrument in Kbr disc. Two milligrams of the sample was mixed with 200 mg KBr (FT-IR grade) and pressed into a pellet. The sample pellet was placed into the sample holder and FT-IR spectra were recorded in the range 4,000–400 cm^{-1} in FT-IR spectroscopy at a resolution of 1 cm^{-1}.

Scanning electron microscopy (SEM)

Scanning electron microscopy (SEM) was used to observe the size, shape and morphology of the resultant nanoparticles. A specimen for SEM sample was made by casting a drop of suspension on a carbon-coated copper grid and the excess solution was removed by tissue paper and allowed to air dry at room temperature. Scanning electron microscopy study was observed on S-3400 2010 (Japan) at an accelerating voltage of 10,000 V and fitted with a CCD camera.

16S rRNA region-based characterization

Genomic DNA Isolation

The freshly cultured cells were pelleted by centrifuging for 2 min at 12,000 rpm to obtain 10–15 mg (wet weight). The cells were resuspended thoroughly in 300 µl of Lysis solution; 20 µl of RNase. A solution was added, mixed and incubated for 2 min at room temperature. About 20 µl of the Proteinase K solution (20 mg/ml) was added to the sample and mixed; the resuspended cells were transferred to Hibead Tube and incubated for 30 min at 55°. The mixture was vortexed for 5–7 min and incubated for 10 min at 95 °C followed by pulse vortexing. Supernatant was collected by centrifuging the tube at 10,000 rpm for 1 min at room temperature. About 200 µl of lysis solution was added, mixed thoroughly by vortexing and incubated at 55 °C for 10 min. To the lysate, 200 µl of ethanol (96–100 %) was added and mixed thoroughly by vortexing for 15 s. The lysate was transferred to new spin column and 500 µl of prewash solution was added to the spin column and centrifuged at 10,000 rpm for 1 min and the supernatant was discarded. The lysate was then washed in 500 µl of wash solution and centrifuged at 10,000 rpm for 3 min. Two hundred micro litres of the elution buffer was pipetted out and added directly into the column without spilling and incubated for 1 min at room temperature. Finally, the DNA was eluted by centrifuging the column at 10,000 rpm for 1 min. (Hipura *Streptomyces* DNA spin kit-MB 527-20pr from Himedia).

Preparation and analysis of 16S rRNA

The primer 27F (51 AGT TTG ATC CTG GCT CAG 31) and 1492R (51 ACG GCT ACC TTG TTA CGA CTT 31) were used to amplify 16S ribosomal sequence from genomic DNA in Thermal Cycler (ep gradient Eppendorf). The cyclic conditions were as follows: initial denaturation at 94 °C for 3 min, 35 cycles of 94 °C for 1 min, 54 °C for 1 min, and 72 °C for 2 min, and final extension of 10 min at 10 min and hold at 4 °C. The PCR products were confirmed by 1 % agarose gel electrophoresis (Farris et al. 2007).

DNA sequence determination

Automated sequencing was carried out according to the dideoxy chain-termination method using Applied Biosystems automated sequencer by synergy Scientific Services (Sanger et al. 1977).

Database searching and nucleotide sequence accession number

The sequence was compared for similarity with the reference species of bacteria contained in genomic database banks, using the NCBI BLAST (Blast'n') tool (http://www.ncbi.nlm.nih.gov/BLAST). The partial 16S rRNA gene sequences of isolate 09 PBT 005 have been deposited in the GenBank. A phylogenetic tree was constructed using the neighbour-joining DNA distance algorithm using software MEGA (version 4.0) (Tamura et al. 2007).

Results and discussion

Isolation and culture characterization

Nanomaterials are the leading substances in the field of nanomedicine and nanobiotechnology. Silver nanoparticles have recently been shown to be promising antibacterial and anticancer material. *Streptomyces* cultures were isolated from the soil sample collected from agricultural field in Vengodu, Thiruvannamalai district, Tamil Nadu, India. This strain 09 PBT 005 was Gram-positive filamentous bacterium. The colour of the substrate mycelia was dark green. The spore's chains were green. These characteristic morphological properties strongly suggested that the isolate belonged to *Streptomyces* genus (Table 1). It showed good growth on medium amended with sodium chloride up to 11 %; no growth was seen at 15 %. The temperature for growth ranged from 25 to 37 °C with optimum of 30 °C and the pH range was 6–10 with normal pH of 7 (Table 2). 09 PBT 005 showed resistance towards ampicillin,

Table 1 Morphological analysis of *Streptomyces* sp. (09 PBT 005) on different media

S. no	Media	Aerial mycelium	Substrate mycelium	Soluble pigment	Reverse side	Growth
1	AIA	Dark green	White	–	Whitish green	+++
2	MHA	–	Slimy yellow	–	Slimy yellow	+
3	SCA	Whitish green	Whitish green	–	Yellow	+++
4	SDA	–	Yellow	–	Whitish yellow	++
5	STP	Dark green	Dark green	–	Yellowish green	+++
6	YPG	White	White	–	White	+++
7	ZMA	–	Yellow	–	Whitish yellow	+
8	ISP-1	Dark green	White	–	Creamy white	+++
9	ISP-2	Dark green	Whitish grey	–	Creamy yellow	+++
10	ISP-3	Creamy white	White	–	Creamy yellow	+++
11	ISP-4	Dark green	Greenish white	–	Creamy white	+++
12	ISP-5	Pale green	White	–	White	+++
13	ISP-6	White	White	–	Creamy white	+++
14	ISP-7	Dark green	White	–	Creamy white	+++

+++ good growth, ++ moderate growth, + weak growth

ceftazidime, ceftazidime/clavulanic acid, co-trimoxazole, oxacillin, penicillin-G, piperacillin and ticarcillin/clavulanic acid (Table 3). The *Streptomyces* isolate was characterized on the basis of colony characteristics and microscopic appearance (Cappuccino et al. 2006) (Fig. 1). The 16S rRNA sequencing showed it to be *Streptomyces* sp. 09 PBT 005 (bases 1–802 linear DNA). The isolate 09 PBT 005 showed 98 % homology to *Streptomyces ghanaensis* 16S ribosomal RNA gene, partial sequence (HE797851) NCBI BLAST available at http://www.ncbi-nlm-nih.gov/. The DNA sequences were aligned and phylogenetic tree was constructed using mega4 software (bootstrap method) (Fig. 2). The sequences were deposited in GenBank (NCBI) with accession number JF710843. *Streptomyces* sp. 09 PBT 005 was further selected for the biosynthesis of silver nanoparticles because there are only very few reports on the synthesis of silver nanoparticles using *Streptomyces* sp. Earlier Faghri and Salouti (2011) (Sathishkumar et al. 2009; Faghri Zonooz. 2011) have reported the biosynthesis of silver nanoparticles using aqueous extract of *Streptomyces* sp.

Table 2 Physiological and biochemical characteristics of *Streptomyces* sp. (09 PBT 005)

Characteristics	Results
Gram staining	Positive
Shape and growth	Filamentous aerial growth
Production of diffusible pigment	–
Range of temperature for growth	25–37 °C
Optimum temperature	30 °C
Range of pH for growth	3–11
Optimum pH	7
Growth in the presence of NaCl	1–11 %
H2S production	–
Amylase	+++
Chitinase	+++
Protease	+
Gelatinase	+
Cellulase	+++
Lipase	++

+++ good growth, ++ moderate growth, + weak growth—no growth

Table 3 Antibiotic sensitivity tests against *Streptomyces* sp. (09 PBT 005)

Standard antibiotics	Resistant	Sensitivity
Amikacin 30 mg/disc	–	S
Ampicillin 10 mg/disc	R	–
Ceftazidime 30 mg/disc	R	–
Ceftazidime/clavulanic acid 30/10 mg/disc	R	–
Ceftriaxone 30 mg/disc	–	S
Chloramphenicol 30 mg/disc	–	S
Cephalothin 30 mg/disc	–	S
Cephotaxime 30 mg/disc	–	S
Cephoxitin 30 mg/disc	–	S
Ciprofloxacin 10 mg/disc	–	S
Co-Trimoxazole 25 mg/disc	R	–
Erythromycin 15 mg/disc	–	S
Gentamicin 10 mg/disc	–	S
Imipnem 1 mg/disc	–	S
Norfloxacin 10 mg/disc	–	S
Oxacillin	R	–
Penicillin—G 100 mg/disc	R	–
Piperacillin	R	–
Polymyxin—B 300 mg/disc	–	S
Rifamycin 30 mg/disc	–	S
Tetracycline 30 mg/disc	–	S
Ticarcillin/clavulanic acid 75/10/disc	R	–
Vencomycin 30 mg/disc	–	S

Microbial synthesis of silver nanoparticles

Synthesis of silver nanoparticles was observed by the addition of selected culture supernatant of *Streptomyces* sp.

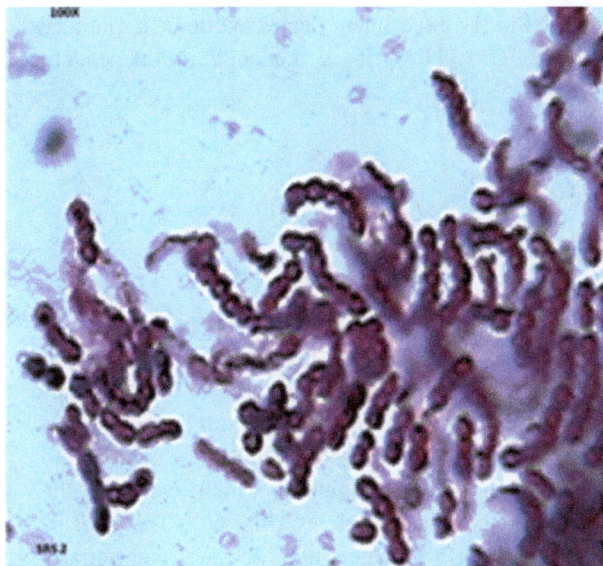

Fig. 1 Microscopic image of S*treptomyces sp.* (09 PBT 005) at ×100

(09 PBT 005) to 0.01, 0.02, 0.03, 0.04 and 0.05 aqueous AgNO3 at room temperature. The *Streptomyces sp.* (09 PBT 005) aqueous filtrate incubated with deionized water (positive control) and the silver nitrate solution (negative control) was observed to retain its original colour, and the silver nitrate-treated supernatant turned dark brown after 72 h due to the deposition of silver nanoparticles (Fig. 3). Generally, the formation of silver can be primarily identified through visible observation in the change of colour solution during the reaction from colourless to pale yellow or dark brown (Sathishkumar et al. 2009). The colour formation is dependent on the excitation of surface plasmon vibrations of silver nanoparticles (Kannan et al. 2011).

Antibacterial activity of synthesized AgNPs

The biologically synthesized AgNPs showed good antimicrobial activity against Gram-positive and Gram-negative bacteria by well diffusion method (Fig. 4a–e). At the 0.02 molar concentrations at pH-7, the *Streptomyces* sp. (09 PBT 005)-based nanosilver particles showed good activity against Gram-negative bacteria such as *S. flexneri* and *E. aerogenes* with maximum zones of inhibition

Fig. 2 The phylogenetic tree of *Streptomyces sp.* (09 PBT 005) (JF710843) was constructed using the neighbour-joining method with aid of MEGA 4.1 program

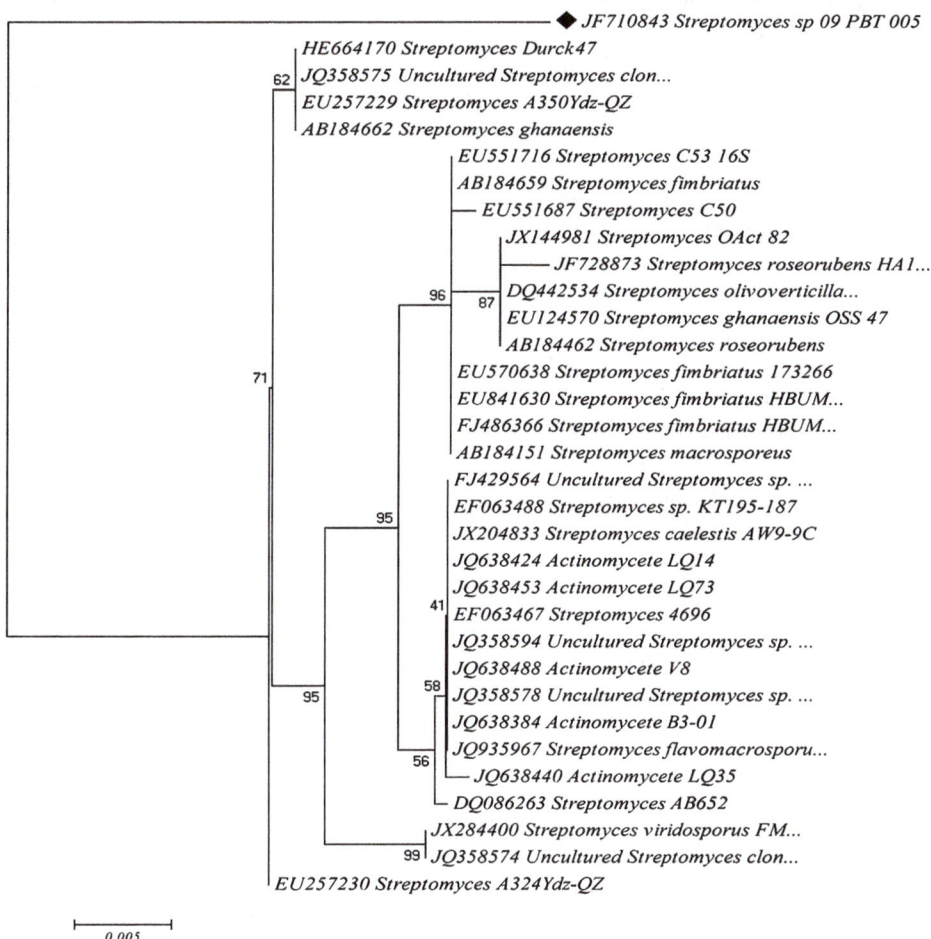

Fig. 3 Biosynthesis of silver nanoparticles—colour change reaction: conical flasks containing the extracellular filtrate of the *Streptomyces sp.* (09 PBT 005) biomass (**a**) and conical flasks are containing the extracellular filtrate of the *Streptomyces* sp. (09 PBT 005) biomass after exposure to AgNO₃ solution (**b**)

(16 mm). *Klebsiella pneumoniae*, *S. typhimurium* and *S. typhi-B* and *P. vulgaris* showed 13 mm zones. A moderate activity was seen in Gram-positive bacteria such as *S. epidermidis* (15 mm), MRSA (13 mm) and *B. subtilis* (12 mm). Nanoparticles synthesized using 0.02 M AgNO₃ was the most efficient to inhibit the tested pathogens when compared to higher concentrations. There was a decrease in antibacterial activity as the concentration of AgNO₃ increased. This shows that a minute quantity can itself act against many pathogenic microorganisms. The antimicrobial activity of colloidal silver particles is influenced by the particle dimensions (Kaviya et al. 2011). Similar results have been reported previously by Panáček et al. (2006) and Augustine et al. (2013). A number of possible mechanisms are proposed for the antibacterial activity of AgNPs. The synthesized AgNPs with smaller size can act drastically on cell membrane and further interact with DNA and cause damage (Morones et al. 2005). In addition, the pitting of the cell membranes by silver nanoparticles causes an increase in permeability and results finally in cell death (Shahverdi et al. 2007). Silver ions have been known to bind with the negatively charged bacterial cell wall resulting in the rupture and consequent denaturation of proteins which leads to cell death (Lin and Vidic et al. 1998). There was not much activity against Gram-positive and Gram-negative bacteria at pH-3, 5, 9 and 11. Based on the bactericidal activity the nanoparticles were taken to assess cytotoxic activity and further characterization.

Cytotoxic properties of synthesized AgNPs

AgNPs showed cytotoxic activity in vitro against A549 adenocarcinoma lung cancer cell line. It showed 83.23 % activity at 100 µl with IC_{50} value of 50 µl. All concentrations used in the experiment decreased the cell viability significantly ($P < 0.05$) in a concentration-dependent manner (Fig. 5).

Characterization

UV–visible spectral analysis

In this study, AgNPs were successfully synthesized by *Streptomyces* sp. (09 PBT 005), when the cell-free extract was subjected to AgNO₃. The biosynthesis reaction started within few minutes and the colour reaction was observed in which clear AgNO₃ solution changed into yellowish-brown coloured solution which indicated the formation of silver nanoparticles (Fig. 6a). The appearance of a yellowish-brown colour in the silver nitrate-treated flask indicated the formation of silver nanoparticles, whereas no colour change was observed in either the culture supernatant without silver nitrate or the silver nitrate control experiments. The UV–visible spectra for the aqueous AgNO₃-culture supernatant of synthesized AgNPs by *Streptomyces* sp (09 PBT 005) alone were recorded. In the UV–visible spectrum, a strong and broad peak was observed at 440 nm, indicating the presence of AgNPs. In agreement with previous reports, the absorption peak at 441 nm is probably due to the excitation of longitudinal plasmon vibrations and formation of quasi-linear superstructures of nanoparticles (Shankar et al. 2003). The optimum time required for the completion of reaction in our study was 72 h; as the duration of reaction and concentration of silver nitrate increase, more silver nanoparticles are formed resulting in large size particle. Different concentrations of silver nitrate solution were used to get maximum antibacterial activity of AgNPs. As the size increases, the peak of plasmon resonance shifts to longer wavelengths and broadens. The nanoparticles in this study are big and have an average diameter from 198 to 595 nm. Similar result was reported by Veerasamy et al. (2011).

FT-IR analysis

Fourier Transform Infrared spectrum showed that the active biomolecules of culture supernatant were responsible for the reduction of Ag+ ions into metallic AgNPS, which revealed distinct peak in the range of

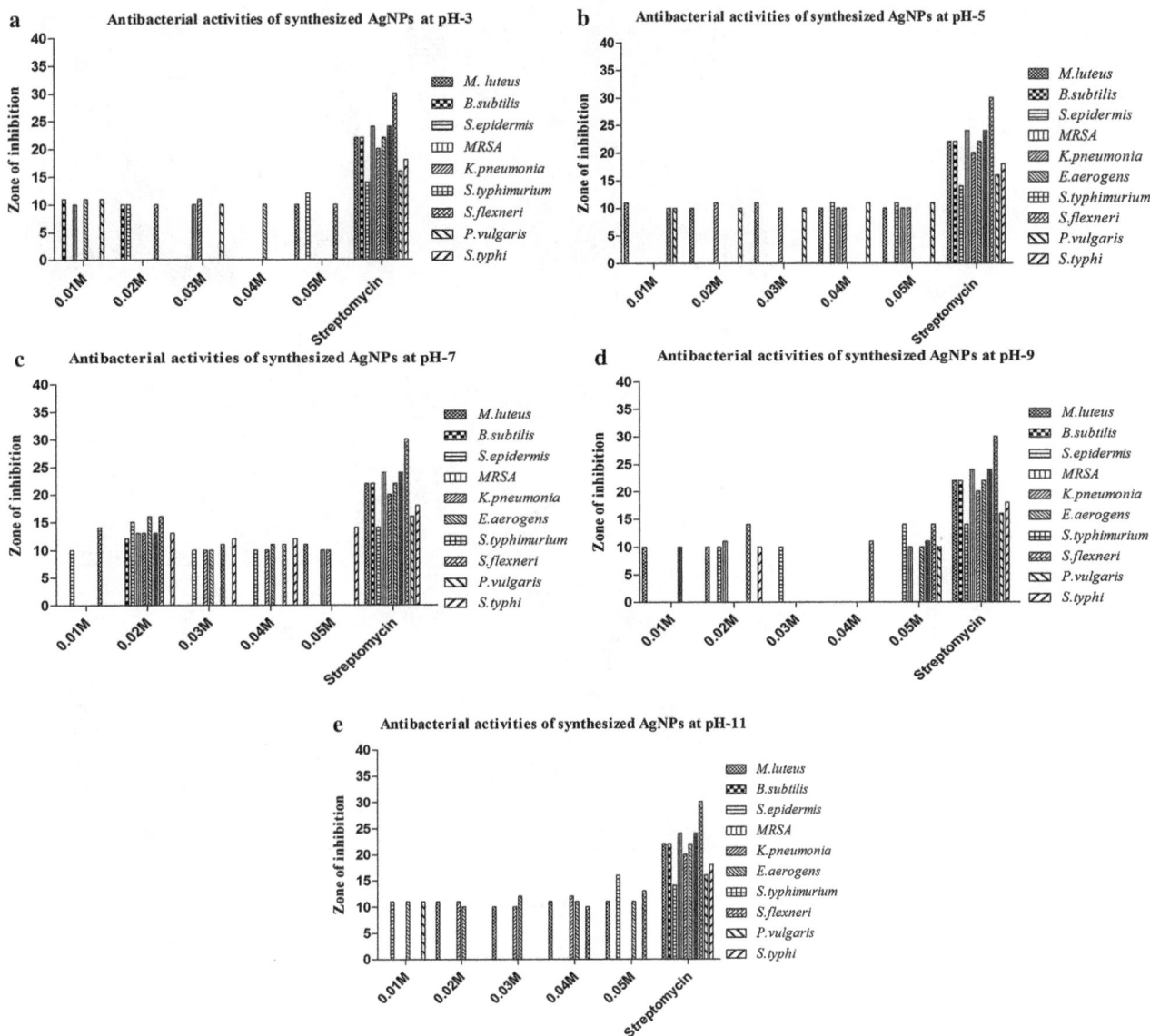

Fig. 4 Antibacterial activities of synthesized AgNPs

$3,586–666$ cm^{-1} (Fig. 6b). The broad peak at $3,586$ cm^{-1} is due to strong stretching vibration of phenolic OH (Gopinath and Mubarkali et al. 2012). The band at $3,547$ cm^{-1} is due to NH stretching of amide group. The band at 3,394 is assigned to the N–H group from peptide linkage present in the cell-free extract of *Streptomyces* sp. (Mubarkali and Thajuddin et al. 2011). The peak at $3,367$ cm^{-1} may be ascribed to NH stretching of primary amine (NH$_2$). The presence of characteristic peak of amide carbonyl is shown by the peak at 1634 (NHCO). IR spectroscopic study has confirmed that the carbonyl group from amino acid residues and peptides of proteins has stronger ability to bind metal, so that the proteins could most pos-

sibly form a coat covering the metal nanoparticles (i.e. capping of AgNPs) to prevent agglomeration of the particles and stabilizing in the medium. This evidence suggests that the biological molecules could possibly perform the function in the formation and stabilization of the AgNP in aqueous medium. It is well known that proteins can bind to AgNPs through free amine groups in the proteins and, therefore, stabilization of the AgNPs by surface-bound proteins is a possibility (Gole and Dash et al. 2001). The peak at $1,634$ cm^{-1} is due to the carbonyl stretch vibrations in the amide linkages of proteins (Basavaraja and Balaji et al. 2008). The carbonyl groups of amino acid residues and peptides have strong ability to bind to silver (Balaji

Fig. 5 Cytotoxic effects of
cancer cell line (A549)
a 100 µl, **b** 75 µl, **c** 50 µl,
d 25 µl, **e** 12.5 µl, **f** 6.25 µl and
g control cells

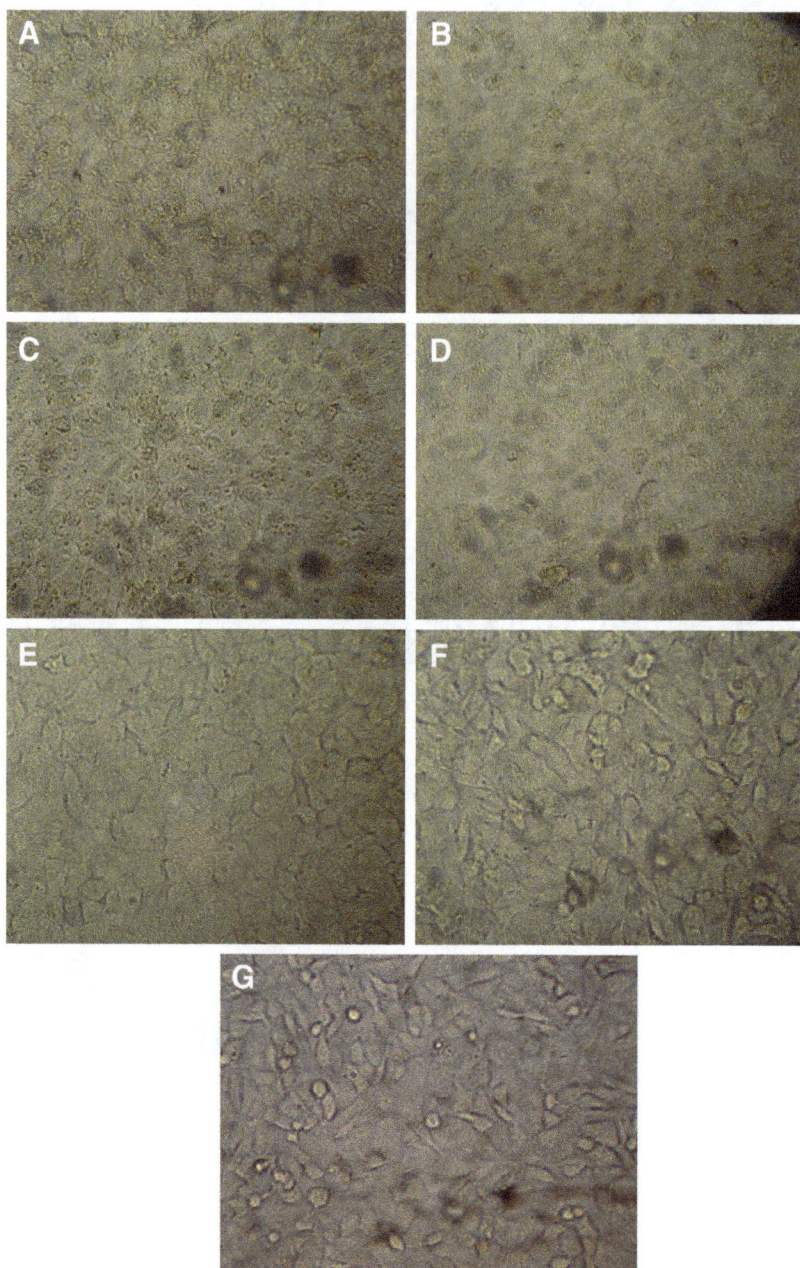

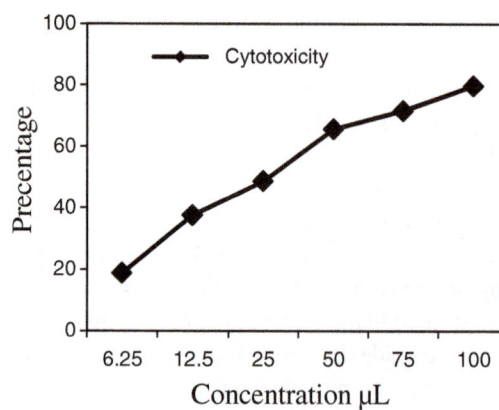

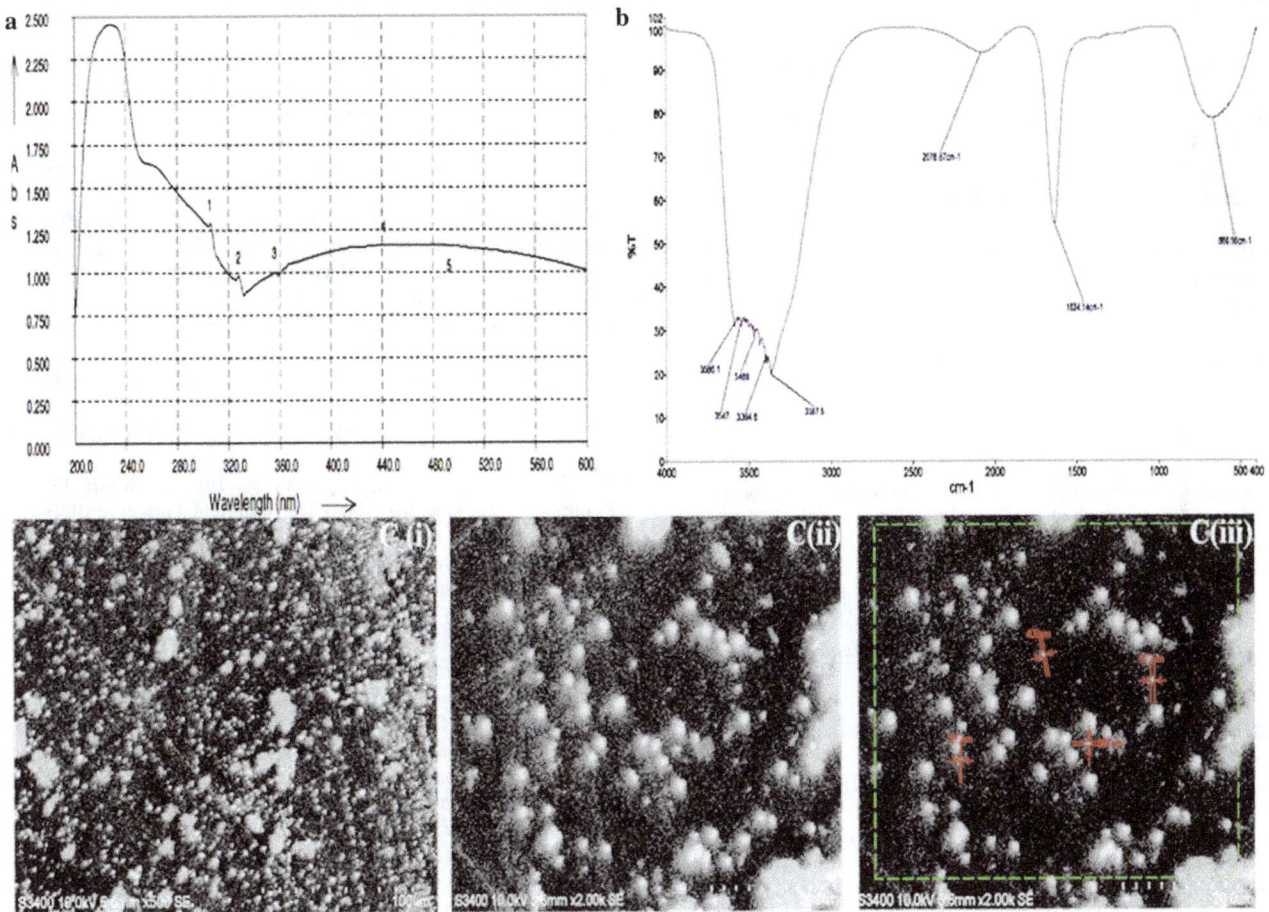

Fig. 6 Characterizations of AgNps. **a** UV–vis absorption spectra of silver nanoparticles synthesized by *Streptomyces* sp. (09 PBT 005). **b** FT-IR analysis of silver nanoparticles biosynthesis using S*treptomyces sp.* (09 PBT 005). **c** Scanning electron microscopy image shows formation of AgNPs by *Streptomyces* sp. (09 PBT 005), **c** (i) agglomeration of AgNPs at 100 μm, **c** (ii) agglomeration of AgNPs at 20 μm and **c** (iii). Size of the AgNPs at 20 μm (198–595 nm)

et al. 2009). It is also reported that proteins can bind to nanoparticles either through free amine or cysteine groups in proteins (Mandal et al. 2005).

SEM

The morphology of the silver nanoparticles was observed using scanning electron microscope. The nanoparticles are polydispersed with a roughly spherical shape Fig. 6c (i, ii), although the exact shape of the nanoparticles was not clearly predicted. Higher magnification showed the average diameter of these spherical nanoparticles to be about 198–595 nm 6c (iii). The SEM analysis of Ag nanoparticles is in agreement with the results of Faghri Zonooz and Salouti (2011) and Sastry et al. 2003. The above results suggested that the silver nanoparticles were synthesized due to the action of *Streptomyces* cell-free extract, which act as good bio-reductant for biosynthesis.

Conclusion

A critical need in the field of nanotechnology is the development of a reliable and eco-friendly process for synthesis of nanoparticles. *Streptomyces* sp. (09 PBT 005) has been effectively used for the synthesis of silver nanoparticles. We have demonstrated the use of a natural, renewable and low-cost bioreducing agent. Biosynthesized silver nanoparticles were confirmed by spectroscopic characterization of UV–visible, FT-IR and SEM. The biosynthesized silver nanoparticles using *Streptomyces* sp. (09 PBT 005) showed good antibacterial and cytotoxic activities.

Acknowledgments The authors are grateful to Entomology Research Institute, Loyola College, Chennai, for financial assistance. We thank the visiting Professorship Program, Deanship of Scientific Research at King Saudi University, Riyadh.

References

Ahmad A, Mukherjee P, Senapati S, Mandal D, Khan MI, Kumar R, Sastry M (2003) Extracellular biosynthesis of silver nanoparticles using the fungus Fusarium oxysporum. Colloids Surf B 28:313–318

Anderson DJ, Moskovits M (2006) A SERS-active system based on silver nanoparticles tethered to a deposited silver film. J Phys Chem B 110:13722–13727

Augustine R, Kalarikkal N, Thomas S (2013) A facile and rapid method for the black pepper leaf mediated green synthesis of silver nanoparticles and the antimicrobial study. Appl Nanosci.

Balachandran C, Duraipandiyan V, Balakrishna K, Ignacimuthu S (2012a) Petroleum and polycyclic aromatic hydrocarbons (PAHs) degradation and naphthalene metabolism in Streptomyces sp. (ERI-CPDA-1) isolated from oil contaminated soil. Bioresour Technol 112:83–90

Balachandran C, Duraipandiyan V, Ignacimuthu S (2012b) Cytotoxic (A549) and antimicrobial effects of Methylobacterium sp. isolate (ERI-135) from Nilgiris forest soil, India. Asian Pac J Trop Biomed 2(9):712–716

Balaji DS, Basavaraja S, Deshpande R, Bedre MD, Prabhakara BK, Venkataraman A (2009) Extracellular biosynthesis of functionalized silver nanoparticles by strains of Cladosporium cladosporioides fungus. Colloids Surf B 68:88–92

Basavaraj U, Praveenkumar N, Sabiha TS, Rupali S, Samprita B (2012) Synthesis and characterization of silver nanoparticles. Int J Pharm Bio Sci 2(3):10–14

Basavaraja SS, Balaji SD, Lagashetty AK, Rajasab AH, Venkataraman A (2008) Extracellular biosynthesis of silver nanoparticles using the fungus Fusarium semitectum. Mater Res Bull 43:1164–1170

Bauer AW, Kirby WMM, Sherris JC, Turck M (1966) Antibiotic susceptibility testing by a standardized single disk method. Am J Clin Pathol 45:493–496

Cappuccino JG, Sherman N (2006) Microbiology: a laboratory manual, Dorling Kindersley (India) Pvt. Ltd., India, 6:237

Cho KH, Park JE, Osaka T, Park SG (2005) The study of antimicrobial activity and preservative effects of nanosilver ingredient. Electrochim Acta 51:956–960

Elliah P, Ramana T, Bapi Raju KVVS, Sujatha P, Uma Sankar AM (2004) Investigation on marine actinomycetes from Bay of Bengal near Karnataka coast of AndhraPradesh. Asian J Microbiol Biotechnol Environ Sci. 6(1):53–56

Faghri Zonooz N, Salouti M (2011) Extracellular biosynthesis of silver nanoparticles using cell filtrate of Streptomyces sp. ERI-3. Scientia Iranica 18(6):1631–1635

Farris MH, Oslon JB (2007) Detection of actinobacteria cultivated from environmental samples reveals bias in universal primers. Lett Appl Microbiol 45:376–381

Gole A, Dash C, Ramachandran V, Sainkar SR, Mandale AB, Rao M, Sastry M (2001) Pepsin-gold colloid conjugates: preparation, characterization, and enzymatic activity. Langmuir 17:1674–1679

Gopinath V, MubarakAli D, Priyadarshini S, Meera PN, Thajuddin N, Velusamy P (2012) Biosynthesis of silver nanoparticles from Tribulus terrestris and its antimicrobial activity: a novel biological approach. Colloids Surf B Biointerfaces 96:69

Hsu H, Huang K, Lu K, Chiou S, Yen J, Chang C, Houng J (2011) Typhonium blumei extract inhibits proliferation of human lung adenocarcinoma A549 cells via induction of cell cycle arrest and apoptosis. J Ethnopharmacol 135:492–500

Kalishwaralal K, Deepak V, Ramkumarpandian S, Nellaiah H, Sangiliyandi G (2008) Extracellular biosynthesis of silver nanoparticles by the culture supernatant of Bacillus licheniformis. Mater Lett 62:4411–4413

Kannan N, Mukunthan KS, Balaji S (2011) A comparative study of morphology, reactivity and stability of synthesized silver nanoparticles using Bacillus subtilis and Catharanthus roseus (L.) G. Don. Colloids Surf B Biointerfaces 86:378–383

Kaviya S, Santhanalakshmi J, Viswanathan B (2011) Green synthesis of silver nanoparticles using Polyalthia longifolia leaf extract along with D-sorbitol: study of antibacterial activity. J Nanotech 2011.

Kim TN, Feng QL, Kim JO, Wu J, Wang H, Chen GC et al (1998) Antimicrobial effects of metal ions (Ag+, Cu2+, Zn2+) in hydroxyapatite. J Mater Sci Mater Med 9:129–134

Lin YE, Vidic RD, Stout JE, McCartney CA, Yu VL (1998) Inactivation of Mycobacterium avium by copper and silver ions. Water Res 32:997–2000

Mandal S, Phadtare S, Sastry M (2005) Interfacing biology with nanoparticle. Curr Appl Phys 5:118–127

Mann S (ed) (1996) Biomimetic materials chemistry. VCH Publishers, New York

Morones JR, Elechiguerra JL, Camacho A, Holt K, Kouri JB, Ramirez JT, Yacaman MJ (2005) The bactericidal effect of silver nanoparticles. Nanotechnology 16:2346–2353

MubaraAli D, Thajuddin N, Jeganathan K, Gunasekaran M (2011) Plant extract mediated synthesis of silver and gold nanoparticles and its antibacterial activity against clinically isolated pathogens. Colloids Surf B Biointerfaces 85:360–365

Nicewarner-Pena SR, Freeman RG, Reiss BD, He L, Pena DJ, Walton ID, Cromer R, Keating CD, Natan MJ (2001) Submicrometer metallic barcodes. Science 294:137–141

Okami Y, Beppu T, Ogawara H (1988) Biology of Actinomycetes. Japan Scientific Societies Press, Tokyo, pp 88–508

Panàček A, Kvitek L, Prucek R, Kolar M, Vecerova R, Pizurova N, Sharma VK, Nevečná T, Zboril R (2006) Silver colloid nanoparticles: synthesis, characterization, and their antibacterial activity. J Phys Chem B 110(33):16248–16253

Rai M, Yadav A, Gade A (2009) Silver nanoparticles as a new generation of antimicrobials. Biotechnol Adv 27:76–83

Rao CNR, Kulkarni GU, Thomas PJ, Edwards PP (2000) Metal nanoparticles and their assemblies. Chem Soc Rev 29:27–35

Sanger F, Nicklen S, Coulson AR (1977) DNA sequencing with chain terminating inhibitors. Proc Natl Acad Sci 74(12):5463–5467

Sasaki T, Yoshida J, Itoh M, Gomi S, Shomura T, Sezaki M (1988) New antibiotics SF2315A and B produced by an Excellospora sp. I. Taxonomy of the strain, isolation and characterization of antibiotics. J Antibiot 41:835–842

Sastry M, Ahmad A, Khan MI, Kumar R (2003) Biosynthesis of metal nanoparticles using fungi and actinomyces. Curr Sci 5:162–170

Sathishkumar M, Sneha K, Won SW, Cho CW, Kim S, Yun YS (2009) Cinnamon zeylanicum bark extract and powder mediated green synthesis of nano-crystalline silver particles and its bactericidal activity. Colloids Surf B Biointerfaces 73:332–338

Shahverdi AR, Fakhimi A, Shahverdi HR, Minaian SA (2007) Synthesis and effect of silver nanoparticles on the antibacterial activity of different antibiotics against Staphylococcus aureus and Escherichia coli. Nanomed Nanotechnol Biol Med 3:168–171

Shaligram NS, Bule M, Bhambure R, Singhal RS, Singh SK, Szakacs G, Pandey A (2009) Biosynthesis of silver nanoparticles using aqueous extract from the compactin producing fungal strain. Process Biochem 44:939–943

Shankar SS, Ahmad A, Sastry M (2003) Geranium leaf assisted biosynthesis of silver nanoparticles. Biotechnol Progr 19:1627–1631

Sharma VK, Ria AY, Lin Y (2009) Silver nanoparticles: green synthesis and their antimicrobial activities. Adv Colloid Interface Sci 145:83–96

Shirling JL, Gottlieb D (1966) Methods for characterization of Streptomyces species. Int J Sys Evol Microbiol 16:313–340

Simkiss K, Wilbur KM (1989) Biomineralization. Cell Biology and Mineral Deposition Academic Press, New York, p 337

Sriram MI, ManiKanth SB, Kalishwaralal K, Gurunathan S (2010) Antitumor activity of silver nanoparticles in Dalton's lymphoma ascites tumor model. Int J Nanomed 5:753–762

Tamura K, Dudley J, Nei M, Kumar S (2007) MEGA4: molecular evolutionary genetics analysis (MEGA) software version 4.0. Mol Biol Evol 24:1596–1599

Valanarasu M, Duraipandiyan V, Agastian P, Ignacimuthu S (2009) In vitro antimicrobial activity of *Streptomyces* spp. ERI-3 isolated from Western Ghats rock soil (India). J Mycol Med 19:22–28

Vaseashta A, Dimova-Malinovska D (2005) Nanostructured and nanoscale devices, sensors and detectors. Sci Technol Adv Mater 6:312–318

Veerasamy R, Xin TZ, Gunasagaran S, Xiang TFW, Yang EFC, Jeyakumar N, Dhanaraj SA (2011) Biosynthesis of silver nanoparticles using mangosteen leaf extract and evaluation of their antimicrobial activities. J Saudi Chem Soc 15:113–120

Verma VC, Kharwar RN, Gange AC (2010) Biosynthesis of antimicrobial silver nanoparticles by the endophytic fungus *Aspergillus clavatus*. Nanomed 5:33–40

Zhong-jie J, Chun-yan L, Lu-wi S (2005) Catalytic properties of silver nanoparticles supported on silica spheres. J Phys Chem B 109:1730–1735

3

Optical, electrochemical and thermal properties of Co^{2+}-doped CdS nanoparticles using polyvinylpyrrolidone

S. Muruganandam · G. Anbalagan ·
G. Murugadoss

Abstract Co^{2+} (1–5 and 10 %)-doped cadmium sulfide nanoparticles were synthesized by the chemical precipitation method using polyvinyl pyrrolidone (PVP) as a surfactant. The X-ray diffraction results showed that Co ions were successfully incorporated into the CdS lattice and the transmission electron microscopy results revealed that the synthesized particles were aligned as rod-like structures. The absorption spectra of all the prepared samples (undoped and doped) were significantly blue shifted (472–504 nm) from the bulk CdS (512 nm). However, the absorption spectra of the doped samples were red shifted (408–504 nm) with respect to the doping concentrations (1–5 and 10 %). Furthermore, a dramatic blue shift absorption is observed at 472 nm for PVP-capped CdS:Co^{2+} (4 %) nanoparticles. In the photoluminescence study, two emission peaks were dominated in the green region at 529 and 545 nm corresponding the CdS:Co^{2+} nanoparticles. By correlating optical and EPR spectral data, the site symmetry of Co^{2+} ion in the host lattice was determined as both octahedral and tetrahedral. The presence of functional groups in the synthesized nanoparticles was identified by Fourier transform infrared spectroscopy. The thermal stability of the Co ions in CdS nanoparticles was studied by TG–DTA. In addition, an electrochemical property of the undoped and doped samples was studied by cyclic voltammetry for electrode applications.

Keywords CdS:Co^{2+} · Nanoparticles · Photoluminescence · Cyclic voltammetry · EPR

Introduction

In recent years, semiconductor nanoparticles exhibit specific properties due to the quantum confinement effects, as a consequence of their size in nanometric range and the special luminescent properties caused by widening the band gap when the spatial dimension is reduced (Brus 1986; Rossetti et al. 1985; Hu and Zhang. 2006). Due to the quantum size effect, semiconductor nanoparticles, especially the II–VI semiconductor nanoparticles, exhibit size-dependent optical properties (Alivisatos 1996). CdS is an n-type semiconductor with a direct band gap of 2.4 eV. Cadmium sulfide can be used as sensitizers in quantum dot-sensitized solar cells. It also has application in nonlinear optics (Liu et al. 2004), light-emitting diodes (Gopal et al. 2009; Kar and Chaudhuri 2006), solar energy conversion (Weller 2003), thin film transistors (Duan et al. 2003), gas detectors (Afify and Battisha 2000), optoelectronics (Hikmet et al. 2003), photo catalysis (Huynh et al. 2002), photovoltaic cells (Uda et al. 1997), X-ray detectors (Frerichs 1950) and as a window material for hetero-junction solar cells because it has usually a high absorption coefficient (Oladeji et al. 2000). Doping with transition metal elements into CdS nanoparticles leads to many interests. Especially, transition metal doped with CdS nanoparticles with good crystal structural, electrical and optical properties has been reported. The optical and electrical properties of CdS are strongly modified by the doping of Co^{2+} because of the sp–d exchange interaction between the localized d electrons of the transition metal magnetic ions and the mobile carriers in the valance band or conduction

S. Muruganandam · G. Anbalagan (✉)
Department of Physics, Presidency College,
Chennai 600 005, India
e-mail: anbu24663@yahoo.co.in

G. Murugadoss
Centre for Nanoscience and Technology, Anna University,
Chennai 600 025, India

band. Among the transition metallic elements, Co^{2+} is an important transition metal element. Because, the ion radius of Co^{2+} is smaller than that of Cd^{2+}, which means that Co^{2+} can easily penetrate into CdS crystal lattice or substitute Cd^{2+} position in crystal (Sathyamoorthy et al. 2010). Currently, tuning the optical absorption of the semiconductor compound by adding dopant is an important issue. In this regard we have successfully doped the Co ions into the CdS lattice for tuning their optical properties for solar cell applications. Further, we tuned the optical absorption of the doped nanoparticles by introducing with different concentrations of the Co^{2+} (1–5, 10 %).

Chemical precipitation method is one of the most popular techniques that are used in industrial applications because of the cheap raw materials, easy handling and large-scale production (Souici et al. 2006). Recently, several methods have been developed to cap the surfaces of the nanoparticles with organic or inorganic groups, so that the nanoparticles not only are stable against agglomeration but also improve some optical properties of the nanoparticles. Some particular passivators used as PVA (Khanna et al. 2005), PAN (Meng et al. 2000), PVP (Wang et al. 2005a, b), and PAA (Xiao et al. 2001) have been investigated. In this research work, polyvinyl pyrrolidone (PVP) is used as a capping agent for synthesis of the Co-doped CdS nanoparticles by chemical method. The polymers may be a good choice as stabilizers as they can interact with the metal ions by complex or ion-pair formation and can be designed for certain physical properties of semiconductor nanoparticles (Murugadoss 2010).

Experimental

Materials

Cadmium acetate (Cd $(CH_3COO)_2 \cdot 2H_2O$), cobalt acetate (Co $(CH_3COO)_2 \cdot 4H_2O$) and sodium sulfide ($Na_2S \cdot xH_2O$) obtained from the Nice chemical company, India were used as precursors. Polyvinylpyrrolidone (PVP—40,000) was received from Aldrich. All chemicals were used as received. Ultrapure water and acetone were used for all dilution and sample preparation.

Synthesis of Co-doped CdS nanoparticles

Cobalt-doped CdS nanoparticles have been synthesized by the aqueous chemical precipitation method. 0.1 M aqueous solution of cadmium acetate dihydrate (Cd $(CH_3COO)_2 \cdot 2H_2O$) and cobalt acetate tetrahydrate (Co $(CH_3COO)_2 \cdot 4H_2O$)) was prepared with deionized water. These two solutions were mixed together and stirred magnetically at 80 °C until a homogeneous solution was obtained. The 0.1 M of sodium sulfide (Na_2S) solution was also prepared with deionized water with the precursor ratio of 1:1. After an hour, aqueous solution of Na_2S was added dropwise into the mixed solution of cadmium acetate and cobalt acetate at room temperature, which resulted in an orange yellowish solution of Co:CdS. The solution was then refluxed with constant stirring at 120 °C for 30 min to attain saturation, which contains Co:CdS nanoparticles. The solution, after attaining the room temperature was added with small quantities of acetone with stirring to precipitate the nanoparticles. It was dried in hot air oven at 80 °C for 2 h. Co^{2+}-doped CdS nanoparticles with six different Co^{2+} concentrations (1, 2, 3, 4, 5 and 10 %) were prepared by the same procedure. In addition, for synthesis of surfactant (PVP)-capped particles different amounts (0.5–2.5 g) of PVP were added in cadmium acetate solution before the addition of cobalt acetate.

Techniques

The synthesized samples were then subjected to powder X-ray diffraction (XRD) analysis for structural characterization using X'pert PRO diffractometer with CuK$_\alpha$ radiation ($\lambda = 1.5406$ Å) in the range of 20°–60° (2θ) at a scanning rate of 0.05° min^{-1}. The infrared spectra of the nanoparticles were recorded at room temperature using NICOLET AVATAR 3330 FT-IR spectrometer by employing a KBr pellet technique. Optical absorption was studied using a Cary 500 UV–Visible diffuse reflectance mode spectrophotometer and luminescence properties were studied by fluorescence spectroscopy using Cary-Eclipse spectrometer with Xenon lamp source of 450 W. Crystallinity and size of the synthesized samples were studied by transmission electron microscope using 100 kV HITACHI (Japan) H-7650. Electron spin resonance (ESR) spectrum of the CdS:Co^{2+} powder was measured on an EPR spectrometer (Bruker EMX Plus), (9.859 GHz/0.6325 mW) at room temperature. Electrochemical measurements were performed using a CHI 660D Biologic instrument. The electrochemical properties of undoped and Co-doped CdS nanoparticles were studied by cyclic voltammetry. Three electrode systems were used consisting of glassy carbon electrode (GCE) with geometric surface area 7.1 mm^2, Ag/AgCl reference electrode (Ag/AgCl$_3$ mol L^{-1} KCl) and Pt counter electrode. The working electrodes were prepared by coating a slurry containing a mixture of the active material (80 wt%), nafion® 117 solution (20 wt%). The coated mesh was dried at 80 °C in vacuum cabinet overnight. The Ag/AgCl electrode with CdS and CdS:Co nanoparticles grown on the surface was characterized by electrochemical measurements. Cyclic voltammetry (CV) measurements were carried out at a scan rate of 20 mV s^{-1}. Thermal analysis including TGA and DTA was carried out using a simultaneous thermal analyzer SDT Q600 V8.3 Build 101 at the heating rate of 20 °C min^{-1} in an air atmosphere.

Results and discussion

X-ray powder diffraction analysis

X-ray powder diffraction pattern of the synthesized nanoparticles (Fig. 1a–g) shows a perfect match with the cubic zinc blende phase of CdS (hawleyite) (JCPDS 10-454). The peaks can be indexed as (111), (220) and (311) which are characteristic peaks of crystal planes for CdS cubic phase. The diffraction pattern for (111) peak position shows a shift towards higher angle of 2θ (lower d value) with increasing cobalt concentration. This clearly implies the lattice compression and thus confirming the dopant incorporation in the synthesized CdS NPs (Hanif et al. 2002; Saravanan et al. 2011). Further the widths of the diffraction peaks are broadened indicating that the cobalt-doped CdS NPs has a nanoscale distribution. Upon doping, no additional reflections were observed up to 5 % Co doping indicating that the cubic phase of CdS structure is not disturbed by the cobalt substitution and there are no impurities present in the sample. However at 10 % of dopant concentration some extra peaks have been observed along with characteristic peaks. The presence of the new peaks at 28.457, 29.829 and 35.824 indicates the hexagonal structure of CdS (Greenockite, JCPDS Card No. 41-1049). It has been observed that the surface capping with PVP molecule does not have any effect on the crystal structure of CdS:Co^{2+} nanorods. The average particle size was calculated using Scherrer formula (Jenkins and Snyder 1996): $D_{XRD} = 0.9\lambda/(\beta\cos\theta)$ where D_{XRD} is the average crystalline size, λ is the wavelength of CuKα, β is the full width at half maximum of the diffraction peak, and θ is the Bragg's angle. The average particle size of uncapped CdS:Co^{2+} (1–5 and 10 %) and PVP-capped CdS:Co^{2+} (4 %) nanoparticles is found to be 7.3, 7.3, 7.8, 9.7, 11.9, 12.6 and 6.4 nm, respectively.

Structural analysis

The morphology of the uncapped and PVP-capped CdS:Co^{2+} (4 %) nanoparticle was studied by TEM techniques. TEM images of CdS:Co^{2+} nanoparticles shown in Fig. 2a indicate that the particles are highly aggregated due to the absence of capping agent. The image of NPs establishes the reasonable uniformity of the particle size, with spherical shape with the average size of ~10 nm, which agrees with the XRD data. Well-shaped nanorods of CdS:Co^{2+} with PVP capped is shown in Fig. 2b, c. This is

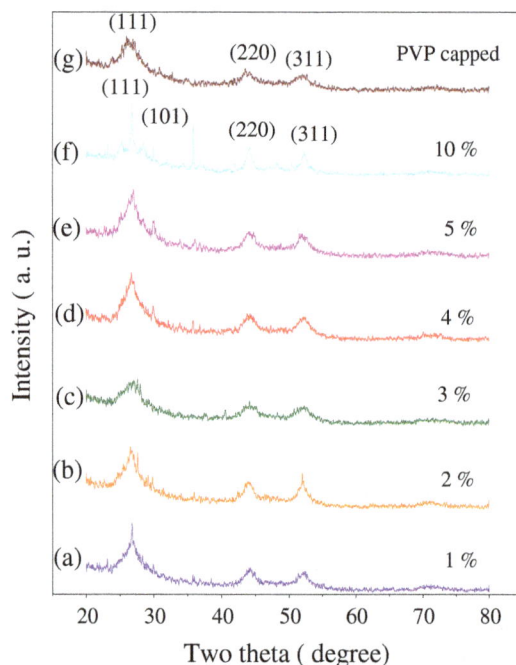

Fig. 1 Powder X-ray diffraction patterns of Co^{2+} CdS nanoparticles

may be due to the assembly of the smaller size nanoparticles (CdS:Co^{2+}) by the carboxylic molecules (PVP). From the TEM photographs, one could find that PVP is not only controlling the particle size, but also assembling the particles as nanorods.

FT-IR spectral analysis

FT-IR spectroscopy gives qualitative information about the way in which the adsorbed surfactant molecules are bound to the surface of CdS:Co^{2+} nanoparticles. The room temperature FT-IR spectra of CdS:Co^{2+} (4 %) and PVP-capped CdS:Co^{2+} (4 %) nanoparticles recorded in the range of 4,000–400 cm^{-1} is shown in Fig. 3. The absorption band observed between 600 and 700 cm^{-1} is due to C–S stretching vibration (Martin and Schaber 1982; Sun et al. 2008). The absorption band appearing at 927 cm^{-1} is assigned to S–O stretching vibration. The additional weak bands observed at 2,343 and 1,638 cm^{-1} are due to microstructure formation of the sample. The absorption band occurring at 1,110 cm^{-1} is due to the C–O stretching bands. Thus, the nanoparticles include a structure containing Cd–S and C–O in all samples Fig. 3a, b. A strong band present at 1,412 cm^{-1} is due to stretching vibration of sulfite. The broad absorption band centered at 3,429 cm^{-1} is attributed to the O–H stretching mode of H$_2$O absorbed on the surface of the samples. The weak and a strong absorption peak centered at 2,360 and 1,560 cm^{-1} were attributed to CO$_2$ adsorbed on the surface of the particles. In fact, adsorbed water and CO$_2$ are common to all powder

Fig. 2 SEM photograph of Co^{2+} doped (4 %):CdS and PVP capped Co^{2+} doped (4 %):CdS

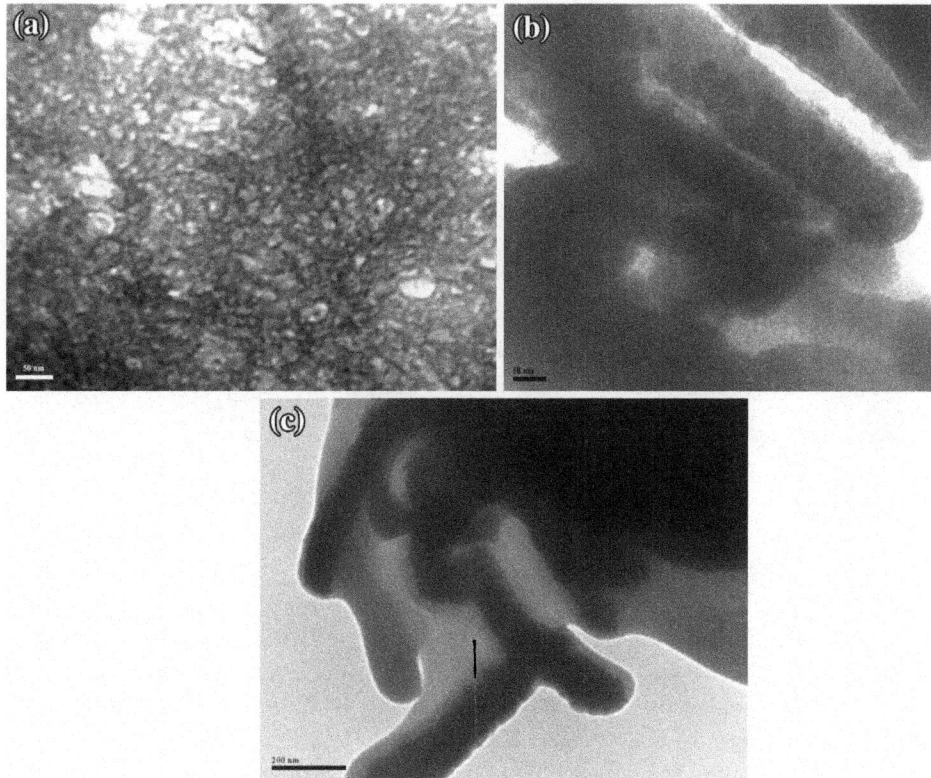

Fig. 3 FT-IR spectrum shows the doped CdS:Co^{2+} (4 %) and CdS:Co/PVP-capped nanoparticles

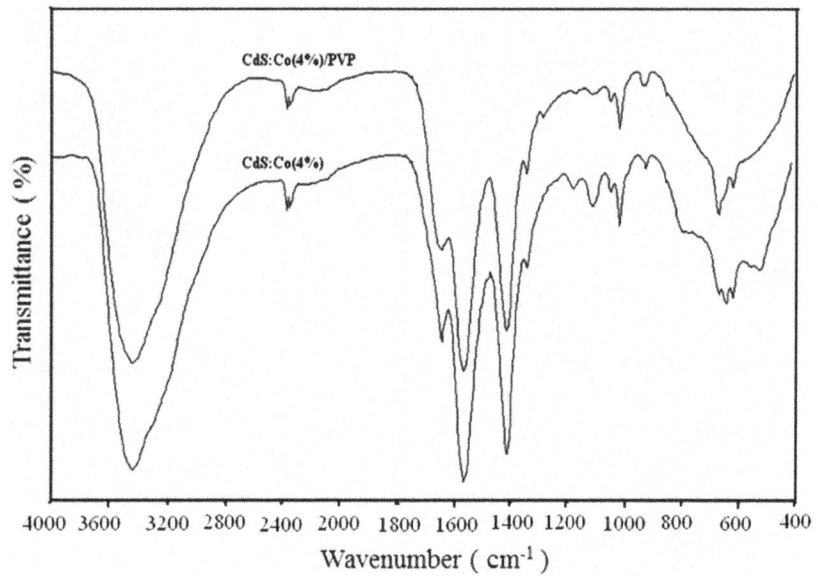

samples exposed to the atmosphere and are more pronounced for nanosized particles with high surface area. The FT-IR spectra suggesting that doped CdS nanoparticles were prepared with a high degree of purity. The presence of absorption peak at 1,290 cm^{-1} is due to C–O which was common in all the PVP-capped nanoparticles of the present study. This clearly confirms that the surface of CdS:Co^{2+} nanoparticles were capped by PVP.

UV–Visible absorption

The absorption spectra of Co^{2+} (1–5 and 10 %)-doped CdS and PVP-capped CdS:Co^{2+} nanoparticles in the range of 200–800 nm are presented in Fig. 4. The bands between 600 and 750 nm represent the tetrahedral coordination of Co^{2+} in CdS. Thus it can be concluded that Co^{2+} exists in a tetrahedral coordination. The presence of absorption band

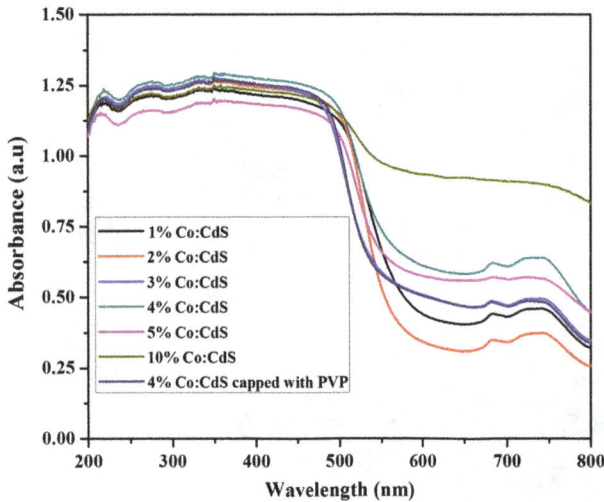

Fig. 4 UV–Vis absorption spectra of CdS:Co Co^{2+} (1–5 and 10 %) and CdS:Co (4 %)/PVP-capped nanoparticles

Table 1 Energy gap, absorption and particle size with respect to the concentrations of Co^{2+}

Doping concentration (wt%)	Band gap (eV)	Absorption wavelength (nm)	Particle size (nm)	
			UV data	PXRD data
CdS:Co^{2+} (1 %)	2.46	504	7.3	7.3
CdS:Co^{2+} (2 %)	2.48	501	7.3	7.3
CdS:Co^{2+} (3 %)	2.51	494	7.8	7.8
CdS:Co^{2+} (4 %)	2.56	484	9.7	9.7
CdS:Co^{2+} (5 %)	2.58	481	11.9	11.9
CdS:Co^{2+} (10 %)	2.58	480	14.6	12.6
CdS:Co^{2+} (4 %)/ PVP	2.63	472	6.5	6.4

around 15,000 cm^{-1} (684 nm) in the visible region corresponds to $^4A_2(F) \rightarrow ^4T_1(P)$ transition of tetrahedral coordination. The absorption bands in 200–350 nm wavelength range can be assigned to the $O^{2-} \rightarrow Co^{2+}$ charge transfer process (He et al. 2005). The absorption values of the prepared samples were noticeably blue shifted from bulk CdS (520 nm). The shift in the band gap with size dominates the spectral changes (Nanda et al. 2000) because the binding energy of the exciton increases with decreasing size due to the increasing columbic overlap enforced by spatial localization of the wave functions. This blue shift in the optical absorption edge indicates the formation of CdS particles in the nanometre region. This result is a direct consequence of the quantum confinement effect associated with smaller particle size. The estimated band gap values for Co^{2+} (1–5 and 10 %)-doped CdS and PVP-capped CdS:Co^{2+} nanoparticles corresponding to the absorption edges are 480 nm (2.58 eV), 481 nm (2.58 eV), 484 nm (2.56 eV), 494 nm (2.51 eV), 501 nm (2.48 eV), 504 nm (2.46 eV), and 472 nm (2.63 eV), respectively, blue shifted compared with the absorption edge of the bulk CdS (512 nm).

The band gap of the nanoparticles (E_{gn}) is calculated using relation $E_{gn} = hc/\lambda$ and the calculated values given in Table 1. Brus (1984) showed that semiconductor nanoparticles with a particle radius significantly smaller than the exciton Bohr radius exhibit strong size-dependent optical properties due to the strong quantum confinement effect (QCE):

$$E_g = E_g^o + \frac{h^2}{8\mu R^2} - \frac{1.8e^2}{4\pi \in R}$$

where E_g^o is the energy band gap of the bulk material, R is the radius of the nanoparticle calculated from XRD data,

$1/\mu = 1/m_e + 1/m_h$ (m_e and m_h being the electron and hole effective masses, respectively), ε is the dielectric constant and e is the electronic charge. Here the electron effective mass (m_e), hole effective mass (m_g) and the dielectric constant (ε) for CdS are 0.19 m_o, 0.8 m_o and 5.7 ε_o, respectively (Ohde et al. 2002). It is clear from Table 1 that there is a decrease in energy band gap values with an increase in cobalt concentration. This red shift of the energy band gap with increasing cobalt concentration is interpreted as mainly due to the sp–d exchange interactions between the band electrons and the localized d electrons of the Co^{2+} ions substituting host ions and is consistent with the reported results (Kumbhoikar et al. 2000), giving an additional evidence of cobalt substitution (Singhal et al. 2010). Also the calculated particle size of the uncapped and PVP-capped CdS:Co^{2+} nanoparticles are in close agreement with the powder XRD result.

Photoluminescence

Figure 5a, b shows the photoluminescence spectra of CdS:Co^{2+} (4 %) and PVP-capped CdS:Co^{2+} (4 %) nanoparticles. The peak position of all the photoluminescence (PL) spectra is nearly same. However, intensity is significantly changed. A maximum PL intensity has been observed at 4 % Co-doped CdS. Further by increasing the doping concentration from 5 to 10 %, the PL intensity is found to be decreased. It indicates that the 4 % of Co^{2+} is an optimum concentration for enhanced PL emission. The emission bands centered at 529 and 545 nm in green emission of Co^{2+}-doped CdS nanoparticles, respectively. The strong PL emission is due to the increased recombination of electrons trapped inside a sulfur vacancy with a hole in the valence band. The present study indicates that the luminescence properties of the capped CdS:Co^{2+} nanoparticles have been attributed to the surface passivation of the nanoparticles with PVP, which can also

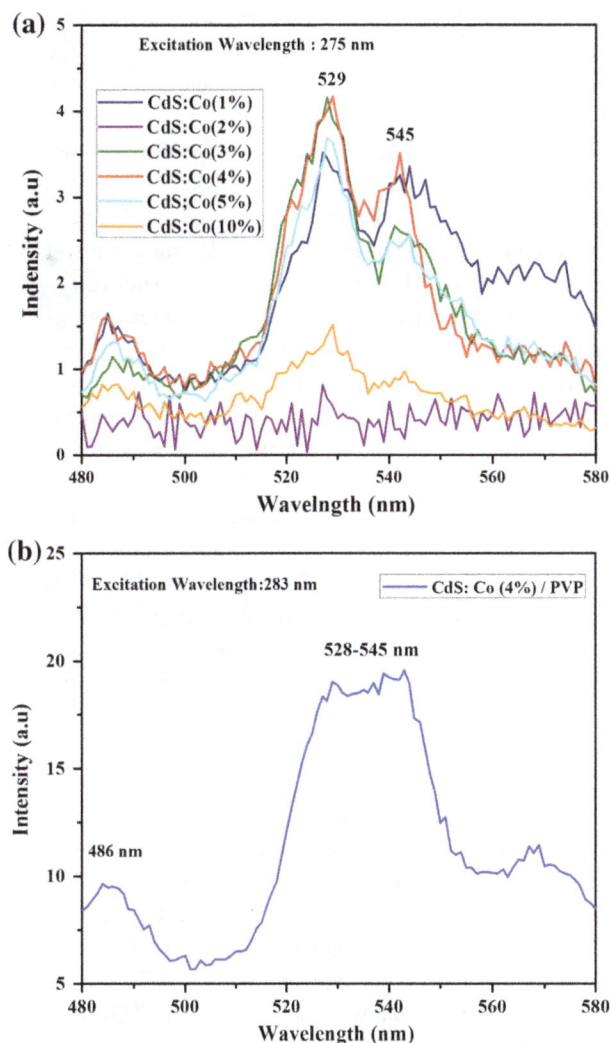

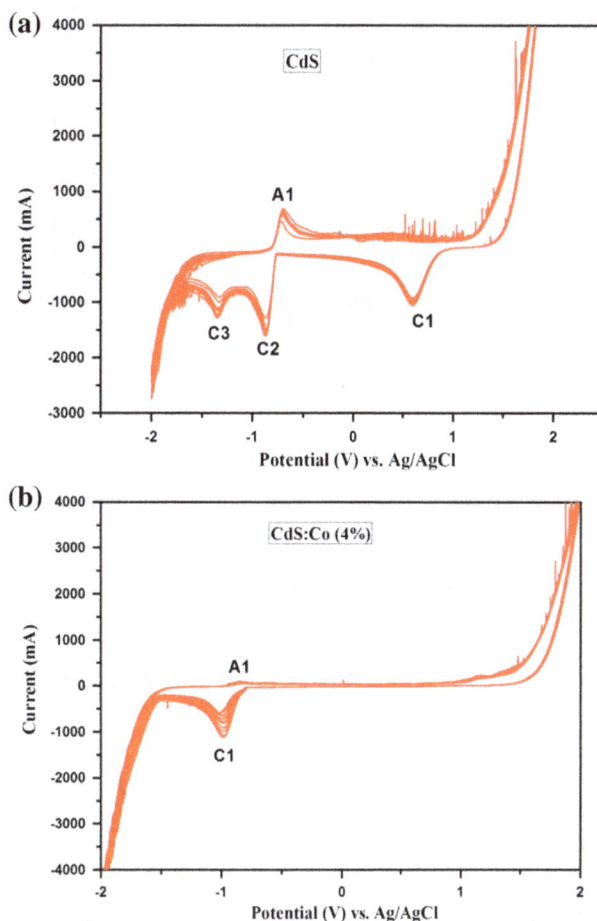

Fig. 6 a, b CV curves of CdS and CdS:Co nanoparticles

Fig. 5 a Photoluminescence spectrum of the CdS:Co^{2+} (1–5 and 10 %) nanoparticles (excitation wavelength = 275 nm), b PL spectrum of CdS:Co (4 %) with PVP-capped nanoparticles (excitation wavelength = 283 nm)

minimize the surface defects and enhance the electron–hole recombination (Paul and Nigel 2001). The present study indicates that the role of PVP is not only used to control the particle size but also reduce the surface defects of CdS (Fig. 5b).

Cyclic voltammetry (CV) studies

Figure 6a, b shows the CVs of CdS and CdS:Co nanoparticles on the glassy carbon with 0.1 M LiClO$_4$ supporting electrolyte at a constant scan rate of 20 mV s^{-1}, respectively. As shown in Fig. 6a, three cathodic peaks (C1, C2 and C3) and one anodic (A1) peak have been observed at 0.60, −0.86, −1.34 and −0.68 V, respectively. Among them C1, corresponding to the reduction of CdS nanoparticle, appears as a modified electrode with CdS:Co

nanoparticles (Fig. 6b). C2 and C3 have been attributed to the reduction of electrolyte. A1 is probably involved in the oxidation of OOH produced by dissolved oxygen (Cui et al. 2004). Figure 6b shows CV curves of CdS:Co nanoparticles which contain one cathodic peak at −0. 98 V (C1) and one anodic peak at −0. 86 V (A1). Compared to the two CV curves, CdS-modified electrode curves showed increasing peak areas with increasing peak current than Co-doped CdS. It dictates the increasing supercapacity behavior of the CdS.

EPR spectral analysis

Electron paramagnetic resonance (EPR) spectroscopy is a method for characterizing structure, dynamics and spatial distribution of paramagnetic ions. Due to the presence of at least one unpaired electron many paramagnetic species are chemically active. EPR spectroscopy is a valuable technique for obtaining detailed information on the geometric and electronic structure of various materials. Therefore, this technique is applied to study the nature of doped Co^{2+} ions in the CdS lattice.

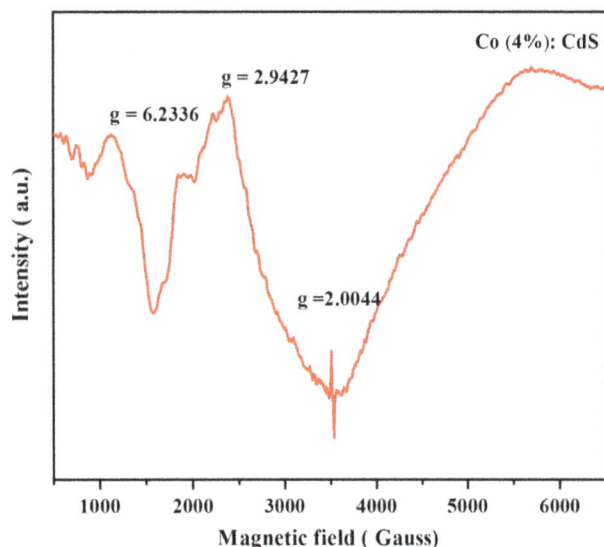

Fig. 7 Room temperature ESR spectrum of free-surfactant 4 % Co:CdS nanoparticles

The EPR spectrum of the Co^{2+}-doped CdS nanoparticles is shown in Fig. 7.

Co^{2+} ions ($3d^7$ configuration with $S = 3/2$, $I = 7/2$) can appear both at high-spin ($S = 3/2$) and at low-spin ($S = 1/2$) configurations and have different characteristics in tetrahedral and octahedral coordination. EPR spectra and spin-Hamiltonian parameters are well understood for high-spin Co^{2+} in ideal high crystal field symmetries and at the tetragonal distortion (Bencini and Gatteschi 1982; Banci et al. 1982). The theory for octahedral field Oh predicts that the EPR spectrum is isotropic with $g = 4.33$ (Abraham and Pryce 1951). This value markedly differs from true g-factors which are close to $g = 2$. This is due to considerable

excited orbital contributions and spin–orbit coupling. As a consequence, the g-values are extremely sensitive to the distortion of the octahedral environment and vary in the range $g = 2$–9.

For high-spin Co^{2+} in the distorted tetrahedral geometry, either the $\pm 1/2$ state or the $\pm 3/2$ state can be lower in energy (Drulis et al. 1985; Pilbrow 1978). The $\pm 1/2$ state is lower in a flattened tetrahedron, and the $\pm 3/2$ state is lower in an elongated tetrahedron of D_{2d} symmetry. The true g-factors are expected in the range 2.2–2.4, whereas the g-factors for the effective spin $S_0 = 1/2$ varies from 2 to 6.

The low-spin Co^{2+} complexes with $S = 1/2$ often appear for the square planar and in pseudo-tetrahedral geometry (Jenkins et al. 2002), and they are easily distinguished from the high-spin complexes. The g-factors are in the range 1.5–3.3, and ground state strongly depends on the crystal field strength, geometry and mixing of configurations (Bencini and Gatteschi 1982; Daul et al. 1979). It is clear from the above that it is possible to distinguish between octahedral and tetrahedral complex geometry and the spin state of Co^{2+} considering the sum of the g-factors only.

TG–DTA study

Thermogravimetric analysis (TGA) and differential thermal analysis (DTA) have been used to study the thermal decomposition of Co^{2+}-doped CdS nanoparticles. The synthesized specimens were heated from room temperature to 800 °C with a heating rate of 20 °C min^{-1} in air atmosphere. Figure 8 shows the combined TG–DTA curves of 4 % Co:CdS nanoparticles. The observed mass loss of TGA curve up to 100 °C mainly corresponds to

Fig. 8 TG and DSC curves of free-surfactant CdS:Co (4 %) nanoparticles

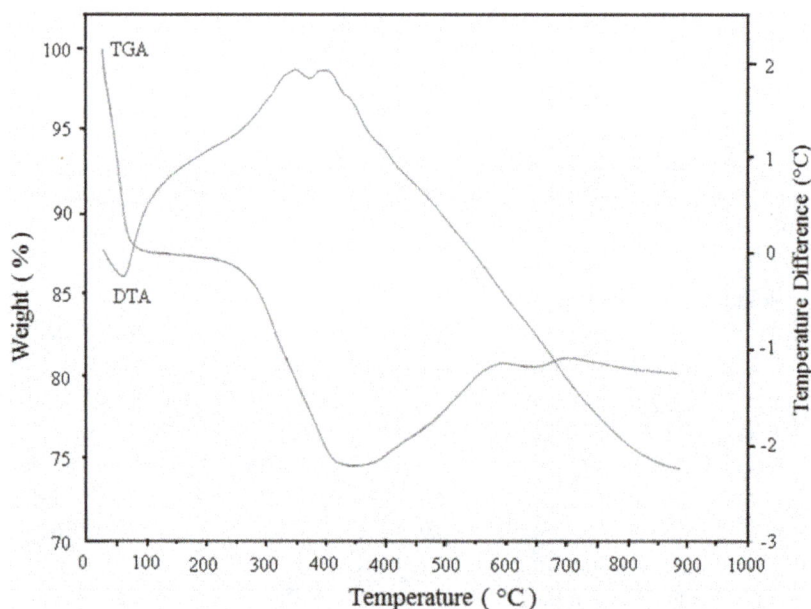

evaporation of water. In addition, the enormous mass loss around 260–400 °C indicates the removal of residual component and organic molecules. The strong exothermic peaks (DTA) at 347 and 399 °C are probably corresponding to the lattice deformation and the improvement of the crystallinity, respectively. Additionally, above 500 °C there is a smooth downward trend is observed in the DTA curve. This may be due to the release of sulfur ions from the sample. Further increase in the temperature from 500 to 700 °C leads to a significant mass gain on the TG curve, due to the oxidation.

Conclusion

Co^{2+} (1–5 and 10 %)-doped CdS nanoparticles were synthesized by the chemical precipitation method and their structural, optical and electrochemical properties were discussed. The broad XRD pattern of all the doped samples shows that the prepared particles were in the nanoscale range (6–14 nm) and that the synthesized rod-like structures of CdS:Co may be used for sensor applications. The doped Co^{2+} ions entered in the host CdS lattice at both octahedral and tetrahedral coordinations without disturbing the hexagonal structure of CdS. Optical absorption and EPR spectra also confirmed that the doped Co^{2+} ions occupied at octahedral and tetrahedral site symmetry. PL spectrum showed characteristic CdS emission bands in UV and blue regions. Optical absorption of all the doped samples was fairly blue shifted from the bulk CdS due to the quantum confinement effect. Thermal stability dopant and phase change were identified by TG–DTA analysis. The well-structured, electrochemical and optical material might be effectively used in some optoelectronic-related fields, for example photocatalytic and photovoltaic devices.

Acknowledgments The author would like to acknowledge the service rendered by CECRI, Karaikudi, Tamilnadu, India, for XRD, UV–Visible, PL, TEM, EPR, TG–DTA studies and CISR, Annamalai University, for recording FT-IR spectra. I would like to gratefully thank Dr. R. Gokul Krishnan, Chairman, Meenakshi Ammal Trust, Chennai, for support.

References

Abraham A, Pryce MHL (1951) The theory of the nuclear hyperfine structure of paramagnetic resonance spectra in the copper Tutton salts. Proc R Soc (Lond) A 206:173–191

Afify HH, Battisha IK (2000) Oxygen interaction with CdS based gas sensors by varying different preparation parameters. J Mater Sci Mater Electron 11:373–377

Alivisatos AP (1996) Semiconductor clusters, nanocrystals, and quantum dots. Science 271:933–937

Banci L, Bencini A, Benelli C, Gatteschi D, Zanchini C (1982) Structures versus special properties. Struct Bond (Berl) 52:38–86

Bencini A, Gatteschi D (1982) The effect of antisymmetric exchange on the EPR spectra of coupled pairs of transition metal ions. Mol Phys 47:161–169

Brus LE (1984) Electron–electron and electron–hole interactions in small semiconductor crystallites: the size dependence of the lowest excited electronic state. J Chem Phys 80:4403–4409

Brus LE (1986) Electronic wave functions in semiconductor clusters: experiment and theory. J Phys Chem 90:2555–2560

Cui H, Xu Y, Zhang ZF (2004) Multi-channel electrochemiluminescence of luminal in neutral and aqueous solutions on a gold nanoparticles self-assembled electrode. Anal Chem 76:4002–4010

Daul C, Schlaepfer CW, von Zalewsky A (1979) The electronic structure of cobalt(II) complexes with schiff bases and related ligands. Struct Bond (Berl) 36:129–171

Drulis H, Dyrek K, Hoffmann KP, Hoffmann SK, Wesełucha-Birczynska A (1985) EPR spectra of low-symmetry tetrahedral high-spin cobalt(II) in a cinchoninium tetrachlorocobaltate(II) dihydrate single crystal. Inorg Chem 24:4009–4012

Duan X, Huang FY, Agarwal R, Lieber CM (2003) Single-nanowire electrically driven lasers. Nature 421:241–245

Frerichs R (1950) The cadmium sulfide X-ray detector. J Appl Phys 21:312–317

Gopal A, Hoshino K, Kim S, Zhang X (2009) Multi-color colloidal quantum dot based light emitting diodes micropatterned on silicon hole transporting layers. Nano Technol 20(23):235201 (pp9)

Hanif KM, Meulenberg RW, Strouse GF (2002) Magnetic ordering in doped $Cd_{1-x}Co_xSe$ diluted magnetic quantum dots. J Am Chem Soc 124:11495–11502

He T, Chen DR, Jiao XL, Wang YL, Duan YZ (2005) Solubility-controlled synthesis of high-quality Co_3O_4 nanocrystals. Chem Mater 17:4023–4030

Hikmet RAM, Talapin V, Weller H (2003) Study of conduction mechanism and electroluminescence in CdS/ZnS quantum dot composites. J Appl Phys 93:3509–3514

Hu H, Zhang W (2006) Synthesis and properties of transition metals and rare-earth metals doped ZnS nanoparticles. J Opt Mater 28:536–550

Huynh WV, Dittmer JJ, Alivisatos AP (2002) Hybrid nanorod–polymer solar cells. Science 295(5564):2425–2427

Jenkins R, Snyder RL (1996) Introduction to X-ray powder diffractometry. John Wiley & Sons, New York

Jenkins DM, Di Bilio AJ, Allen MJ, Betley TA, Peters JC (2002) Elucidation of a low spin cobalt(II) system in a distorted tetrahedral geometry. J Am Chem Soc 124:15336–15350

Kar S, Chaudhuri S (2006) Cadmium sulfide one-dimensional nanostructures: synthesis, characterization and application. Synth React Inorg Met Org Nano Met Chem 36:289–312

Khanna PK, Gokhale RR, Subbarao VVVS, Singh N, Jun KW, Das BK (2005) Synthesis and optical properties of CdS/PVA nanocomposites. Mater Chem Phys 94:454–459

Kumbhoikar N, Nikesh VV, Kshirsagar A, Mahamuni S (2000) Photophysical properties of ZnS nanoclusters. J Appl Phys 88:6260–6264

Liu YK, Zapien JA, Geng CY, Shan YY, Lee CS, Lifshitz Y, Lee ST (2004) High-quality CdS nanoribbons with lasing cavity. Appl Phys Lett 85:3241–3243

Martin TP, Schaber H (1982) Matrix isolated II–VI molecules: sulfides of Mg, Ca, Sr, Zn and Cd. Spectrochem Acta A 38:655–660

Meng C, Yi X, Chen HY, Qiao ZP, Zhu YJ, Qian YT (2000) Templated synthesis of Cds/PAN composite nanowires under ambient conditions. J Colloid Interface Sci 229:217–221

Murugadoss G (2010) Synthesis and optical characterization of PVP and SHMP-encapsulated Mn^{2+}-doped ZnS nanocrystals. J Lumin 130:2207–2214

Nanda J, Sapra S, Srama DD, Chandrasekharan N, Hodes G (2000) Size-selected zinc sulfide nanocrystallites: synthesis, structure, and optical studies. Chem Mater 12:1018–1024

Ohde H, Ohde M, Bailey F, Kim H, Wai CM (2002) Water-in-CO_2 microemulsions as nanoreactors for synthesizing CdS and ZnS nanoparticles in supercritical CO_2. Nano Lett 2:721–724

Oladeji IO, Chow L, Ferekides CS, Viswanathan V, Zhao Z (2000) Metal/CdTe/CdS/$Cd_{1-x}Zn_xS$/TCO/glass: a new CdTe thin film solar cell structure. Sol Energy Mater Sol Cells 61(2):203–211

Paul OB, Nigel LP (2001) Nanocrystalline semiconductors: synthesis, properties and perspectives. Chem Mater 13:3843–3858

Pilbrow JR (1978) Effective g values for $S = 3/2$ and $S = 5/2$. J Magn Reson 31:479–490

Rossetti R, Hull R, Gibson JM, Brus LE (1985) Excited electronic states and optical spectra of ZnS and CdS crystallites in the 15 to 50 A size range: evolution from molecular to bulk semiconducting properties. J Chem Phys 82:552–559

Saravanan L, Pandurangan A, Jayavel R (2011) Synthesis of cobalt-doped cadmium sulphide nanocrystals and their optical and maganetic properties. J Nanoparticle Res 13:1621–1628

Sathyamoorthy R, Sudhagar P, Balerna A, Balasubramanian C, Bellucci S, Popov AI, Asokan K (2010) Surfactant-assisted synthesis of $Cd_{1-x}Co_xS$ nanocluster alloys and their structural, optical and magnetic properties. J Alloys Compd 493:240–245

Singhal S, Chawla AK, Gupta HO, Chandra R (2010) Influence of cobalt doping on the physical properties of $Zn_{0.9}Cd_{0.1}S$ nanoparticles. Nanoscale Res Lett 5:323–331

Souici AH, Keghouche N, Delaire JA, Remita H, Mostafavi M (2006) Radiolytic synthesis and optical properties of ultra-small stabilized ZnS nanoparticles. Chem Phys Lett 422:25–29

Sun ZB, Dong XZ, Chen WQ, Shoji S, Duan XM, Kawata S (2008) Two- and three-dimensional micro/nanostructure patterning of CdS–polymer nanocomposites with a laser interference technique and in situ synthesis. J Nanotechnol 19(3):035610–035611

Uda H, Sonomura H, Ikegami S (1997) Screen printed CdS/CdTe cells for visible-light-radiation sensor. Meas Sci Technol 8:86–91

Wang C, Lp KM, Hark SK, Li Q (2005a) Structure control of CdS nanobelts and their luminescence properties. J Appl Phys 97:054303–054306

Wang QQ, Zhao GL, Han GR, Wang QQ, Zhao GL, Han GR (2005b) Synthesis of single crystalline CdS nanorods by a PVP-assisted solvothermal method. Mater Lett 59:2625–2629

Weller H (2003) Synthesis and self-assembly of colloidal nanoparticles. Philos Trans R Soc A Math Phys Eng Sci 361:229–240

Xiao M, Wang CY, You M, Zhu YR, Chen ZY, Hu Y (2001) A novel ultraviolet-irradiation route to CdS nanocrystallites with different morphologies. Mater Res Bull 36:2277–2282

4

Size effects in magnetotransport in sol–gel grown nanostructured manganites

N. A. Shah · P. S. Solanki · Ashish Ravalia ·
D. G. Kuberkar

Abstract We report the results of the studies on polycrystalline nanostructured La$_{0.7}$Pb$_{0.3}$MnO$_3$ (LPMO) manganites synthesized using sol–gel method employing metal acetate precursor route. Interestingly, it is observed that crystallite size decreases with increase in sintering temperature while microscopic investigations reveal the second grain growth in all the samples. A correlation between the grain morphology and secondary grain growth with the transport and magnetotransport in LPMO manganites has been established. Observation of large temperature sensitivity (~ -28.29 %/K @ 0 T; >300 K) and field sensitivity (~ -48.70 %/T @ 0.2 T; 5 K) in the samples sintered at higher temperature (~ 1150 °C) has been understood in the light of observed secondary grain growth in the form of nanosized grains over the surface of primary grains.

Keywords Size effect · Nanostructure · Manganites · Field sensitivity

Introduction

Several 3d-metal oxide-based materials exhibit variety of interesting and interrelated phenomena including high-T_C superconductivity (HTSC) in cuprates, colossal magnetoresistance (CMR) in manganites, high spin polarizability of conduction electrons in CrO$_2$, Fe$_3$O$_4$, etc. and ferromagnetism in diluted magnetic semiconductors (DMS). All of these compounds exhibit strong correlations between spin, charge, orbital and lattice degrees of freedom (Urban et al. 2004; Rivas et al. 2000; Yuan et al. 2001). It is well established that a well-defined linkage between the theoretical predictions and experimental findings exists in all of them leading to well-known CMR effect in manganites (Jin et al. 1994), giant thermal expansion (Ibarra et al. 1995), isotopic effect (Zhao et al. 1996) and charge ordering in manganites (Doshi et al. 2011). Various theoretical aspects such as double exchange mechanism (Zener 1951), spin–phonon coupling (Millis et al. 1995) and percolated phase segregated regions (Uehara et al. 1999) have been extensively studied in these compounds. Studies on mixed valent manganites in various form such as polycrystalline bulk (Doshi et al. 2009), nanostructures (Kuberkar et al. 2012), bulk composites (Stoyanova-Ivanova et al. 2011), thin films (Solanki et al. 2011), multilayers (Vachhani et al. 2011), thin film composites (Cheng and Wang 2007) and heterostructures (Khachar et al. 2012) have been reported which shows their application potential in various devices.

Nanophasic La$_{0.7}$Sr$_{0.3}$MnO$_3$ (LSMO) manganites synthesized by sol–gel have been studied for the modification in transport and magnetic behavior (Gaur and Varma 2006). Also, intrinsic and extrinsic magnetoresistance (MR) behavior of nanophasic La$_{0.7}$Ca$_{0.3}$MnO$_3$ (LCMO) has been found to depend strongly on particle size (Siwach et al. 2006). Among various chemical methods of synthesizing nanostructured manganites, namely, co-precipitation (Solanki et al. 2010), pechini method (Pechini 1967), urea gel complex method (Vazquez-Vazquez et al. 1998), citric acid–ethylene diamine gel route (Mahesh et al. 1996), molten alkali metal nitrate flux (Luo et al. 2003), amorphous citrate method (Courty et al. 1973) and PVA-based

N. A. Shah (✉)
Department of Electronics, Saurashtra University,
Rajkot 360005, India
e-mail: snikesh@yahoo.com

P. S. Solanki · A. Ravalia · D. G. Kuberkar
Department of Physics, Saurashtra University,
Rajkot 360005, India

chemical synthesis route (Pandya et al. 2001), it is seen that sol–gel (Kuberkar et al. 2012) is the most simple and cost-effective method for obtaining uniformly distributed manganite nanoparticles.

In order to understand various physical properties of nanostructured manganites and their thin films, few proposed mechanisms and theories have been reported. Lopez-Quintela et al. (2003) have studied the intergranular magnetoresistance (IMR) in sol–gel-grown nanostructured $La_{2/3}Ca_{1/3}MnO_3$ manganites. They have explained that magnetization of nanoparticles decreases with increase in surface/volume ratio in the context of core–shell model wherein nanoparticles composed of an inner core and particle boundaries are considered as outer shells. They argued that with increase in particle size, intrinsic magnetism remains unchanged in inner core, while in outer shells different magnetic states are expected mainly due to oxygen vacancies and superficial stress (Lopez-Quintela et al. 2003). They have also used electrostatic blockade model for carriers between grains to explain the unexpected low-temperature resistivity upturn using which authors have found the blocking energy values for charge carriers of nanostructured manganites (Lopez-Quintela et al. 2003). Similar low-temperature resistivity behavior, observed in $La_{0.5}Pr_{0.2}Ba_{0.3}MnO_3$ manganite films, has been discussed by Rana et al. (2005) which has been attributed to the electron–electron scattering mechanism due to coulombic interactions between charge carriers. Similar theory has been discussed for low-temperature resistivity behavior for the $La_{0.7}Pb_{0.3}MnO_3$ manganite-based chemically grown thin films (Solanki et al. 2011). Other reports are also available on the various theoretical aspects considered to understand the transport properties of manganites (Rana et al. 2004). Recently, Cossu et al. (2013) have performed the first principle calculations to investigate $LaMnO_3$/$SrTiO_3$ superlattice.

Till date, several research reports are available on the particle size-dependent changes in the transport and magnetic properties of mixed valent manganites (Gaur and Varma 2006; Siwach et al. 2006) but very few studies are reported on the temperature and field sensitivity in manganite-based thin films (Markna et al. 2006; Parmar et al. 2006; Kataria et al. 2013).

From application point of view, it is necessary to investigate the manganite materials having appreciably large MR at or near room temperature. Occurrence of phase transitions and magnetic ordering at low temperatures under relatively large applied fields has become a bottle neck for the suitability of manganites for device applications. LPMO is a unique substituted rare earth manganite having large MR under relatively low applied field and ferromagnetic ground state at room temperature (Mahendiran et al. 1995). In this communication, we report the results of the studies on the magnetotransport properties and field and temperature sensitivity in nanostructured $La_{0.7}Pb_{0.3}MnO_3$ (LPMO) manganites. Also, the grain morphology and grain size affect the field and temperature sensitivity in LPMO manganites which has been discussed in the light of size and surface effects.

Experimental details

Nanostructured polycrystalline bulk $La_{0.7}Pb_{0.3}MnO_3$ (LPMO) manganites were synthesized using low-cost and simple sol–gel method using metal acetate precursor route. Mixing of all the metal acetate precursors (and Pb in carbonate form) was carried out in a solution of acetic acid and double distilled water having 1:1 volume ratio. Continuous mixing, stirring, heating and drying of the appropriate stoichiometric quantities of the metal acetates of La and Mn and Pb-carbonate resulted in clear xerogel form of material. Grinding of the xerogel for 30 min resulted in a brown-colored powder which was then calcined at 750 °C for hours. Resultant black powder was subsequently palletized and sintered at different temperatures (1050, 1100 and 1150 °C) in order to obtain the samples with different grain sizes. Hereafter, the samples sintered at 1050, 1100 and 1150 °C temperatures are referred as 1050, 1100 and 1150, respectively. XRD measurements were performed to know the structural phases present while AFM micrographs were obtained to study the effect of sintering temperature on the grain morphology and grain size. Resistance and magnetoresistance (MR) measurements performed using dc four probe method in the temperature range from 2 to 380 K under 0–9 T applied magnetic field.

Results and discussion

XRD patterns of all the LPMO samples sintered at different temperatures in Fig. 1 show the single phasic nature of samples crystallizing in rhombohedral structure having R-3C space group (no. 167). Figure 2 shows an enlarged view of most intense *(110)* and *(104)* peaks of all the samples indicating higher angle shifting of the peaks with increase in sintering temperature suggesting the decrease in unit cell parameters and cell volume. Obtained cell volume [using Rietveld refinement of the XRD patterns using FULLPROF program (Rodriguez-Carvajal 1990)] (Rietveld fitted XRD pattern of 1050 sample is shown as an inset of Fig. 2) decreases from 0.4158 nm^3 (1050) to 0.4093 nm^3 (1150). Interestingly, it is observed that with increase in sintering temperature, XRD peak intensity decreases while peak width increases (Figs. 1, 2). This results in the decrease in crystallite size (CS) with increase in sintering temperature, which is contrary to the earlier reported studies (Gaur

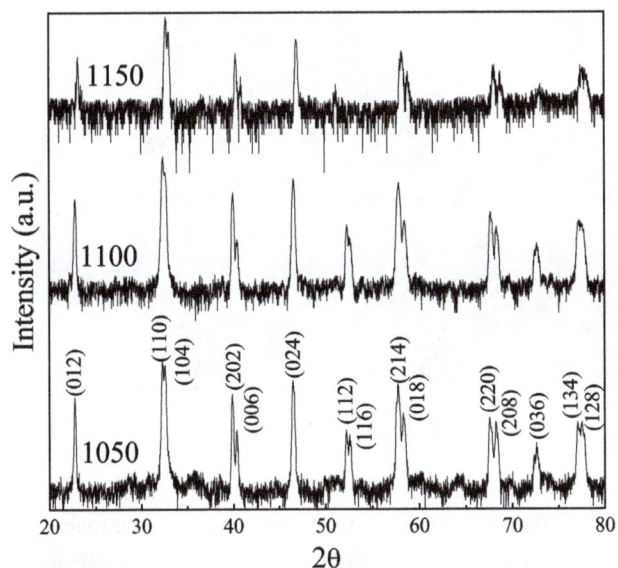

Fig. 1 XRD patterns of $La_{0.7}Pb_{0.3}MnO_3$ manganites sintered at different temperatures. *Inset* Rietveld fitted XRD pattern of $La_{0.7}Pb_{0.3}MnO_3$ manganite sintered at 1050 °C

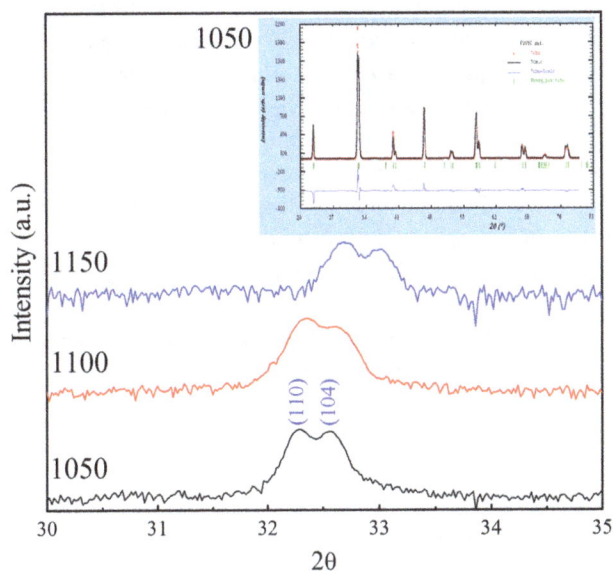

Fig. 2 Enlarged view of *(110)* and *(104)* peaks at $2\theta \sim 32.5°$ for $La_{0.7}Pb_{0.3}MnO_3$ manganites sintered at different temperatures

Table 1 Values of Mn–O–Mn bond angles, Mn–O bond lengths and differences between the values of average basal and epical Mn–O–Mn bond angles (Δ) and Mn–O bond lengths (D) for nanostructured $La_{0.7}Pb_{0.3}MnO_3$ manganites sintered at different temperatures

Parameters	LaMnO$_3$ (Alonso 1998)	1050	1100	1150
Mn–O$_1$ (°)	1.9789 (6)	1.969 (3)	1.963 (2)	1.958 (3)
Average Mn–O$_2$ (°)	1.9745 (2)	1.965 (2)	1.961 (4)	1.957 (4)
Δ (°)	–	0.004	0.002	0.001
Mn–O$_1$–Mn (Å)	161.49 (1)	162.13 (5)	163.82 (1)	164.28 (4)
Mn–O$_2$–Mn (Å)	–	174.47 (2)	174.98 (3)	175.05 (3)
D (Å)	–	12.34	11.16	10.77

and Varma 2006; Siwach et al. 2006). Calculated values of CS, using Scherer's formula: CS = 0.9 λ/B cosθ, where λ is the X-ray wavelength used, B is the peak width (FWHM) and θ is the Bragg angle, are 20.01 nm (1050), 15.29 nm (1100) and 14.88 nm (1150). The values of Rietveld-refined Mn–O–Mn bond angles and Mn–O bond lengths for all the LPMO samples are listed in Table 1. In Table 1, reported values of Mn–O–Mn bond angle and Mn–O bond lengths are also listed for reference purpose (Alonso 1998).

For understanding the sintering temperature-dependent modifications in the grain size and morphology, AFM micrographs were obtained for all the nanostructured LPMO samples. Figure 3 shows clear granular nature of the samples evident from the micrographs with grain size increases from 1.0 μm (1050) to 1.3 μm (1100) and 1.9 μm (1150) resulting in the decrease in grain boundary density. It can be seen that, in the samples sintered at 1050 °C, there is a growth of smaller secondary grains over the surface of micron-sized primary grains (Fig. 3). The size of secondary grains decreases from 150 nm (1050) to 90 nm (1150) with sintering temperature. It is clear that the secondary grain density and hence connectivity between them increase with the sintering temperature which can be correlated with the decrease in crystallinity and CS (Figs. 1, 2).

The simultaneous effect of grain growth and secondary grain size on the transport properties of sol–gel-grown nanostructured LPMO manganites was understood by performing temperature-dependent resistivity measurements under various applied magnetic fields as shown in Fig. 4. All the samples show metal (dρ/dT > 0) to insulator (dρ/dT < 0) transition at T_P. Resistivity decreases with increase in sintering temperature which can be understood as—with increase in sintering temperature, the primary grain size increases from 1.0 to 1.9 μm resulting in the decrease in grain boundary density which consequences in the decrease in scattering of the charge carriers across the grain boundaries and hence decrease in resistivity. Also, resistivity gets suppressed with applied magnetic field leading to negative MR in all the samples which can be ascribed to the field-induced reduction in the magnetic spin scattering of the charge carriers at the grain boundaries and improved magnetic order of the grain boundaries. In addition, T_P increases with increase in sintering temperature as well as with applied field, which can be due to the sintering temperature-induced improved grain boundary nature and field-induced reduction in the non-magnetic phase fraction which significantly support the zener double exchange (ZDE) mechanism and hence enhance the T_P.

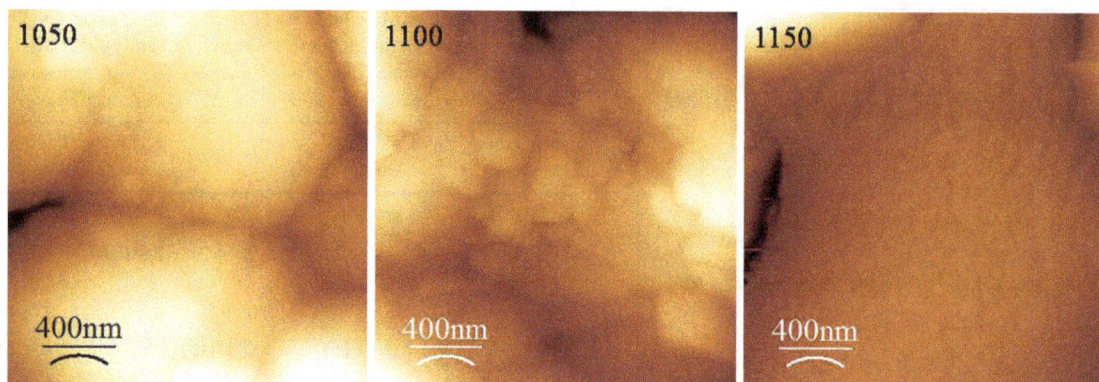

Fig. 3 Microstructural (AFM) images of nanostructured $La_{0.7}Pb_{0.3}MnO_3$ manganites sintered at different temperatures

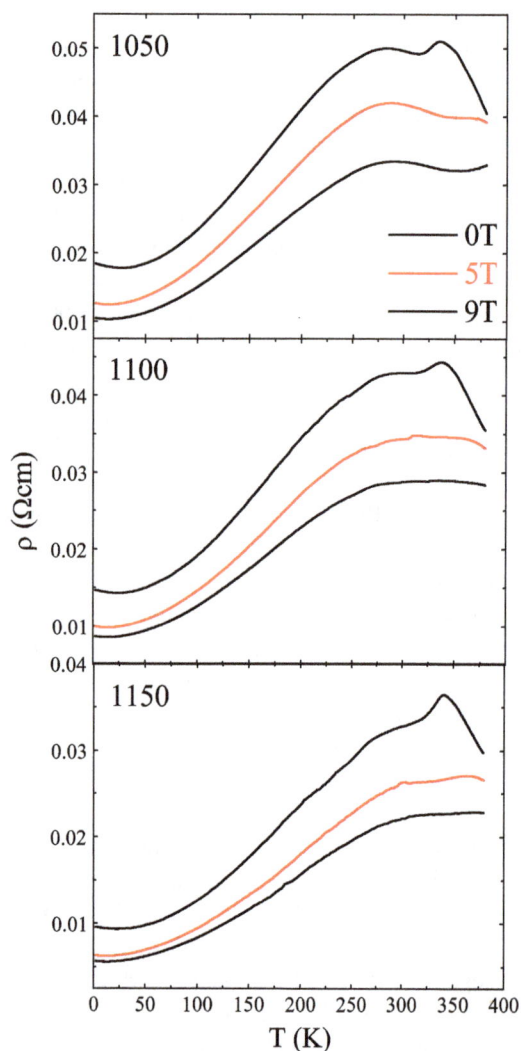

Fig. 4 Resistivity (ρ) vs. temperature (T) plots under 0, 5 and 9 T applied magnetic fields for nanostructured $La_{0.7}Pb_{0.3}MnO_3$ manganites sintered at different temperatures

The effect of decrease in the secondary grain size on the transport in LPMO can be understood as—smaller secondary grain size results in better connectivity between them in higher sintered samples having sharp and ordered grain boundaries which in turn results in the improved charge transport. In addition, the observation of a small hump-like behavior observed at ~ 300 K in ρ–T plots of LPMO may be due to the secondary grain growth which can be modified under applied field.

From Table 1, it can be seen that upon the substitution of Pb^{2+} at La^{3+} site in $LaMnO_3$, Mn–O bond lengths (both, basal and epical) get reduced while Mn–O–Mn bond angles get increased. With increase in sintering temperature, Mn–O bond length decreases while Mn–O–Mn bond angles increase which in turn improve the transfer integral of charge carriers and hence decrease in resistivity and increase in T_P. This can also be supported by the MnO_6 octahedral distortion which can be quantified by the differences between the Rietveld-refined values of average basal and epical Mn–O–Mn bond angles (Δ) and Mn–O bond lengths (D). As listed in Table 1, values of Δ and D decrease from 1050 to 1150 indicating the decrease in MnO_6 octahedral distortion and improved transport.

MR isotherms (MR = $[\{(\rho_H - \rho_0)/\rho_0\} \times 100]$) were recorded at 5 K (to understand the extrinsic contribution to MR) and 300 K (to study the intrinsic contribution to MR) for all the samples studied. Extrinsic MR is the low-temperature, low-field (≤ 1 T) MR exhibited by the manganites mainly governed by external parameters such as grain size, grain boundary density, grain boundary nature, temperature history, final sintering/annealing temperature, etc. Intrinsic MR is the MR observed at $\sim T_P/T_C$ under high magnetic field (>1 T), originating mainly due to the zener double exchange mechanism. Figure 5 shows the MR vs. H isotherms of all the nanostructured LPMO manganites studied. All the samples exhibit negative MR at 5 and 300 K which increases with field. At 5 K, on increasing the field, MR continuously increases with large slope (dMR/dH) below 1 T and smaller slope at higher fields. Low-field, low-temperature MR (extrinsic MR) can be ascribed to the spin-polarized tunneling (Hwang et al. 1996) or spin-dependent scattering (Gupta and Sun 1999) across the grain

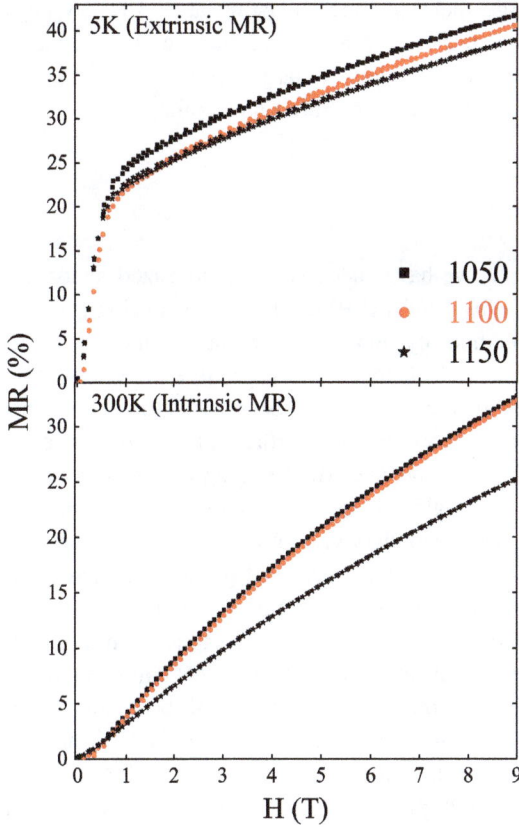

Fig. 5 Field-dependent MR isotherms collected at 5 and 300 K for nanostructured $La_{0.7}Pb_{0.3}MnO_3$ manganites sintered at different temperatures

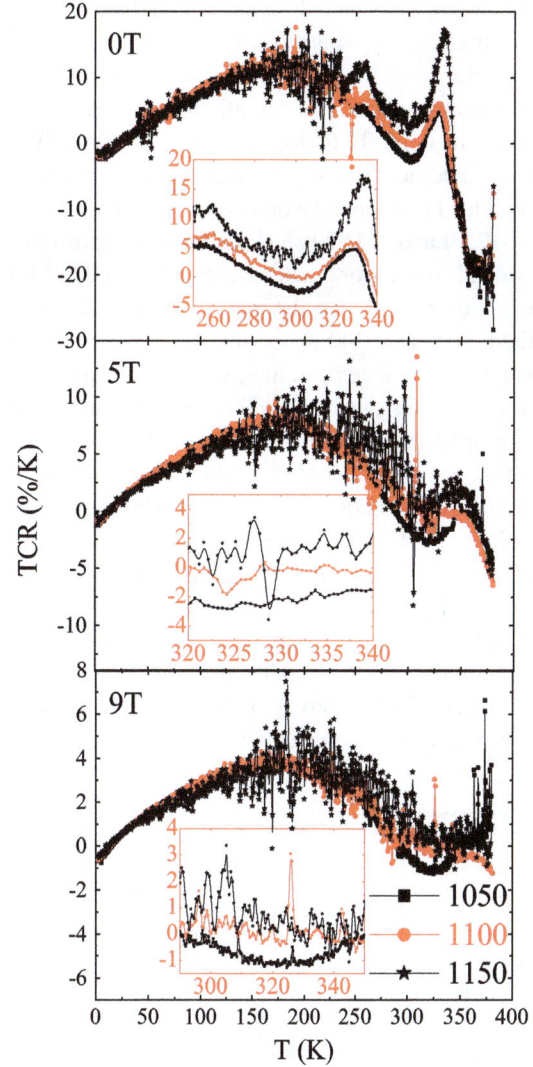

Fig. 6 Variation in temperature sensitivity (TCR) with temperature under 0, 5 and 9 T applied magnetic field for nanostructured $La_{0.7}Pb_{0.3}MnO_3$ manganites sintered at different temperatures. *Inset* enlarged view of TCR vs. *T* plots (where fluctuations are less) under 0, 5 and 9 T fields for the clarity purpose

boundaries while high-field, low-temperature MR depends on the reorientation and stiffness of the spins and connectivity between the micron-sized grains and nanostructured secondary grains over the primary grains. At 5 K, large low-field extrinsic MR ~ 20 to 25 % is observed which decreases with sintering temperature throughout the field range studied. Similarly, at 300 K, MR decreases from 32.70 % (1050) to 25.26 % (1150) under 9 T. Reduction in MR at 5 and 300 K with sintering temperature can be understood as—with increase in sintering temperature, grain size increases, grain boundary density decreases and secondary grain connectivity increases leading to the decrease in non-magnetic disordered phase resulting into the total reduction in the field-induced suppression in scattering of the charge carriers and magnetic disorder (i.e. reduction in total field effect) consequence in the decrease in MR with sintering temperature.

To explore the application potential of sol–gel-grown nanostructured LPMO samples, we have calculated the temperature and field sensitivity of resistivity quantified by temperature coefficient of resistance [TCR = $(1/R) \times (dR/dT) \times 100$ (%/K)] and field coefficient of resistance [FCR = $(1/R) \times (dR/dH) \times 100$ (%/T)], respectively.

Few reports are available on the field and temperature sensitivities of manganite-based thin films (Parmar et al. 2006; Markna et al. 2006; Kataria et al. 2013) but, for the first time, we report the TCR and FCR studies on nanostructured manganites to understand the effect of surface and nanostructured secondary grain growth. Figure 6 shows the variation in TCR with temperature under various applied fields. All the samples show large fluctuations in sensing the temperature in terms of resistance which may be attributed to the disordered structure of the nanophasic samples (Khachar et al. 2013) and thermal fluctuations of the spins. Overall, it can be seen that with increase in sintering temperature, TCR increases which can be attributed to the large surface area (due to smaller secondary grain size) in higher sintered samples and hence large reactivity of the dangling spins of the charge carriers. This also supports the comparatively larger fluctuations in TCR in higher sintered samples, under all the applied fields. With increase in field, TCR decreases mainly due to the field-induced modifications in the surface spin structure. Under 0 T field, 1150 sample exhibits maximum values of TCR $\sim +17.30$ and -28.29 %/K well above room temperature which is useful for practical applications of LPMO manganites. Figure 7 shows the variation in FCR with applied field at 5 and 300 K for all the nanostructured LPMO samples. At low temperature, where spins are most stable, favoring ferromagnetism and hence able to sense the field change appreciably, large FCR values are observed. No fluctuation has been recorded at low temperature with maximum FCR ~ -42.60 %/T (1050) under relatively lower applied field ~ 0.2 T which increases up to ~ -48.70 %/T (1150) under the same field (inset of Fig. 7). Increased field sensitivity of the samples sintered at higher temperatures can be ascribed to the large surface area of the samples having smaller-sized secondary grains. With insignificant fluctuations in FCR at 300 K, overall FCR gets suppressed as compared to that at 5 K, mainly due to large thermal effects existing at 300 K. By engineering the surface area and designing the grains in these materials, one can achieve large temperature and field sensitivity of the nanostructured manganites.

Conclusions

In summary, we have successfully synthesized nanostructured $La_{0.7}Pb_{0.3}MnO_3$ (LPMO) manganites using low-cost sol–gel method by employing metal acetate precursor route. Observed decrease in crystallite size has been understood in the context of secondary growth of nanostructured grains over the surface of the micron-sized grains. The transport and MR behaviors have been understood in the light of grain morphology and observed nanostructured secondary grain growth. Improved transport has been correlated with increased grain size, decrease in grain boundary density and compactness and connectivity between the nanostructured grains. MR has been discussed on the basis of intrinsic and extrinsic MR components and its dependence on the microstructural behavior of the samples. Field and temperature sensitivity and its variation with sintering temperature have been ascribed to the large surface area in higher sintered samples due to the presence of nanostructured secondary grain growth over the primary grains.

Acknowledgments Authors thankfully acknowledge Dr. V. Ganesan, UGC-DAE CSR, Indore for providing the experimental facilities for microstructural and transport measurements. Mr. Mohan Gangrade is thankfully acknowledged for his valuable support in carrying out AFM micrographs and specially detecting secondary grain growth. PSS is thankful to DST, New Delhi for the award of Fast Track Young Scientist (File No. SR/FTP/PS-138/2010).

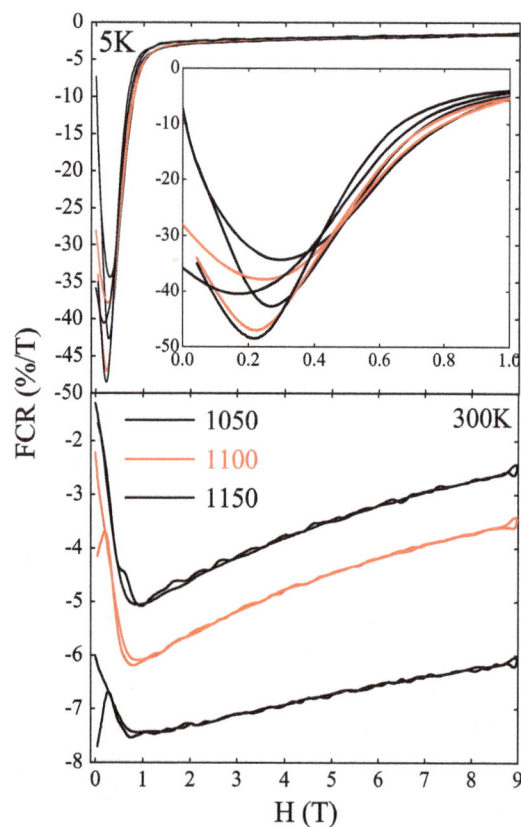

Fig. 7 Variation in field sensitivity (FCR) with field at 5 and 300 K for nanostructured $La_{0.7}Pb_{0.3}MnO_3$ manganites sintered at different temperatures. *Inset* enlarge view of FCT vs. H plots at 5 K (field range 0–1 T) for clarity purpose

References

Alonso JA (1998) Non-stoichiometry and properties of mixed valent manganites. Philos Trans R Soc Lond 356:1617–1634

Cheng Z, Wang X (2007) Room temperature magnetic-field manipulation of electrical polarization in multiferroic thin film composite $BiFeO_3/La_{2/3}Ca_{1/3}MnO_3$. Phys Rev B 75(172406): 1–4

Cossu F, Singh N, Schwingenschlogl U (2013) High mobility half metallicity in the $(LaMnO_3)_2/(SrTiO_3)_8$ superlattice. Appl Phys Lett 102:042401

Courty P, Ajot H, Marcilly C (1973) Mixed oxides or a solid solution form very divided obtained by thermal decomposition of precursor amorphous. Powder Technol 7:21–38

Doshi RR, Solanki PS, Krishna PSR, Das A, Kuberkar DG (2009) Magnetic phase coexistence in Tb^{3+} - and Sr^{2+}—doped

La$_{0.7}$Ca$_{0.3}$MnO$_3$ manganite: a temperature-dependent neutron diffraction. J Magn Magn Mater 321:3285–3289

Doshi RR, Solanki PS, Khachar U, Kuberkar DG, Krishna PSR, Banerjee A, Chaddah P (2011) First order paramagnetic–ferromagnetic phase transition in Tb^{3+} doped La$_{0.5}$Ca$_{0.5}$MnO$_3$ manganite. Phys B 406:4031–4034

Gaur A, Varma GD (2006) Sintering temperature effect on electrical transport and magnetoresistance of nanophasic La$_{0.7}$Sr$_{0.3}$MnO$_3$. J Phys Condens Matter 18:8837–8846

Gupta A, Sun JZ (1999) Spin polarized transport and magnetoresistance in magnetic oxides. J Magn Magn Mater 200:24–43

Hwang HY, Cheong SW, Ong ON, Batlogg B (1996) Spin-polarized intergrain tunneling in La$_2$/3Sr$_1$/3MnO$_3$. Phys Rev Lett 77:2041–2044

Ibarra MR, Algarabel PA, Marquina C, Blasco J, Garcia J (1995) Large magnetovolume effect in yttrium doped La–Ca–Mn–O perovskite. Phys Rev Lett 75:3541–3544

Jin S, Tiefel TH, McCormack M, Fastnacht RA, Ramesh R, Chen LH (1994) Thousandfold change in resistivity in magnetoresistive La–Ca–Mn–O films. Science 264:413–415

Kataria B, Solanki PS, Khachar U, Vagadia M, Ravalia A, Keshvani MJ, Trivedi P, Venkateshwarlu D, Ganesan V, Asokan K, Shah NA, Kuberkar DG (2013) Role of strain and microstructure in chemical solution deposited La$_{0.7}$Pb$_{0.3}$MnO$_3$ manganite films: thickness dependent swift heavy ions irradiation studies. Radiat Phys Chem 85:173–178

Khachar U, Solanki PS, Choudhary RJ, Phase DM, Ganesan V, Kuberkar DG (2012) Current–voltage characteristics of PLD grown manganite based ZnO/La$_{0.5}$Pr$_{0.2}$Sr$_{0.3}$MnO$_3$/SrNb$_{0.002}$Ti$_{0.998}$O$_3$ thin film heterostructure. Solid State Commun 152:34–37

Khachar U, Solanki PS, Choudhary RJ, Phase DM, Kuberkar DG (2013) Positive MR and large temperature-field sensitivity in manganite based heterostructures. J Mater Sci Technol 29:989–994

Kuberkar DG, Doshi RR, Solanki PS, Khachar U, Vagadia M, Ravalia A, Ganesan V (2012) Grain morphology and size disorder effect on the transport and magnetotransport in sol–gel grown nanostructured manganites. Appl Surf Sci 258:9041–9046

Lopez-Quintela MA, Hueso LE, Rivas J, Rivadulla F (2003) Intergranular magnetoresistance in nanomanganites. Nanotechnology 14:212–219

Luo F, Huang YH, Yan CH, Jiang S, Li XH, Wang ZM, Liao CS (2003) Molten alkali metal nitrate flux to well-crystallized and homogeneous La$_{0.7}$Sr$_{0.3}$MnO$_3$ nanocrystallites. J Magn Magn Mater 260:173–180

Mahendiran R, Mahesh R, Raychaudhuri AK, Rao CNR (1995) Room temperature giant magnetoresistance in La$_{1-x}$Pb$_x$MnO$_3$. J Phys D Appl Phys 28:1743–1745

Mahesh R, Mahendiran R, Raychaudhuri AK, Rao CNR (1996) Effect of particle size on the giant magnetoresistance of La$_{0.7}$Ca$_{0.3}$MnO$_3$. Appl Phys Lett 68:2291–2293

Markna JH, Parmar RN, Kuberkar DG, Kumar R, Rana DS, Malik SK (2006) Thickness dependent swift heavy ion irradiation effects on electronic transport of (La$_{0.5}$Pr$_{0.2}$)Ba$_{0.3}$MnO$_3$ thin films. Appl Phys Lett 88(152503):1–3

Millis AJ, Littlewood PB, Shraiman BI (1995) Double exchange alone does not explain the resistivity of La$_{1-x}$Sr$_x$MnO$_3$. Phys Rev Lett 74:5144–5147

Pandya DK, Kashyap SC, Pattanaik GR (2001) Magnetoresistive behavior of La$_{0.67}$(Ca$_{0.33-x}$Pb$_x$)MnO$_3$ nanopowders prepared by lower temperature. J Alloys Compd 326:255–259

Parmar RN, Markna JH, Kuberkar DG, Kumar R, Rana DS, Bagve VC, Malik SK (2006) Swift-heavy-ion-irradiation-induced

enhancement in electrical conductivity of chemical solution deposited La$_{0.7}$Ba$_{0.3}$MnO$_3$ thin films. Appl Phys Lett 89(202506):1–3

Pechini MP (1967) Method of preparing lead and alkaline earth titanates and niobates and coating method using the same to form a capacitor. US Pat. No. 3330697

Rana DS, Thaker CM, Mavani KR, Kuberkar DG, Kundaliya DC, Malik SK (2004) Magnetic and transport properties of (La$_{0.7-2x}$Eu$_x$)(Ca$_{0.3}$Sr$_x$)MnO$_3$: effect of simultaneous size disorder and carrier density. J Appl Phys 95:4934–4940

Rana DS, Markna JH, Parmar RN, Kuberkar DG, Raychaudhuri P, John J, Malik SK (2005) Low temperature transport anomaly in the magnetoresistive compound (La$_{0.5}$Pr$_{0.2}$)Ba$_{0.3}$MnO$_3$. Phys Rev B 71:212404

Rivas J, Hueso LE, Fondado A, Rivadulla F, Lopez-Quintela MA (2000) Low field magnetoresistance effects in fine particles of La$_{0.67}$Ca$_{0.33}$MnO$_3$ perovskites. J Magn Magn Mater 221:57–62

Rodriguez-Carvajal J (1990) FULLPROF: a program for rietveld refinement and pattern matching analysis. In: Abstracts of the satellite meeting on powder diffraction of the XV congress of the IUCr, Toulouse, France, p 127

Siwach PK, Goutam UK, Srivastava P, Singh HK, Tiwari RS, Srivastava ON (2006) Colossal magnetoresistance study in nanophasic La$_{0.7}$Ca$_{0.3}$MnO$_3$ manganite. J Phys D Appl Phys 39:14–20

Solanki PS, Doshi RR, Khachar UD, Vagadia MV, Ravalia AB, Kuberkar DG, Shah NA (2010) Structural, microstructural, transport and magnetotransport properties of nanostructured La$_{0.7}$Sr$_{0.3}$MnO$_3$ manganites synthesized by coprecipitation. J Mater Res 25:1799–1802

Solanki PS, Doshi RR, Khachar UD, Choudhary RJ, Kuberkar DG (2011) Thickness dependent transport and magnetotransport in CSD Grown La$_{0.7}$Pb$_{0.3}$MnO$_3$ manganite films. Mater Res Bull 46:1118–1123

Stoyanova-Ivanova AK, Staneva AD, Shoumarova JM, Blagoev BS, Zaleski AJ, Mikli V, Dimitriev YB (2011) Microstructure and superconductivity of bulk BPSCCO/LPMO composite. Philos Mag Lett 91:190–199

Uehara M, Mori S, Chen CH, Cheong SW (1999) Percolative phase separation underlies colossal magnetoresistance in mixed valent manganites. Nature (London) 399:560–563

Urban JJ, Ouyang L, Jo MH, Wang DS, Park H (2004) Synthesis of single crystalline La$_{1-x}$Ba$_x$MnO$_3$ nanocubes with adjustable doping levels. Nano Lett 4:1547–1550

Vachhani PS, Solanki PS, Doshi RR, Shah NA, Rayaprol S, Kuberkar DG (2011) Substrate dependent transport and magnetotransport in manganite multilayer. Phys B 406:2270–2272

Vazquez-Vazquez C, Blanco MC, Lopez-Quintela MA, Sanchez RD, Rivas J, Oseroff SB (1998) Characterization of La$_{0.67}$Ca$_{0.33}$MnO$_3$ particles prepared by the sol–gel route. J Mater Chem 8:991–1000

Yuan SL, Zhang GQ, Peng G, Tu F, Zeng XY, Liu J, Yang YP, Jiang Y, Tang CQ (2001) Electrical transport and low field magnetoresistance in the series of mixed polycrystals (1-m)La$_{2/3}$Ca$_{1/3}$MnO$_3$ + mLa$_{2/3}$Ca$_{1/3}$MnO$_3$. J Phys Condens Matter 13:5691–5698

Zener C (1951) Interaction between the d-shells in the transition metals. II. Ferromagnetic compounds of manganese with perovskite structure. Phys Rev 82:403–405

Zhao G, Conder K, Keller H, Muller KA (1996) Giant oxygen isotope shift in the magnetoresistive perovskite La$_{1-x}$Ca$_x$MnO$_{3+y}$. Nature (London) 381:676–678

5

Lantana camara Linn leaf extract mediated green synthesis of gold nanoparticles and study of its catalytic activity

Shib Shankar Dash · Braja Gopal Bag · Poulami Hota

Abstract A facile one-step green synthesis of stable gold nanoparticles (AuNPs) has been described using chloroauric acid ($HAuCl_4$) and the leaf extract of *Lantana camara* Linn (Verbenaceae family) at room temperature. The leaf extract enriched in various types of plant secondary metabolites is highly efficient for the reduction of chloroaurate ions into metallic gold and stabilizes the synthesized AuNPs without any additional stabilizing or capping agents. Detailed characterizations of the synthesized gold nanoparticles were carried out by surface plasmon resonance spectroscopy, transmission electron microscopy, dynamic light scattering, Zeta potential, X-ray diffraction and Fourier transform-infrared spectroscopy studies. The synthesized AuNPs have been utilized as a catalyst for the sodium borohydride reduction of 4-nitrophenol to 4-aminophenol in water at room temperature under mild reaction condition. The kinetics of the reduction reaction has been studied spectrophotometrically.

Keywords *Lantana camara* Linn · Gold nanoparticles · Catalytic reduction · Green synthesis · Phytochemicals

Introduction

Creation of nanoscale objects having at least one of their dimensions in the size range of 1–100 nm and their

S. S. Dash · B. G. Bag (✉) · P. Hota
Department of Chemistry and Chemical Technology, Vidyasagar University, Midnapore 721 102, West Bengal, India
e-mail: braja@mail.vidyasagar.ac.in

utilizations in various facets of science and technology have become an area of tremendous research interest in recent years because the physicochemical properties of the materials at this scale are significantly different compared with their bulk scale (Alkilany et al. 2013; Pan et al. 2013; Titoo et al. 2014). Among various noble metal nanoparticles, gold nanoparticles (AuNPs) having unique optical, electronic and magnetic properties have found applications in pharmacology, biodiagnostics, medicine, drug-delivery, catalysis, etc. (Gong and Mullins 2009; Zhang et al. 2012; Murphy et al. 2008; Laura and Alberto 2014; Thomas and Kamat 2003). Depending upon their size, shape and degree of aggregation, AuNPs exhibit different colors (Weisbecker et al. 1996; Fujiwara et al. 1999; Aslan and Perez-Luna 2002; Mie 1908). Though the colloidal gold particles have been used since fifth to fourth century B.C., the scientific method for the reductive synthesis of colloidal gold can be traced back to 1857 when Michael Faraday reported a reductive synthesis of gold hydrosols from an aqueous solution of chloroaurate using phosphorus dissolved in carbon disulfide (Daniel and Astruc 2004; Faraday 1857). AuNPs can be synthesized using a number of routinely used chemical and physical methods. But, most of these methods employ toxic chemicals and nonpolar solvents during synthesis followed by addition of synthetic additives or capping agents as stabilizers thereby limiting their applications in clinical and biomedical fields. Therefore, there is a growing need for the development of eco-friendly, benign, biocompatible, reliable and synthetic methods to avoid any undesired environmental and health effects (De et al. 2008). The plant extract-based reductive method, involving the reduction of Au(III) to Au(0) by the phytochemicals, has gained profound significance in recent years for the development of a clean, reliable, biocompatible, benign, cost-effective and eco-friendly process

(Jain et al. 2011). As the phytochemicals present in the plant extracts act as stabilizers for the synthesized gold nanoparticles and no additional stabilizers or capping agents are needed, the method is advantageous over other synthetic methods. Syntheses of AuNPs from the extracts of *Acacia nilotica* leaves (Majumdar and Bag 2013), *Punica granatum* juice (Dash and Bag 2014), *Saraca indica* bark (Dash et al. 2014), *Ananas comosus* L. (Basavegowda et al. 2013), *Terminalia arjuna* bark (Majumdar and Bag 2012), *Ocimum sanctum* stem (Paul and Bag 2013), etc. have recently been reported. As newer applications of nanoparticles are emerging rapidly, there is an ever growing need for the development of newer methods for the synthesis of metal nanoparticles utilizing the rich diversity of plant resources as renewables.

Lantana camara Linn (*L. camara*) is a popular ornamental and garden plant growing up to 2–4 meter in height, with a number of flower colors viz. yellow, red, pink and white. It also grows naturally at road or river sides up to elevations of 2,000 meters in tropical and subtropical temperature regions. The leaves of the plant are used in the treatment of tumors, tetanus, rheumatism, malaria, etc., and its antiseptic and carminative properties have also been reported (Raju 2000; Ganjewala et al. 2009; Ghisalberti 2000). During our investigations on the utilization of triterpenoids (C30 s) as renewable functional nanoentities (Bag and Dash 2011; Bag and Majumdar 2012; Bag and Paul 2012; Bag et al. 2012, 2013), it occurred to us that the medicinally important leaf extract of *L. camara*, rich in polyphenolic compounds, can be utilized for the synthesis of AuNPs from HAuCl$_4$ (Mittal et al. 2013). Herein, we report a very mild and environment friendly method for the synthesis of AuNPs from the leaf extract of *L. camara* without any additional capping or stabilizing agents. The AuNPs were characterized by surface plasmon resonance (SPR) spectroscopy, high-resolution transmission electron microscopy (HRTEM), X-ray diffraction, energy dispersive X-ray (EDX) and Fourier transform-infrared spectroscopy (FTIR) studies. *Flower-like assemblies* of AuNPs were observed at higher concentration of the leaf extract. Catalytic application of the synthesized AuNPs has been demonstrated for the sodium borohydride reduction of 4-nitrophenol to 4-aminophenol, and the reduction kinetics have been investigated spectrophotometrically.

Materials and method

Plant materials

The leaves of *L. camara* were collected from the campus of Vidyasagar University, Midnapore, West Bengal, India, identified at the Department of Botany and Forestry of this University, and the specimen was deposited in our laboratory.

Chemicals

All chemicals used in the experiment were analytical reagent grade. Chloroauric acid (HAuCl$_4$·H$_2$O) was purchased from SRL and used without further purification. 4-Nitrophenol and sodium borohydride were purchased from Merck. Double distilled water was used for the experiments.

Synthesis of AuNPs

HAuCl$_4$ (36.5 mg) was dissolved in distilled water (10 mL) to obtain Au(III) (10.7 mM) stock solution. The purified *L. camara* leaf extract (see supporting information, 2 0 mg) was dissolved in 10 mL of distilled water and sonicated for 10 min to afford a yellowish stock solution (2,000 mg L^{-1}). Aliquots of Au(III) solution (0.2 mL and 10.7 mM) were added drop wise to the leaf extract to prepare a series of stabilized AuNPs, where the concentration varied from 100 to 500 mg L^{-1}, keeping the concentration of Au(III) ion fixed at 0.54 mM. UV–visible spectroscopy of the gold colloids was carried out after 7 h of mixing HAuCl$_4$, and the purified leaf extract and a band in the vicinity of 530 nm in the UV–visible spectrum confirmed the formation of AuNPs.

Reduction of 4-nitrophenol to 4-aminophenol

The sodium borohydride reduction of 4-nitrophenol to 4-aminophenol in the presence of AuNPs as catalyst was carried out as follows: an aliquot of 4-nitrophenol (0.2 mL and 1 mM) was treated with freshly prepared sodium borohydride solution (3.6 mL and 16.5 mM) and 0.2 mL freshly prepared colloidal AuNPs (synthesized with 100 mg L^{-1} plant extract) in a 10-mm quartz cuvette. Then, the reaction mixture was shaken thoroughly, and the UV–visible spectrum was recorded at room temperature (23–27 °C). The progress of the reaction was monitored by recording the absorption intensity of 4-nitrophenolate ion with certain time intervals, and the apparent rate constant was calculated. The experiment was repeated with 0.4-mL colloidal AuNPs (synthesized with 100 mg L^{-1} plant extract), and the apparent rate constant was measured.

Characterization

The morphology of the nanoparticles was analyzed from the high-resolution images obtained with a TECNAI G^2 20 transmission electron microscope at an accelerating voltage of 200 kV. The X-ray diffraction (XRD) patterns of the stabilized AuNPs were acquired by PANalytical X'Pert PRO diffractometer with Cu-Kα radiation ($\lambda = 1.542$ Å). Mass spectra of the purified leaf extract were recorded in Shimadzu GCMS QP 2100 Plus instrument. UV–visible spectrum was carried out in Shimadzu 1601 spectrophotometer using 2-mm optical path quartz cuvette. FTIR spectra of the samples (with KBr pellet) were recorded using a Perkin Elmer spectrum 2 instrument where eight scans were accumulated to record the data. Particle size in the bulk and surface charge of gold nanoparticles were studied by dynamic light scattering (DLS) measurement and Zeta potential analysis using Malvern Nano-ZS90 instrument.

Results and discussions

The presence of a wide range of plant secondary metabolites such as triterpenoids, flavanoids as well as β-sitosteryl-3-O-β-D-glucoside and a mixture of campesterol, stigmasterol and β-sitosterols have been reported in the leaf extract of *L. camara*. Mass spectral analysis of the leaf extract carried out by us has revealed the presence of most of the above compounds (supporting information Fig. S1 and S2). Presence of phenolic compounds was obtained by positive ferric chloride test (see supporting information). Hence, it occurred to us that the easily oxidizable phytochemicals present in the leaf extract of *L. camara* can be utilized for the reduction of chloroaurate ions having very high reduction potential to atomic gold. Collision of the gold atoms will lead to the formation of nano-sized particles and get stabilized by the polyphenols, quinones and other coordinating phytochemicals. To test this, we treated aqueous HAuCl$_4$ solutions with increasing concentration of the leaf extract of *L. camara*. Interestingly, we observed the appearance of pinkish red coloration within a 5 min, indicating the formation of AuNPs. The intensity of the color increased with time (Fig. 1).

UV–visible absorption and TEM Studies

On addition of HAuCl$_4$ solution to an aqueous leaf extract of *L. camara*, the color of the mixture changed from light yellow to reddish brown within five minutes, indicating the formation of AuNPs. UV–visible spectroscopy of HAuCl$_4$ solution and the mixtures containing HAuCl$_4$ and varied amounts of the leaf extract were carried out (Fig. 1) to find out the shifts in the absorption maxima. The strong absorption maximum at 222 nm and a shoulder peak at 296 nm appeared in the UV–visible spectrum of HAuCl$_4$ are due to charge transfer interactions between metal and chloro ligands (Fig. 1a). However, in the UV–visible spectrum of stabilized AuNPs, the intensities of these peaks

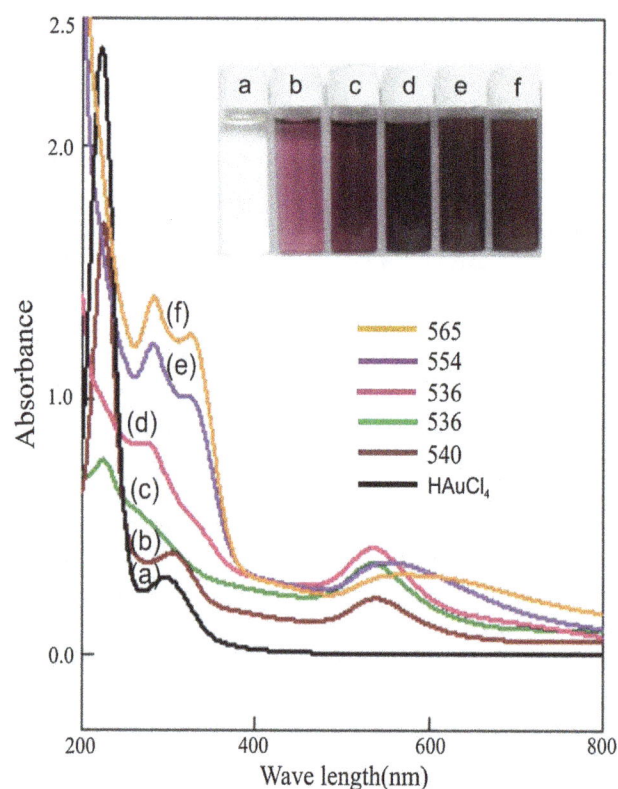

Fig. 1 UV–visible absorption spectra of gold alone and colloidal AuNPs at different concentrations of *L. camara* leaf extract: **a** solution of HAuCl$_4$ (0.54 mM); **b** 100 mg L^{-1}; **c** 200 mg L^{-1}; **d** 300 mg L^{-1}; **e** 400 mg L^{-1} and **f** 500 mg L^{-1}. The wavelengths of the SPR bands of the respective AuNP colloids are shown in color. *Inset* Photograph of vials containing AuNPs of various concentrations

reduced, and a new peak in the region of 540–565 nm due to surface plasmon resonance phenomenon of AuNPs. At lower concentrations of plant extract (100 and 200 mg L^{-1}), the SPR bands appeared at 540 and 536 nm, respectively (Fig. 1b, c). Interestingly, on further increasing the concentration of the leaf extract (from 300 to 500 mg L^{-1}), broadening of SPR band took place with concomitant red shift, suggesting the formation of *assembly* of nanoparticles (Fig. 1d–f) (Ghosh and Pal 2007).

High-resolution transmission electron microscopy was carried out to investigate the size, shape and morphology of the synthesized AuNPs. Mostly spherical-shaped AuNPs were formed at 300 and 500 mg L^{-1} concentration of the leaf extract having average size of 6–7 nm (Fig. 2a–f). Poly-shaped AuNPs were obtained at 100 mg L^{-1} concentration of the leaf extract (Fig. 2g–i). Interestingly, at higher concentration of the leaf extract (500 mg L^{-1}), *flower-like assembly of nanoparticles* of nearly 100 nm diameter embedded in organic matrix was observed along with spherical-shaped particles (Fig. 2c). This result is in agreement with the pattern of SPR band obtained at this concentration (Fig. 1f). The fringe spacings measured to be 0.21 nm from the HRTEM

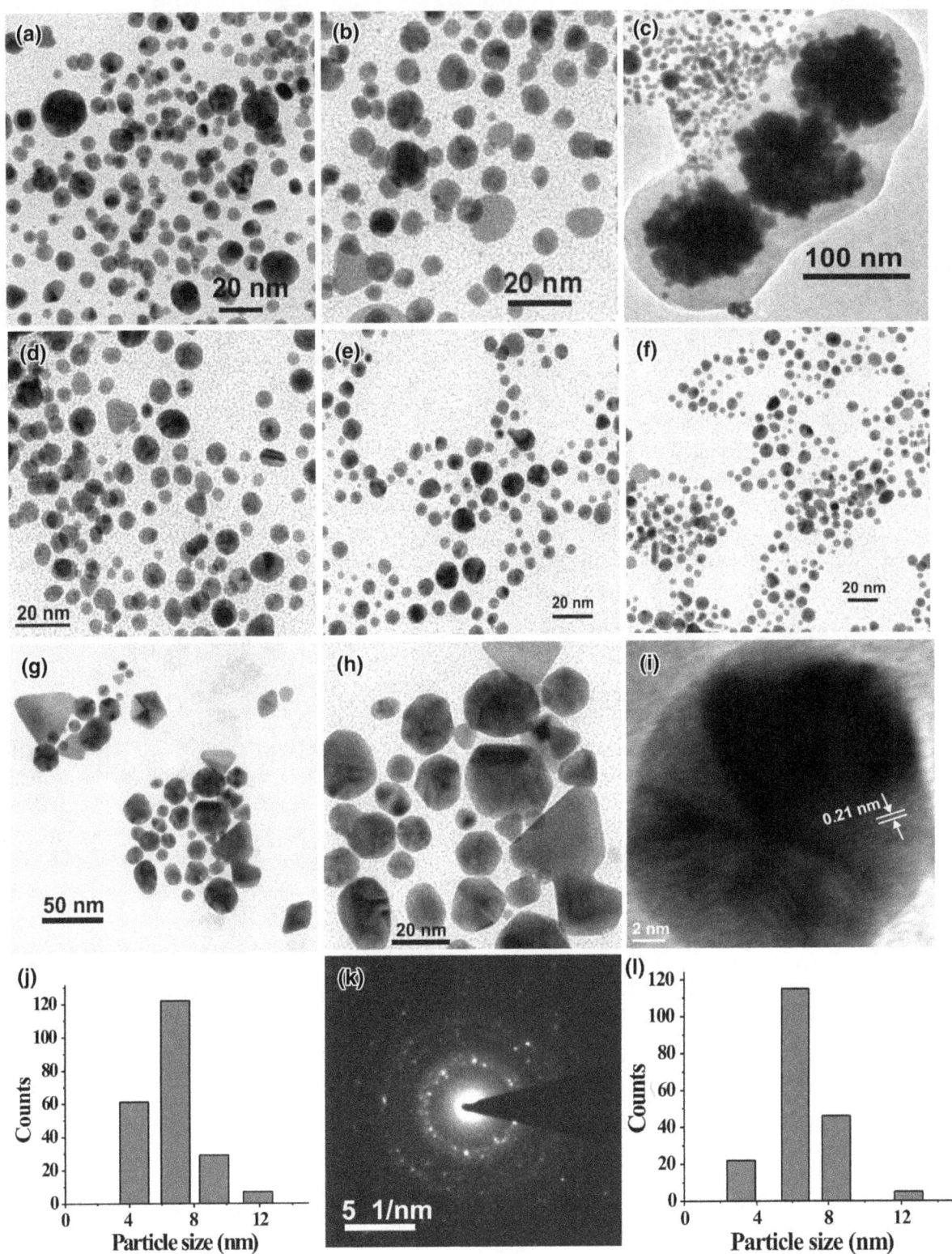

Fig. 2 HRTEM images of colloidal Au nanoparticles at various concentration of *L. camara* leaf extract: **a–c** at 500 mg L^{-1}; **d–f** 300 mg L^{-1}; **g–i** 100 mg L^{-1}; **j, l** histograms at 500 and 300 mg L^{-1} concentration and (k) SAED pattern of AuNP

images (Fig. 2i) were in agreement with the spacing between two (111) planes of crystalline face-centered cubic (fcc) gold (0.235 nm) (Kanan and John 2008). The selected area electron diffraction (SAED) pattern revealed that the nanoparticles were crystalline in nature (Fig. 2k). The nanoparticles were so stable that no further aggregation of the AuNPs took place on

keeping the colloids for several months at 23–27 °C. Elemental analysis of the AuNPs carried out using EDX analysis
(supporting information Fig. S3) suggested that the nanoparticles were composed of pure crystalline gold. A strong band
of carbon indicated the presence of biomolecules on the surface of AuNPs that might have played a role in the reduction of
Au(III) and stabilization of AuNPs. This is in conformity with
the HRTEM images (Fig. 2c).

Dynamic light scattering (DLS) and Zeta potential measurement studies

Dynamic light scattering studies of the colloidal AuNPs
were carried out to investigate the average size of the
particles in the bulk. The hydrodynamic diameters of the
particles increased consistently as the concentration of leaf
extract increased from 100 to 500 mg L^{-1} (supporting
information Fig. S4 and S5). At each concentration of the
leaf extract, the particle size obtained by DLS study was
larger than the particle size obtained by HRTEM study.
This is due to the ligands attached to the surface of AuNPs
(Gangula et al. 2011; Chandran et al. 2012). The surface
charges (Zeta potential) of the AuNPs synthesized at 100
and 200 mg L^{-1} concentration of the leaf extract were
−27.8 and −37.4 mV, respectively (see supporting information Fig. S6). High negative value of Zeta potential
indicated high electrical charge on the surface of the
AuNPs. This can cause strong repulsive forces among the
AuNPs and prevent their aggregation, resulting in higher
stability of colloidal AuNPs.

X-ray diffraction (XRD) and FTIR studies

The XRD pattern of the synthesized AuNPs is shown in
Fig. 3. Five sharp diffraction peaks at 38.3°, 44.5°, 64.8°,
77.7° and 81.6° can be indexed as (111), (200), (220), (311)
and (222) Bragg reflections, respectively, confirming the
fcc nature of colloidal AuNPs (JCPDS file no. 04-0784).
The intensity of the peak due to (111) plane at 38.3° was
much larger than the other peaks, indicating predominant
orientation of the (111) plane.

Fourier transform-infrared spectroscopy studies of the
leaf extract and the stabilized AuNPs were carried out to
investigate the various functional groups involved in the
reduction and capping of gold nanoparticles (Fig. 4). The
appearance of a broad peak approximately around
3,314 cm^{-1} region in the FTIR spectrum of *L. camara* leaf
extract indicated the presence of –OH/N–H group containing compounds (Fig. 4a). The broadening of the peak
was due to strong intermolecular H-bonding between these
groups. The peak at 2,932.4 cm^{-1} is due to the stretching
vibration of aliphatic –CH– group containing compounds
present in the leaf extract. The molecules present in the leaf

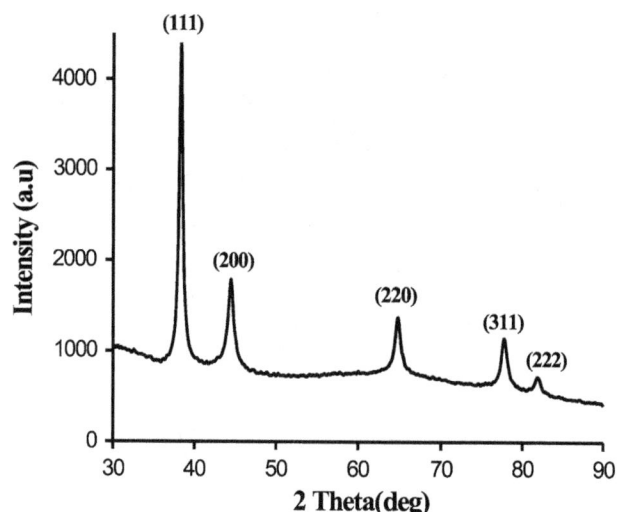

Fig. 3 X-ray diffraction pattern of stabilized AuNPs

extract containing –C=O group were evident from the peak
at 1,704 cm^{-1} region. The presence of aromatic rings
containing compounds in the leaf extract was confirmed
from the peaks at 1,611 and 1,522 cm^{-1} arising due to the
stretching vibration of C=C bond. The peaks at 1,270 and
1,374 cm^{-1} can be assigned to stretching vibrations of C–
N bond. In case of colloidal AuNPs, the peaks due to –OH/
N–H groups became slightly narrower, suggesting the
interaction between these groups and the AuNPs (Fig. 4b).
The peak due to carbonyl group became significantly
weaker in the FTIR spectrum of stabilized AuNPs, indicating the interaction of carbonyl groups with the AuNPs.

Reaction mechanism

The leaf extract of *L. camara* is rich in different types of
plant secondary metabolites such as terpenoids, steroids
and polyphenols including flavanoids (Ghisalberti 2000).
Mass spectral analysis carried out by us indicated the
presence of different types of polyphenols such as quer-
cetin, gallic acid, caffeic acid and chlorogenic acid. (supporting information Fig. S1). A positive ferric chloride test
(see supporting information) supported the presence of the
phenolic compounds in the leaf extract. A schematic representation for the formation of AuNPs is given in Fig. 5.
Phenolic compounds and other easily oxidizable phytochemicals can reduce Au(III) to Au (0). The electron-rich
o-dihydroxy compounds present in the leaf extract can
easily form a five-membered chelate ring with the Au(III)
ions. In this case, the redox reaction can take place in the
chelated complex where the o-dihydroxy compounds can
be oxidized to corresponding quinones with concomitant
reduction of Au(III) ions to Au (0). Collision of the
neighboring Au (0) atoms with each other may lead to the

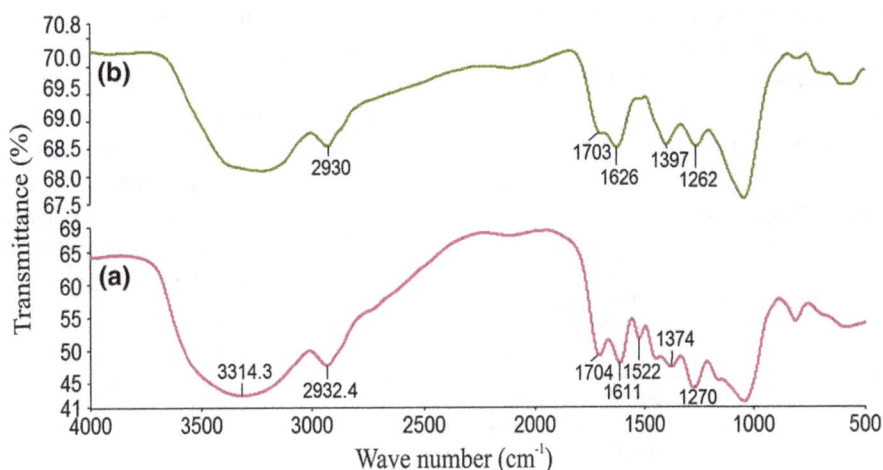

Fig. 4 FTIR spectrum of **a** *L. camara* leaf extract and **b** stable AuNPs

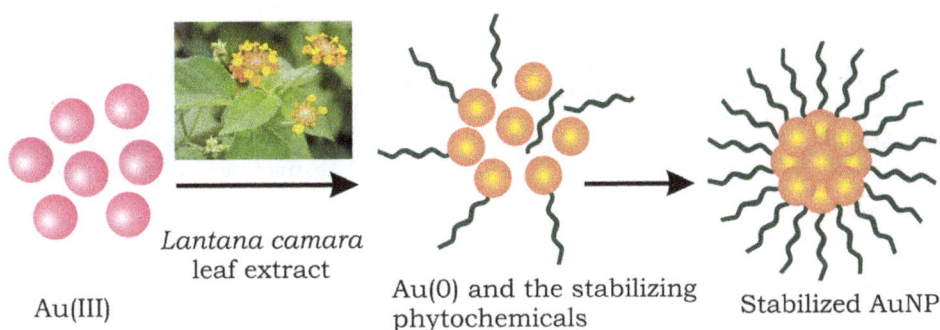

Fig. 5 Mechanism of the formation and stabilization of AuNPs by the phytochemicals present in the leaf extract of *L. camara*

formation of the nano-sized gold particles. The AuNPs thus formed can be stabilized by the quinones, phenolic compounds as well as other coordinating phytochemicals present in the leaf extract. Further aggregation of the AuNPs is prevented by the stabilizing ligands surrounding the AuNPs.

Application of AuNPs in catalytic reduction

Nanoscale materials have drawn considerable attention in recent years because of their unique application as catalyst for some chemical transformations, which are normally restricted due to the large kinetic barrier of the reaction (Kim et al. 2009). Whereas bulk gold at the macro scale is inactive as a catalyst, the AuNPs are useful as a catalyst for a number chemical transformation (Pan et al. 2013; Gong and Mullins 2009). The electrochemical potential (E_0) value suggests that the reduction of 4-nitrophenol to 4-aminophenol in the presence of sodium borohydride is a thermodynamically favorable reaction (E_0 for 4-nitrophenol/4-aminophenol −0.76 and for H_3BO_3/BH_4^- -1.33 V). However, on treatment of an aqueous solution of 4-nitrophenol (0.05 mM) with a freshly prepared aqueous solution of sodium borohydride (16.5 mM), the absorption maxima at 319 nm shifted to 401.5 nm due to the formation of more

stable 4-nitrophenolate ion in alkaline medium (Fig. 6a). In the absence of AuNPs or in the presence of the leaf extract alone, no reduction of the nitro group to amino group took place on standing the reaction mixture for several days due to a large kinetic barrier for the reduction reaction. Interestingly, on addition of *L. camara* leaf extract stabilized AuNPs, the decolorization of yellowish color (of 4-nitrophenolate solution) was observed. The progress of the reduction reaction was monitored by UV–visible spectroscopy at various time intervals. After 30 s, the intensity of the peak at 401.5 nm reduced, and concomitantly a new peak appeared around 300 nm, suggesting the conversion of 4-nitrophenol to 4-aminophenol. The peak due to 4-nitrophenolate ion completely disappeared after 20 min, indicating the completion of reduction reaction. The absorption intensity (A) of 4-nitrophenolate ion at various intervals allowed us to calculate the apparent rate constant (k_{app}) for the reduction reaction. As the concentration of 4-nitrophenol largely exceeded (300 times) the concentration of sodium borohydride, the reaction was assumed to be pseudo-first order, and a good linear correlation was obtained between ln A vs time (t) plot. From the slope of the linear plot, the apparent rate constant (k_{app}) was measured to be 0.26 and 0.29 min^{-1} using 0.2 and 0.4 mL stabilized colloidal AuNPs, respectively (Table 1, supporting information TS1, TS2 and Fig. S7 and S8). The small increase in the apparent catalytic rate constant (k_{app}) with higher volume of colloidal AuNPs (keeping other parameters constant) was perhaps due to increase in the number of reaction centers (Gangula et al. 2011). These rate constant values were comparable to the recently reported values on related systems (Gangula et al. 2011; Dash et al. 2014) (Fig. 6).

Table 1 Catalytic reduction of 4-nitrophenol (4-NP) at 25 °C using different amounts of colloidal AuNPs (synthesized with 100 mg L^{-1} of leaf extract)

Entry	Conc. of 4-NP (mM)	Conc. of NaBH$_4$ (mM)	Colloidal AuNPs used (mL)	Time for completion of reaction (min)	Apparent Rate constant (k_{app}/min^{-1})
1	0.05	16.5	0.2	7.5	0.26
2	0.05	16.5	0.4	6	0.29
3	0.05	0	0.2	No reduction	NA
4	0.05	16.5	0.0	No reduction	NA

NA Not applicable

Conclusions

One-step synthesis of colloidal gold nanoparticles has been reported by utilizing the leaf extract of *L. camara* under very mild reaction condition. According to our knowledge, this is the first report for the synthesis of AuNPs utilizing the leaf extract of *L. camara*. The method described here is very simple, cost-effective and nontoxic in nature. Various phytochemicals present in the leaf extract are highly efficient to reduce Au(III) to Au (0) and simultaneously stabilize the synthesized AuNPs. The synthesized AuNPs were of 6–7 nm in size and mostly spherical shaped as evident from HRTEM studies. *Flower-like assemblies* of AuNPs of approximately 100 nm diameter were observed at a higher concentration of the leaf extract along with discrete AuNPs. The synthesized AuNPs were utilized for the sodium borohydride reduction of 4-nitrophenol to 4-aminophenol at room temperature, and the apparent catalytic rate constant for the reduction reaction was measured spectrophotometrically. As *L. camara* leaf extract has various medicinal applications, the results

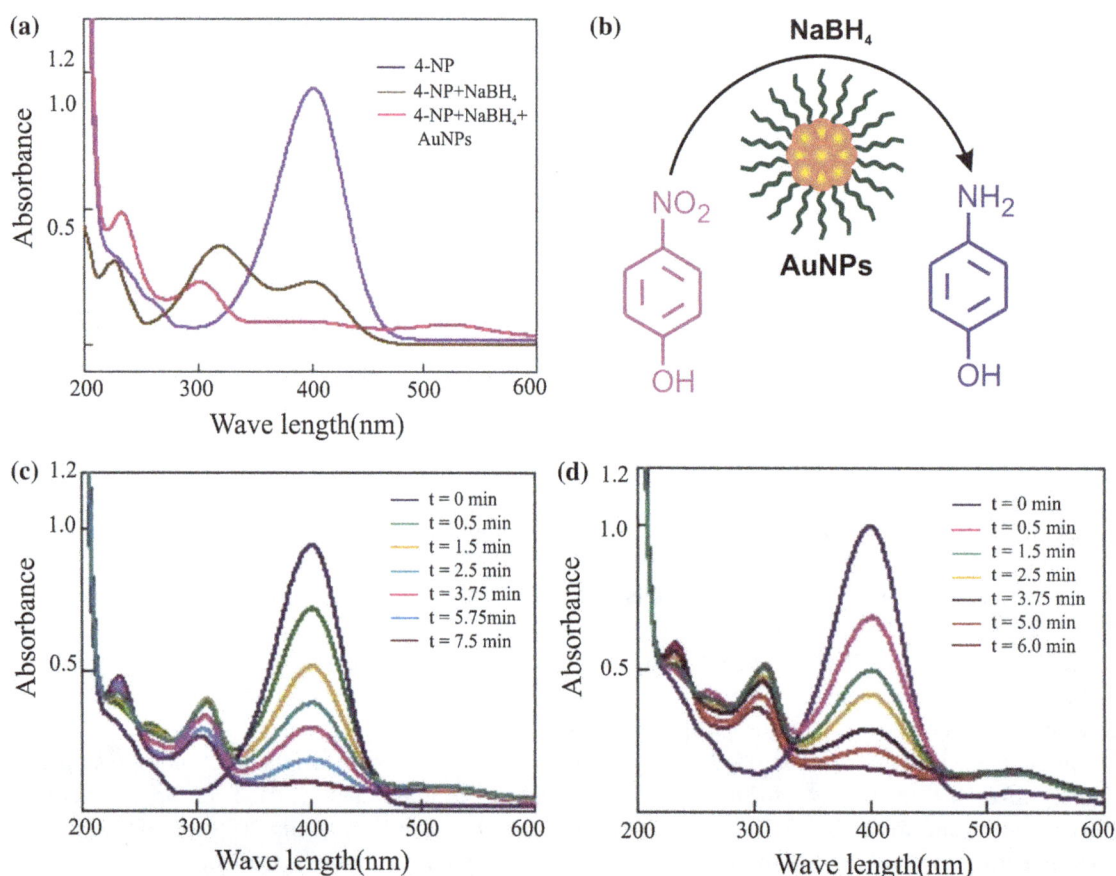

Fig. 6 **a** UV–visible spectrum of 4-nitrophenol, 4-nitrophenolate ion and 4-amino phenol; **b** a schematic representation of catalytic reduction of 4-nitrophenolate ion to 4-aminophenolate ion; **c** and **d** overlay of UV–visible spectrum at various time intervals using 0.2 and 0.4 mL stabilized AuNPs, respectively, during catalytic reduction

described here open up its use in biomedical applications as well as nanoscience and nanotechnology.

Acknowledgments BGB thanks CSIR, New Delhi for a research Grant. SSD thanks CSIR, New Delhi for a research fellowship.

References

Alkilany AM, Lohse SE, Murphy CJ (2013) The gold standard: gold nanoparticle libraries to understand Hydrogenation Catalyst. Acc Chem Res 46:650–661

Aslan K, Perez-Luna VH (2002) Surface modification of colloidal gold by chemisorption of alkanethiols in the presence of a nonionic surfactant. Langmuir 18:6059–6065

Bag BG, Dash SS (2011) First self-assembly study of betulinic acid, a renewable nano-sized, 6-6-6-6-5 pentacyclic monohydroxy triterpenic acid. Nanoscale 3:4564–4566

Bag BG, Majumdar R (2012) Self-assembly of a renewable nano-sized triterpenoid 18β-glycyrrhetinic acid. RSC Adv 2:8623–8626

Bag BG, Paul K (2012) Vesicular and fibrillar gels by self-assembly of nanosized oleanolic acid. Asian J Org Chem 1:150–154

Bag BG, Garai C, Majumdar R, Laguerre M (2012) Natural triterpenoids as renewable nanos. Struct Chem 23:393–398

Bag BG, Majumdar R, Dinda SK, Dey PP, Maity GC, Mallia AV, Weiss RG (2013) Self-assembly of ketals of arjunolic acid into vesicles and fibers yielding gel-like dispersions. Langmuir 29:1766–1778

Basavegowda N, Sobczak-Kupiec A, Malina D, Yathirajan HS, Keerthi VR, Chandrashekar N, Dinkar S, Liny P (2013) Plant mediated synthesis of gold nanoparticles using fruit extracts of Ananas comosus L. (Pineapple) and evaluation of biological activities. Adv Mater Lett 4:332–337

Chandran PR, Naseer M, Udupa N, Sandhyarani N (2012) Size controlled synthesis of biocompatible gold nanoparticles and their activity in the oxidation of NADH. Nanotechnology 23:15602–15609

Daniel MC, Astruc D (2004) Gold nanoparticles: assembly, supra-molecular chemistry, quantum-size-related properties, and applications toward biology, catalysis, and nanotechnology. Chem Rev 104:293–346

Dash SS, Bag BG (2014) Synthesis of gold nanoparticles using renewable Punica granatum juice and study of its catalytic activity. Appl Nanosci 4:55–59

Dash SS, Majumdar R, Sikder AK, Bag BG, Patra BK (2014) Saraca indica bark extract mediated green synthesis of polyshaped gold nanoparticles and its application in catalytic reduction. Appl Nanosci 4:485–490

De M, Ghosh PS, Rotello VM (2008) Applications of nanoparticles in biology. Adv Mater 20:4225–4241

Faraday M (1857) Experimental relations of gold (and other metals) to light. Philos Trans R Soc Lond 147:145–181

Fujiwara H, Yanagida S, Kamat PV (1999) Visible laser induced fusion and fragmentation of thionicotinamide-capped gold nanoparticles. J Phys Chem B 103:2589–2591

Gangula A, Podila R, Ramakrishna M, Karanam L, Janardhana C, Rao AM (2011) Catalytic reduction of 4-nitrophenol using biogenic gold and silver nanoparticles derived from Breynia rhamnoides. Langmuir 27:15268–15274

Ganjewala D, Sam S, Khan KH (2009) Biochemical compositions and antibacterial activities of Lantana camara plants with yellow, lavender, red and white flowers. EurAsian J BioSci 3:69–77

Ghisalberti EL (2000) Lantana camara L. (Verbenaceae). Fitoterapia 71:467–486

Ghosh SK, Pal T (2007) Interparticle coupling effect on the surface plasmon resonance of gold nanoparticles: from theory to applications. Chem Rev 107:4797–4862

Gong J, Mullins CB (2009) Surface science investigations of oxidative chemistry on gold. Acc Chem Res 42:1063–1073

Jain N, Bhargava A, Majumdar S, Panwar J (2011) Extracellular biosynthesis and characterization of silver nanoparticles using Aspergillus flavus NJP08: a mechanism prospective. Nanoscale 3:635–641

Kanan P, John SA (2008) Synthesis of mercaptothiadiazole-functionalized gold nanoparticles and their self-assembly on Au substrates. Nanotechnology 19:85602–85611

Kim S, Sang WB, Lee JS, Park J (2009) Recyclable gold nanoparticle catalyst for the aerobic alcohol oxidation and C–C bond forming reaction between primary alcohols and ketones under ambient conditions. Tetrahedron 65:1461–1466

Laura P, Alberto V (2014) Gold colloids: from quasi-homogeneous to heterogeneous catalytic systems. Acc Chem Res. 47:855–863

Majumdar R, Bag BG (2012) Terminalia arjuna bark extract mediated size controlled synthesis of polyshaped gold nanoparticles and its application in catalysis. Int J Res Chem Environ 2:338–344

Majumdar R, Bag BG (2013) Acacia nilotica (Babool) leaf extract mediated size controlled rapid synthesis of gold nanoparticles and study of its catalytic activity. Int Nano Lett.

Mie G (1908) Contributions to the optics of cloudy media, particularly of colloidal metal solutions. Ann Phys 330:377–445

Mittal AK, Chisti Y, Banerjee UC (2013) Synthesis of metallic nanoparticles using plant extracts. Biotechnol Adv 31:346–356

Murphy CJ, Gole AM, Stone JW, Sisco PN, Alkilany AM, Goldsmith EC, Baxter SC (2008) Gold nanoparticles in biology: beyond toxicity to cellular imaging. Acc Chem Res 41:1721–1730

Pan M, Gong J, Dong G, Mullins CB (2013) Model studies with gold: a versatile oxidation and the nano bio interface. Acc Chem Res 46:650–661

Paul K, Bag BG (2013) Ocimum sanctum (Tulasi) stem extract mediated size controlled green synthesis of polyshaped gold nanoparticles and its application in catalysis. Int J Res Chem Environ 3:128–135

Raju A (2000) Wild plants of Indian sub-continent and their economic uses. CBS Pub. & Distribution, New Delhi 65

Thomas KG, Kamat PV (2003) Chromophore-functionalized gold nanoparticles. Acc Chem Res 36:888–898

Titoo J, Qingxin T, Thomas B, Kasper N (2014) Wet chemical synthesis of soluble gold nanogaps. Acc Chem Res.

Weisbecker CS, Merritt MV, Whitesides GM (1996) Molecular self-assembly of aliphatic thiols on gold colloids. Langmuir 12:3763–3772

Zhang Y, Cui X, Shi F, Deng Y (2012) Nano-gold catalysis in fine chemical synthesis. Chem Rev 112:2467–2505

6

Green synthesis, characterization and catalytic activity of palladium nanoparticles by xanthan gum

Amrutham Santoshi kumari · Maragoni Venkatesham ·
Dasari Ayodhya · Guttena Veerabhadram

Abstract Here, we report the synthesis, characterization and catalytic evaluation of palladium nanoparticles (PdNPs) using xanthan gum, acting as both reducing and stabilizing agent without using any synthetic reagent. The uniqueness of our method lies in its fast synthesis rates using hydrothermal method in autoclave at a pressure of 15 psi and at 120 °C temperature by 10 min time. The formation and size of the PdNPs were characterized by UV–visible spectroscopy, X-ray diffraction, Fourier transform infrared spectroscopy and transmission electron microscopy. The catalytic activity of PdNPs was evaluated on the reduction of 4-nitrophenol to 4-aminophenol by sodium borohydride using spectrophotometry.

Keywords Autoclave · Catalytic activity · 4-Nitrophenol · Palladium nanoparticle · Xanthan gum

Introduction

In the last decade, the importance of understanding of science at the nanometre scale has attracted the interest of number of groups all over the world and has led to the emergence of a new interdisciplinary field called nanoscience, because of the size-dependent optical, electronic, and catalytic properties of nanoparticles (Prashant et al. 2006), and even in biological and medical science applications (Salata 2004). There are enormous changes in physical and chemical properties, both quantitatively and qualitatively, from those of bulk materials as these materials are derived by manipulation at the atomic or molecular levels and because of surface to volume ratio and quantum size effect (Roucoux et al. 2002; Rosi and Mirkin 2005; Yang et al. 2006). Palladium nanoparticles are of great interest owing to their application both in heterogeneous and homogeneous catalyses, their high surface-to-volume ratio and high surface energy (Narayanan and El-Sayed 2005). In addition, Pd is exploited as a catalyst in various coupling reactions like Heck coupling (Karimi and Enders 2006), Suzuki coupling (Klingensmith and Leadbeater 2003), and hydrogenation of allyl alcohols (Wilson et al. 2006). Till now, production of PdNPs involved different reducing chemicals such as $NaBH_4$ (Jana et al. 2000; Domenech et al. 2011), N_2H_4 (Yonezawa et al. 2001; Szilvia et al. 2007), ascorbic acid (Sun et al. 2007) and PEG (Luo et al. 2005; El-Houta et al. 2012). Over the past decade there has been an increased interest in the green chemistry (Roucoux et al. 2002; Raveendran et al. 2003). In the synthesis of metal nanoparticles by reduction of the corresponding metal ion salt solutions, there are three areas of opportunity to engage in green chemistry: (1) choice of solvent, (2) the reducing agent, and (3) the capping agent. There has also been increasing interest in identifying environmental friendly materials that are multifunctional (Nadagouda and Rajender 2008). Previously PdNPs were prepared through green methods using annona squamosa L peel extract (Roopana et al. 2012), banana peel extract (Ahok et al. 2010), cinnamom zeylanicum bark (Sathishkumar et al. 2009), broth of cinnamom camphora leaf (Yang et al. 2010) and gum acacia (Keerthi et al. 2011).

Xanthan gum (XG) is an extracellular polysaccharide secreted by the fermentation of the bacterium *Xanthomonas campestris* (Barrére et al. 1986). It is composed of pentasaccharide repeat units, comprising glucose,

A. Santoshi kumari · M. Venkatesham · D. Ayodhya ·
G. Veerabhadram (✉)
Department of Chemistry, University College of Science,
Osmania University, Hyderabad 500007, India
e-mail: gvbhadram@gmail.com

mannose, and glucuronic acid in the molar ratio 2.0:2.0:1.0 (Garcia-Ochoa et al. 2000). XG is used as a stabilizer, thickener and emulsifier. Due to its nontoxic and biocompatible properties, XG is widely used in food and pharmaceutical industries. In this paper, we report green synthesis of PdNPs using XG as both reducing and stabilizing agent without using any toxic chemicals. This reaction is carried out in an autoclave at a pressure of 15 psi and at 120 °C temperature for 10 min. This method is more efficient and rapid synthesis process compared to previously reported green methods mentioned above (Roopana et al. 2012; Sathishkumar et al. 2009; Yang et al. 2010) for the synthesis of Pd nanoparticles.

Nitro phenols are important chemical materials which are widely used to manufacture explosives, drugs, insecticides and dyes, and also used as corrosion inhibitors of woods and rubber chemicals (Zhang et al. 2011). In addition, 4-nitrophenol (4-NP) is an intermediate in the synthesis of many organic compounds. Since 4-amino phenol (4-AP) is a potent industrial intermediate in the manufacturing of many analgesic (Sandip et al. 2010). Thus, being a common precursor material for 4-AP, a newer and cheaper method for catalytic hydrogenation of 4-NP is always in demand. The catalytic activities of PdNPs are tested on the reduction reaction of 4-NP by $NaBH_4$. The reaction was studied by spectrophotometric methods. In the present study, we report the catalytic activity of green synthesized PdNPs towards 4-NP reduction.

Experimental

Materials

Palladium (II) chloride, HCl, 4-NP, xanthan gum and sodium borohydride were purchased from S D Fine-chem Limited, Mumbai, India.

Preparation of PdNPs

Solutions were prepared with double distilled water. 0.0356 g of $PdCl_2$ was accurately weighed and dissolved in 100 ml of HCl (0.000413 M) to form an H_2PdCl_4 aqueous solution. An aliquot of 5 ml of the aqueous H_2PdCl_4 solution was mixed with 5 ml of a 0.2 % aqueous solution of XG in a boiling tube. This reaction is carried out in an autoclave at a pressure of 15 psi and at 120 °C temperature by 10 min time.

Catalytic activity

To monitor the homogeneous catalysis of 4-NP reduction by PdNPs, UNICAM UV–3600 UV–visible Spectrophotometer (UV–Vis) (Thermo Spectronic) was used. To a 3-ml cuvette containing freshly prepared sodium borohydrate (1 ml, 15 mM) solution, 2 ml of 4-NP (0.2 mM) solution was added. After adding PdNPs stabilized in XG (10 μl) solution, cuvette was shaken vigorously for mixing and kept in UV–Vis spectrophotometer to examine the reaction.

Characterization

To study the formation of palladium nanoparticles, the UV–Vis absorption spectra of the prepared solutions were recorded using an UNICAM UV-3600 spectrophotometer (Thermo Spectronic). Fourier transform infrared (FTIR) spectra of PdNPs stabilized in XG and XG alone were recorded in KBr pellets using FTIR spectrophotometer (Bruker optics, Germany) and the scan was performed in the range of 400–4,500 cm^{-1}. X-ray diffraction (XRD) measurement of PdNPs stabilized in XG was carried out on X'pert Pro X-ray diffractometer (Panalytical B.V., Netherlands) operating at 40 kV and a current of 30 mA at a scan rate of 0.388 min^{-1}. The size and morphology of the nanoparticles were determined by TEM using TechnaiG2 under 200 kv.

Results and discussion

UV–visible analysis

Present studies focus on the synthesis of PdNPs using XG as reducing as well as stabilizing agent without using any toxic chemicals, which is purely green approach. The reaction was carried out in an autoclave at a pressure of 15 psi and at 120 °C temperature. The yellow colour reaction mixture was converted to the characteristic black colour after the autoclaving. The appearance of black colour indicates the formation of PdNPs nanoparticles (Keerthi et al. 2011). Figure 1 shows the absorption peak at 412 nm, corresponding to mixture of H_2PdCl_4 and XG. After autoclaving formation of PdNPs the peak at 412 nm disappears confirming the reduction of Pd (II) ions to Pd (0).

XRD analysis

The crystalline nature of resulting PdNPs synthesized can be seen in Fig. 2, in which all the peaks are clearly distinguishable. The broad peak at 39.73 is characteristics peak of the (111) indices of Pd (0) which is a face-centred cubic structure. Three other peaks at 2θ values of 46.09, 67.82, 81.52, 86.87 were observed corresponding to the major reflections of the (200), (220), (311), (222) crystal planes, respectively (Ramesh et al. 2012). Broadening of the diffraction peaks was observed

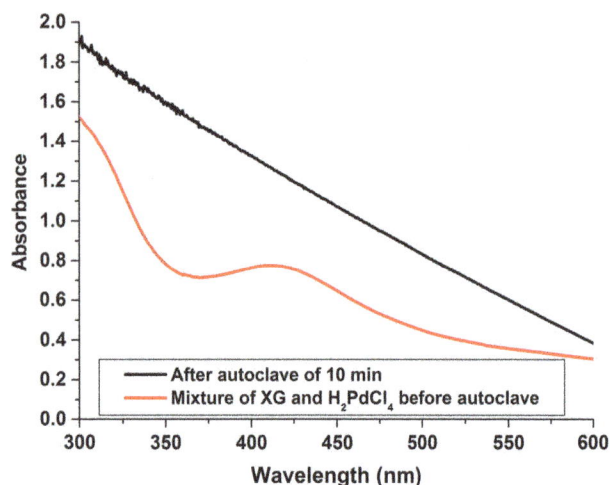

Fig. 1 UV–visible spectra of reaction mixture of H$_2$PdCl$_4$ and xanthan gum solution before and after autoclaving

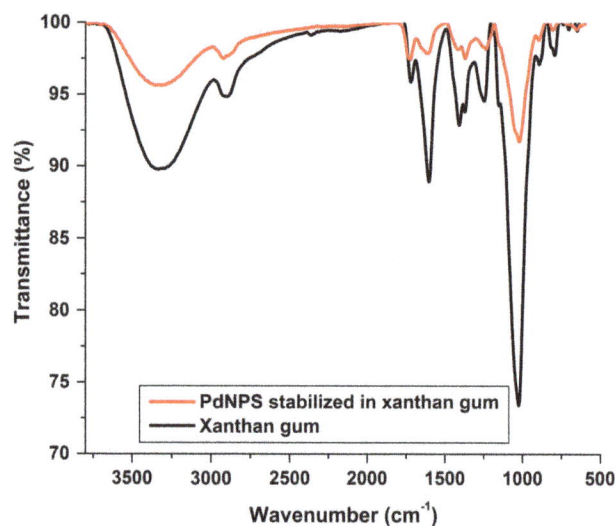

Fig. 2 Typical XRD pattern of the palladium nanoparticles stabilized in xanthan gum

owing to the effect of the nano-sized particles. The crystallite size of palladium nanoparticles is 10 nm which was calculated using peak broadening profile of (111) peak at 40° using Sherrer's formula given below

$$d = \frac{0.94\,(\lambda)}{\beta \cos \theta}$$

where λ is wavelength (1.5418 Å) and β is full-width half-maximum (FWHM) of corresponding peak. The calculated crystallite size of the synthesized palladium nanoparticles is 10 nm.

FTIR spectra analysis

FTIR analysis was used to identify the role of XG for the reduction and the capping of NPs surfaces. Figure 3 shows

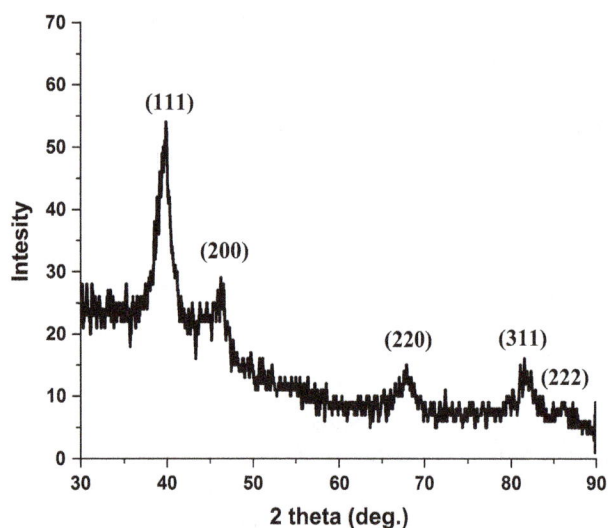

Fig. 3 FTIR spectra of palladium nanoparticles stabilized in xanthan gum and pure xanthan gum

FTIR spectrum of XG and PdNPs stabilized in XG. The major absorbance bands present in the spectrum of XG were at 3,334, 2,906, 1,719 and 1,603 cm^{-1}. The broad bands observed at 3,334 and 2,906 cm^{-1} could be assigned to the stretching vibrations of O–H groups and –CH2,–CH3 aliphatic groups in XG. The bands found at 1,719 and 1,603 cm^{-1} could be due to the characteristic asymmetrical stretch of carboxylate group and carbonyl group. The peak at 1,024 cm^{-1} is due to the C–O stretching vibration of alcoholic groups. The band at 3,334 cm^{-1} shifted to 3,320 cm^{-1} in the presence of PdNPs. These observations clearly show the interaction of Pd with the OH group of XG. The interactions among the resultant Pd nanoparticles and oxygen atoms of O–H, –COO^{-} and –CO become stronger. This can lead to corresponding changes both in the positions and in the strengths of FTIR spectra of XG. The variations in the shape and peak positions of the –OH stretching vibration, –COO^{-} group, –CO group and –OH bending vibration at 3,334, 1,719, 1,603 and 1,024 cm^{-1}, respectively, are observed, because of the contribution of XG towards the reduction and stabilization process (Venkatesham et al. 2014).

TEM analysis

The size distribution, shape and morphology of the PdNPs stabilized in XG were studied by high-resolution transmission electron microscopy. The TEM image of PdNPs stabilized in XG is shown in Fig. 4a. The TEM image shows that the PdNPs are spherical and are well distributed in the gum polymer matrix. To obtain size distributions of PdNPs, approximately 38 particles were counted and then converted into histograms. Figure 4b presents a histogram

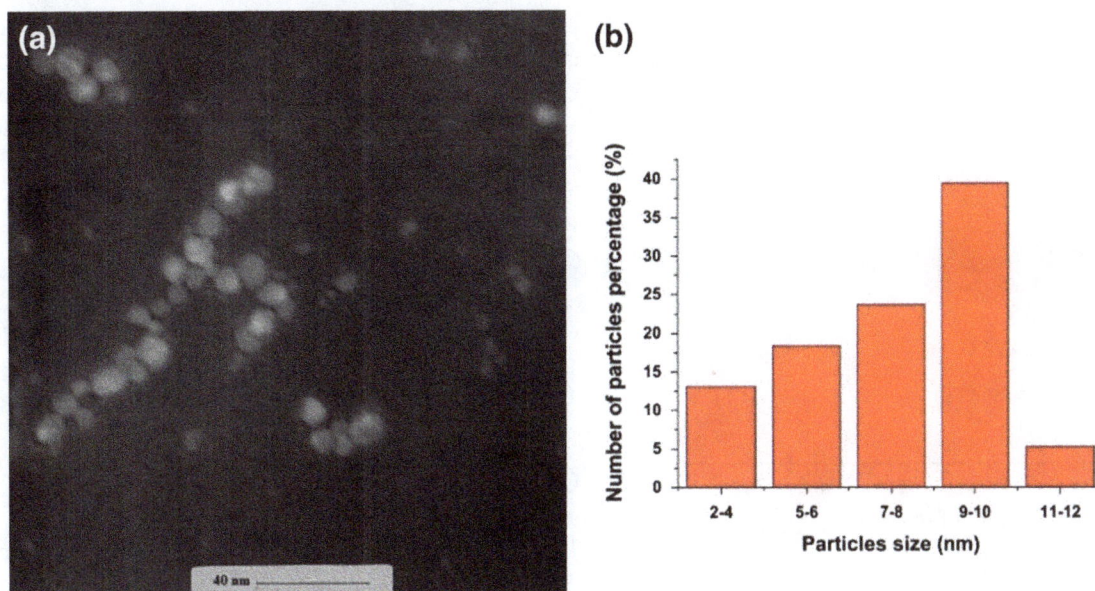

Fig. 4 a Typical TEM image of PdNPs in aqueous system using xanthan gum as reducing and stabilizing agent. b Histogram showing the size distribution of PdNPs

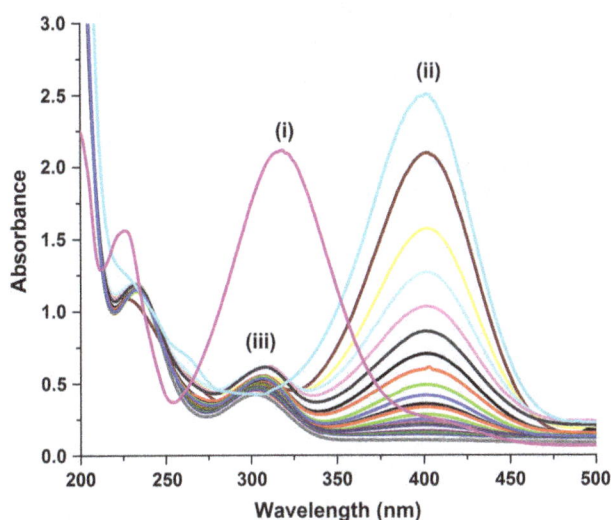

Fig. 5 UV–visible spectra of *i* 4-nitrophenol, *ii* successive reduction of nitrophenolate ion with a time interval of 1 min, and *iii* 4-aminophenol

Fig. 6 The conversion percentage of 4-nitrophenol to 4-aminophenol with time

of the particle size distribution of PdNPs. Most of the particles were in the size around 10 nm.

Catalytic reduction of 4-nitro phenol

As Pd is recognized to be an excellent catalyst for hydrogenation reactions, we tried the PdNPs stabilized in XG for the reduction of 4-NP to 4-AP using borohydride. In a typical catalytic reaction, freshly prepared NaBH$_4$ (1 ml, 15 mM) solution, taken in a 3-ml cuvette, was added to the aqueous 4-NP solution (2 ml, 0.2 mM). The 4-NP shows

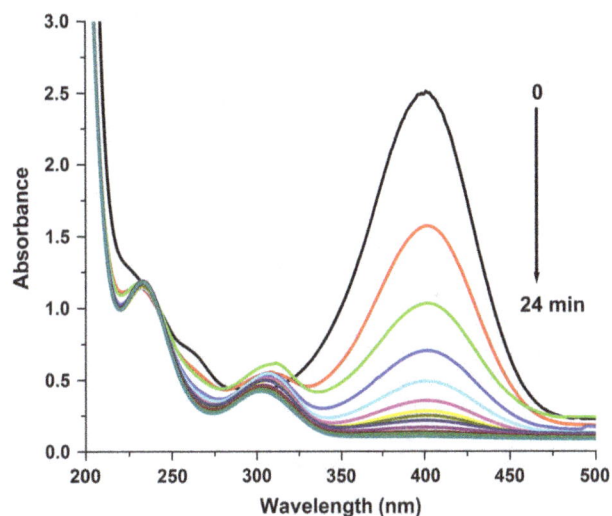

an absorbance peak at 317 nm, which shifts to 400 nm (red shift) in the presence of NaBH$_4$ due to the formation of 4-nitrophenolate ion (Fig. 5). In the absence of catalyst, the 4-nitrophenolate ions cannot be reduced further, even in the presence of strong reducing agent like NaBH$_4$ (Venkatesham et al. 2014). However, immediately after adding 10 µl of PdNPs to the reaction mixture, the intensity of the absorbance band at 400 nm successively decreased with increasing reaction time. At the same time a new band with increased absorbance intensity appears at 300 nm, which is known to be due to absorption of 4-AP.

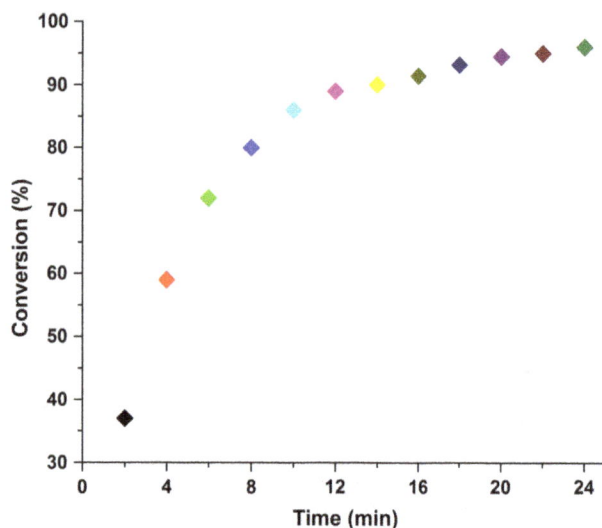

Fig. 7 Reduction of UV–Vis spectra of 4-nitrophenol using PdNPs stabilized in xanthan gum with time of interval 2 min

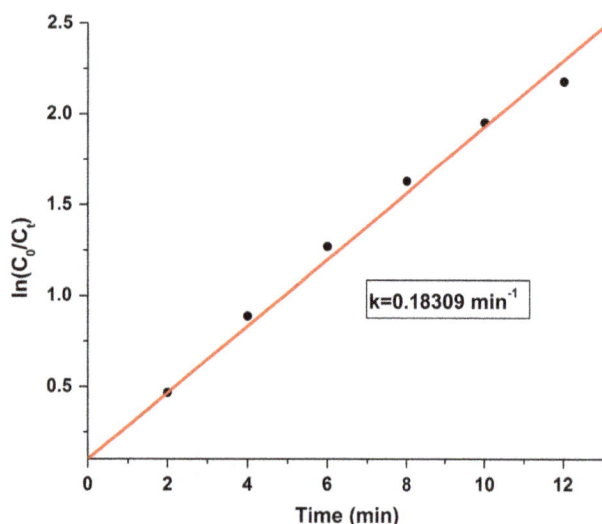

Fig. 8 The plot of $\ln C_0/C_t$ versus time for the reduction of nitrophenol to aminophenol

The absorbance at time $t = 0$ (A_0) and at t (A_t) is proportional to the initial concentration C_0 and concentration at time t (C_t) of 4-NP, respectively. The conversion percentage (α) of 4-NP to 4-aminophenol was calculated by the formula:

$$\alpha = \frac{(C_0 - C_t)}{C_0} 100.$$

The conversion percentage of 4-NP to 4-AP is shown in Fig. 6.

It has been found that the reduction of 4-NP to 4-AP by sodium borohydride in the presence of PdNPs as catalyst follows pseudo first-order rate equation with respect to 4-NP; the concentration of sodium borohydride was too high

as compared to 4-NP. So, concentration of sodium borohydride was considered constant throughout the reaction. UV–Vis spectra of successive reduction of 4-NP to 4-aminophenol using PdNPs stabilized in XG with time interval of 2 min are shown in Fig. 7. The rate equation can be written as

$$k = \frac{\ln (C_0/C_t)}{t}.$$

Figure 8 shows a good linear correlation of $\ln (C_0/C_t)$ versus time and the rate constant of the reaction is obtained as 0.18309 min^{-1} for PdNPs stabilized in XG.

Conclusions

The present study reports the green the synthesis, characterization and catalytic evaluation of PdNPs from aqueous H_2PdCl_4 solution using XG. The adapted method is compatible with green chemistry principles as the XG serves as a matrix for both reduction and stabilization of the PdNPs synthesized. The PdNPs stabilized in XG exhibited a very good catalytic activity and the kinetics of the reaction was found to be pseudo first order with respect to the 4-NP.

Acknowledgments The authors wish to thank the Coordinator, DBT-OU-ISLARE, Instrumentation Laboratory (Funded by UGC), Osmania University for providing facilities.

References

Ahok B, Bhagyashree J, Ameeta RK, Smita Z (2010) Banana peel extract mediated novel route for the synthesis of palladium nanoparticles. Mat Lett 64:1951–1953

Barrére GC, Barber CE, Daniels MJ (1986) Molecular cloning of genes involved in the production of the extracellular polysaccharide xanthan by *Xanthomonas campestris*. Intern J Bio Macromol 8:372–374

Domenech B, Munoz M, Muraviev DN, Macanas J (2011) Polymer-stabilized palladium nanoparticles for catalytic membranes: *ad hoc* polymer fabrication. Nanoscale Res Lett 6:406

El-Houta SE, Killab HM, Ibrahima IA, Harraz FA (2012) Palladium nanoparticles stabilized by polyethylene glycol: efficient, recyclable catalyst for hydrogenation of styrene and nitrobenzene. J Catal 286:184–192

Garcia-Ochoa F, Santos VE, Casas JA, Gomez E (2000) Xanthan gum: production, recovery, and properties. Biotech Adv 18:549–579

Jana NR, Wang ZL, Pal T (2000) Redox catalytic properties of palladium nanoparticles: surfactant and electron donor–acceptor effects. Langmuir 16:2457–2463

Karimi B, Enders D (2006) New N-heterocyclic carbene palladium complex/ionic liquid matrix immobilized on silica: application as recoverable catalyst for the heck reaction. Org Lett 8:1237–1240

Keerthi DD, Veera PSV, Harith R, Samba SK, Radhika P, Sreedhar B (2011) Gum acacia as a facile reducing, stabilizing, and templating agent for palladium nanoparticles. J Appl Polym Sci 121:1765–1773

Klingensmith LM, Leadbeater NE (2003) Ligand-free palladium catalysis of aryl coupling reactions facilitated by grinding. Tetrahedron Lett 44:765–768

Luo C, Zhang Y, Wang Y (2005) Palladium nanoparticles in poly (ethylene glycol): the efficient and recyclable catalyst for heck reaction. J Mol Catal A 229:7–12

Nadagouda MN, Rajender SV (2008) Green synthesis of silver and palladium nanoparticles at room temperature using coffee and tea extract. Green Chem 10:859–862

Narayanan R, El-Sayed MA (2005) FTIR study of the mode of binding of the reactants on the Pd nanoparticle surface during the catalysis of the Suzuki reaction. J Phys Chem B 109:4357–4360

Prashant KJ, Kyeong SL, El-Sayed IH, El-Sayed MA (2006) Calculated absorption and scattering properties of gold nanoparticles of different size, shape, and composition: applications in biological imaging and biomedicine. J Phys Chem B 110:7238–7248

Ramesh KP, Vivekanandhan S, Manjusri M, Amar KM, Satyanarayana N (2012) Soybean (Glycine max) leaf extract based green synthesis of palladium nanoparticles. J Biomater Nanobiotechnol 3:14–19

Raveendran P, Fu J, Wallen SL (2003) Completely "Green" synthesis and stabilization of metal nanoparticles. J Am Chem Soc 125:13940–13941

Roopana SM, Bharathi A, Rajendran K, Khanna VG, Prabhakarn A (2012) Acaricidal, insecticidal, and larvicidal efficacy of aqueous extract of Annona squamosa L peel as biomaterial for the reduction of palladium salts into nanoparticles. Collid Surf B Biointerfaces 92:209–212

Rosi NL, Mirkin CA (2005) Nanostructures in biodiagnostics. Chem Rev 105:1547–1562

Roucoux A, Schulz J, Patin H (2002) Reduced transition metal colloids: a novel family of reusable catalysts? Chem Rev 102:3757–3778

Salata OV (2004) Applications of nanoparticles in biology and medicine. J Nanobiotechnol 2:3

Sandip S, Anjaliv P, Subrata K, Soumen B, Tarasankar P (2010) Photochemical green synthesis of calcium-alginate-stabilized Ag and Au nanoparticles and their catalytic application to 4-nitrophenol reduction. Langmuir 26:2885

Sathishkumar M, Sneha K, Kwak IS, Mao J, Tripathy SJ, Yun YS (2009) Phyto-crystallization of palladium through reduction process using cinnamom zeylanicum bark extract. J Hazard Mater 171:400–404

Sun Y, Zhang LH, Zhou H, Zhu Y, Sutter E, Ji Y, Rafailovich MH, Sokolov JC (2007) Seedless and templateless synthesis of rectangular palladium nanoparticles. Chem Mater 19:2065–2070

Szilvia P, Pi Rita, Imre D (2007) Formation and stabilization of noble metal nanoparticles croat. Chem Acta 80:493–502

Venkatesham M, Ayodhya D, Madhusudhan A, Veera BN, Veerabhadram G (2014) A novel green one-step synthesis of silver nanoparticles using chitosan: catalytic activity and antimicrobial studies. Appl Nanosci 4:113–119

Wilson OM, Knecht MR, Garcia-Martinez JC, Crooks RM (2006) Effect of Pd nanoparticle size on the catalytic hydrogenation of allyl alcohol. J Am Chem Soc 128:4510–4511

Yang M, Yang Y, Liu Y, Shen G, Yu R (2006) Platinum nanoparticles-doped sol–gel/carbon nanotubes composite electrochemical sensors and biosensors. Biosens Bioelectron 21:1125–1131

Yang X, Li Q, Wang H, Huang J, Lin L, Wang W, Sun D, Su Y, Opiyo JB, Hong L, Wang Y, He N, Jia L (2010) Green synthesis of palladium nanoparticles using broth of Cinnamomum camphora leaf. J Nano par Res 12:1589–1598

Yonezawa T, Imamura K, Kimizuka N (2001) Direct preparation and size control of palladium nanoparticle hydrosols by water-soluble isocyanide ligands. Langmuir 17:4701–4703

Zhang P, Shao C, Zhang Z, Zhang M, Mu J, Guo Z, Liu Y (2011) In situ assembly of well-dispersed Ag nanoparticles (AgNPs) on electrospun carbon nanofibers (CNFs) for catalytic reduction of 4-nitrophenol. Nanoscale 3:3357

7

Linear and non-linear optical properties of GaAs nanowires

Satyendra Singh · Pankaj Srivastava

Abstract The linear and non-linear optical properties of different geometrical structures of gallium arsenide (GaAs) nanowires have been studied by employing ab initio method. We have calculated the optical response of four different GaAs nanowires, viz., two-atom linear wire, two-atom zigzag wire, four-atom square wire and six-atom hexagonal wire. We have investigated imaginary part of the zz component of the linear dielectric tensor and second-order susceptibility for different structures along with bulk material. We revealed that the strongest absorption occurs for four-atom square nanowire configuration.

Keywords GaAs nanowires · Linear optical properties · Non-linear optical properties

Introduction

Gallium arsenide (GaAs) semiconducting nanowires with diameters ranging from 1 to 400 nm and length of up to hundreds of micrometers have shown remarkable optical properties and hence they will prove to be the versatile building blocks for optoelectronic circuits at a nanoscale. GaAs semiconducting nanowires also offer a unique approach for the bottom-up assembly of electronic and photonic devices with a potential for on-chip integration of non-silicon-based photonic devices with silicon nanoelectronics-based devices (Lieber 2001, 2003; Law et al. 2004; Xia et al. 2003; Dekker 1999; Collins and Avouris 2000). GaAs nanowires are also very promising candidates for the development of a number of new nanoscale optical devices a few to be named as nanowire-based photodetectors, single nanowire laser, etc. (Duan et al. 2000, 2001).

The existence of the exciton has a strong influence on the electronic properties of the semiconductor and its optical absorption. The second-order response or non-linear property is a two-photon process where the excited electron absorbs another photon of the same frequency and makes a transition to another allowed state at higher energy. When this electron is falling back to its original state, it emits a photon of frequency, which is twice the frequency of that of the incident light (Srivastava et al. 2008; Srivastava and Singh 2008; Singh and Srivastava 2013).

Electronic and optoelectronic applications demand straight, uniformly aligned nanowires with uniform diameters and excellent crystallographic, electronic and optical properties; currently, the usefulness of GaAs nanowires in these devices is limited due to the presence of certain defects like crystallographic twin defects (Ihn and Song 2007), nanowire kinking (Zou et al. 2007), intrinsic doping and surface states (Joyce et al. 2007). Imaging and optical properties of single core–shell GaAs–AlGaAs nanowires were studied by Hoang et al. (Titova et al. 2006). Duan et al. (2000) synthesized and studied the optical properties of gallium arsenide nanowires. Electronic and optical properties of InAs/GaAs nanowires were studied by Niquet (2006). Redliski and Peters (2008) studied the optical properties of free-standing GaAs semiconducting nanowires and their dependence on the growth direction. Growth, structural and optical properties of GaAs/AlGaAs nanowires with and without quantum well shells were

S. Singh (✉)
Department of Physics, Shri Ram College of Engineering and Management, Banmore, Morena 476444, Madhya Pradesh, India
e-mail: satyendra7171@yahoo.co.in

P. Srivastava
Nanomaterials Research Group, Indian Institute of Information Technology and Management (ABV-IIITM), Gwalior 474010, Madhya Pradesh, India

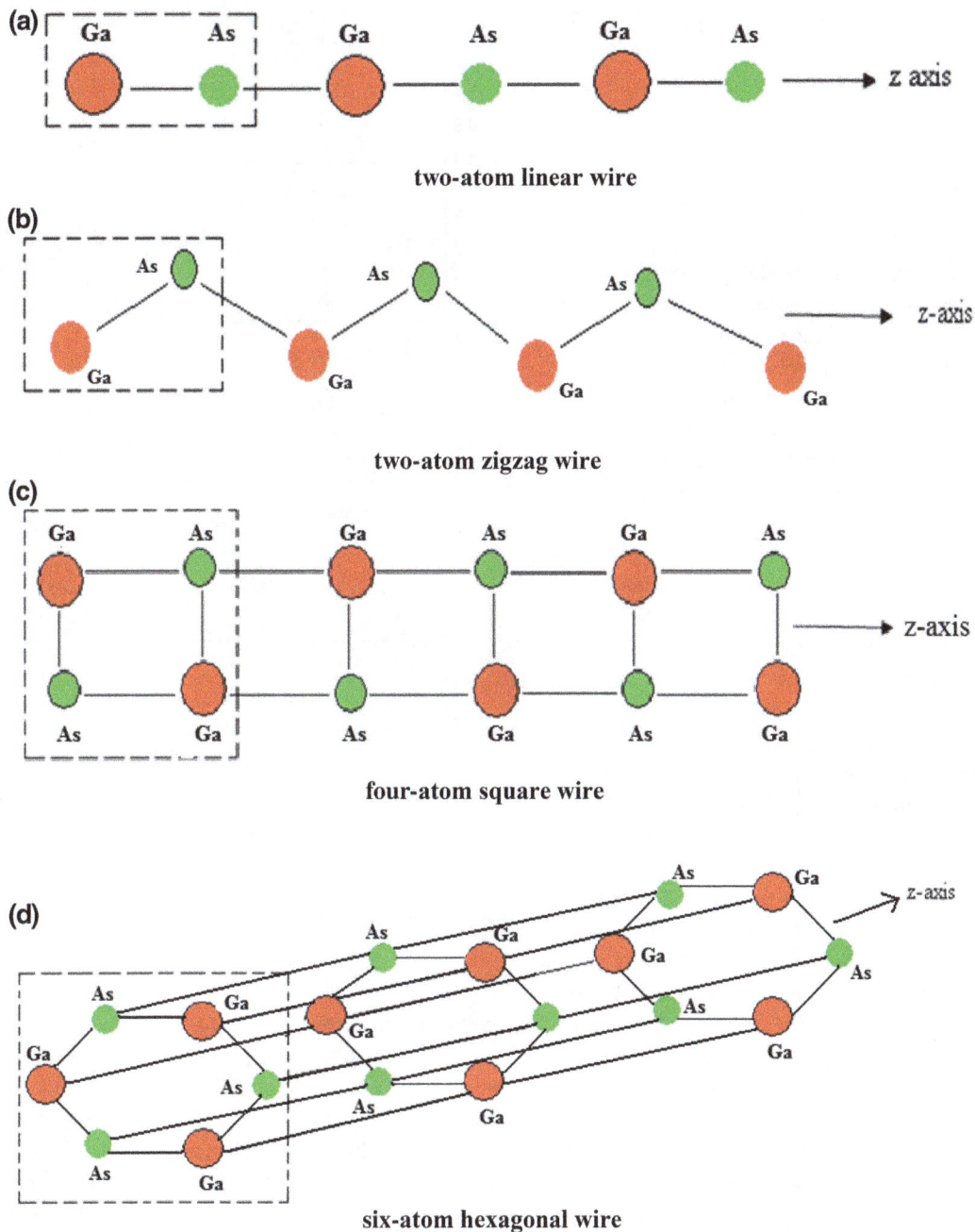

Fig. 1 Optimized structures of GaAs nanowires: **a** two-atom linear wire, **b** two-atom zigzag wire, **c** four-atom square wire, **d** six-atom hexagonal wire

demonstrated by Joyce et al. (2006). GaAs nanowires for optoelectronics were studied by Kim et al. (2006). Hoang et al. (2009) studied the optical properties of single wurtzite GaAs nanowires and GaAs nanowires with GaAsSb inserts. Zhang et al. (2009) synthesized the GaAs nanowires with very small diameters and studied their optical properties with the radial quantum-confinement effect. The study of effect of structure, surface passivation and doping on the electronic and optical properties of GaAs

nanowires by a first principle study was done by Khare et al. (2009).

Four different geometrical structures of GaAs nanowires are considered and the linear and non-linear optical properties of considered structures are being presented in this paper. The linear and non-linear spectra for considered structures are investigated and finally the optical transition in different energy regions is discussed. The details about the geometrical structures have been given in our earlier

published papers (Srivastava et al. 2008; Singh et al. 2009a,
b; Srivastava and Singh 2008, 2011; Srivastava et al. 2011).
The considered geometrical structures of GaAs nanowires
are shown in Fig. 1.

Methodology

ABINIT code (Hohonberg and Kohn 1964; Kohn and
Sham 1965) has been used for the computational work. The
ab initio DFT calculations (Martin 2009; Gonze et al.
2002) are employed within the plane-wave pseudopotential
method to investigate the linear and non-linear optical
properties of GaAs nanowires. As is evident from the
above literature that the pseudopotential method has been
very successful in exploring the structural, electronic and
optical properties of various materials (Martin 2009); in
this calculation, the generalized gradient approximation
and the exchange–correlation functional of Perdew et al.
(1996) were applied. The exchange correlation potential of
Troullier and Martins (1991) has been used and these
pseudopotentials were taken from ABINIT web page
(Gonze et al. 2002). The potentials were tested by performing calculations on bulk GaAs material in which the
results were found to be consistent with the experimental
ones. All the calculations were performed in a self-consistent manner. The studied structures were optimized for
Hellmann–Feynman forces as small as 10^{-3} eV/Å on each
atom and the calculations were performed with a kinetic
energy cut-off of 816.3418 eV. The wires were positioned
in a supercell of side 10.583 Å along the x and y directions;
the axis of the wire was taken along the z direction and the
periodic boundary conditions were applied. The Monkhorst–Pack method with 15 k-points sampling along the
z direction was used in the integration of the Brillouin
zone; all atoms were allowed to relax without any imposed
constraint. In order to check the self-consistent calculations, we have determined the self-consistent optimized
value for the lattice parameter of bulk GaAs materials, the
magnitude of atomic relaxation depends on the plane cut-
off energy and one has to obtain the convergence with
respect to cut-off energy too.

Results and discussion

The imaginary parts of the zz component of the linear
dielectric tensor or linear susceptibility for various gallium
arsenide (GaAs) nanowires along with bulk are presented
in Fig. 2. The linear optical response of bulk is quite
smooth and the linear susceptibility gradually increases
with energy, for two-atom linear wire, it is observed that
there are two strong peaks around 4.5 and 5.8 eV with

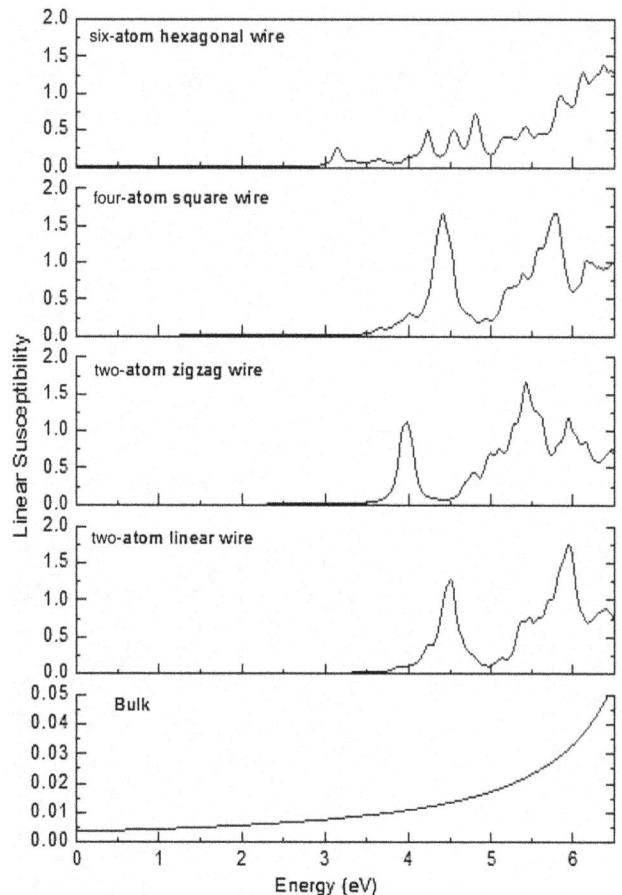

Fig. 2 Calculated imaginary part of the zz component of the linear
dielectric tensor for different GaAs nanowires along with bulk GaAs
material

small peaks towards higher energy side, but in the case of a
two-atom zigzag wire, strong absorption is observed with
three major peaks at 4.0, 5.5 and 6.0 eV. Beside this there
are other peaks also present. The four-atom square wire
linear spectra has strongest peak around 4.2 and 5.8 eV,
which exhibit strong absorption in such type of geometrical
configuration. The other weak absorption peaks at higher
energy region are also seen. For a six-atom hexagonal wire
cross-section, peaks occur in between 4.0 and 6.5 eV.

Thus, from linear absorption spectra of all the configurations, it is predicted that the strongest absorption is
shown by four-atom square wire. In the study of electronic
properties, it can be presented that same structure having
the highest stability (Singh et al. 2009a, b), thus it can be
predicted that the same structure that has the highest stability, must also have the strongest absorption which is
evident from our investigation.

The non-linear optical spectra of different configuration
of GaAs nanowires are analyzed and various contributions
to the imaginary part of χ_{zzz} (2ω, ω, ω) are presented in
Figs. 3, 4, 5, 6, 7.

Fig. 3 Calculated second-order susceptibility I_m [χ_{zzz} (2ω, ω, ω)] and different contributions for the bulk GaAs material

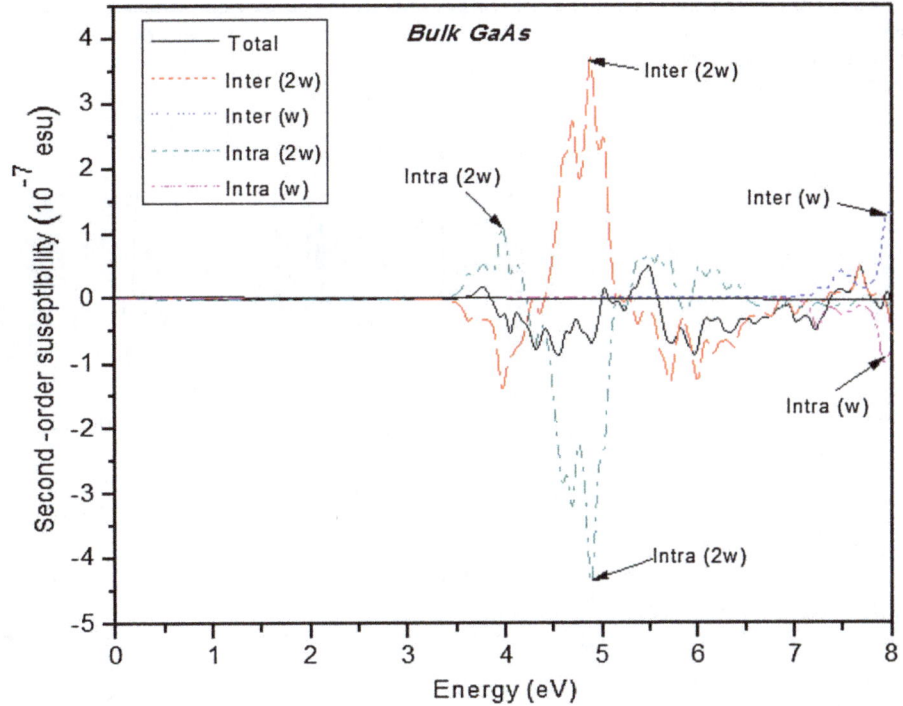

Fig. 4 Calculated second-order susceptibility I_m [χ_{zzz} (2ω, ω, ω)] and different contributions for the two-atom GaAs linear wire

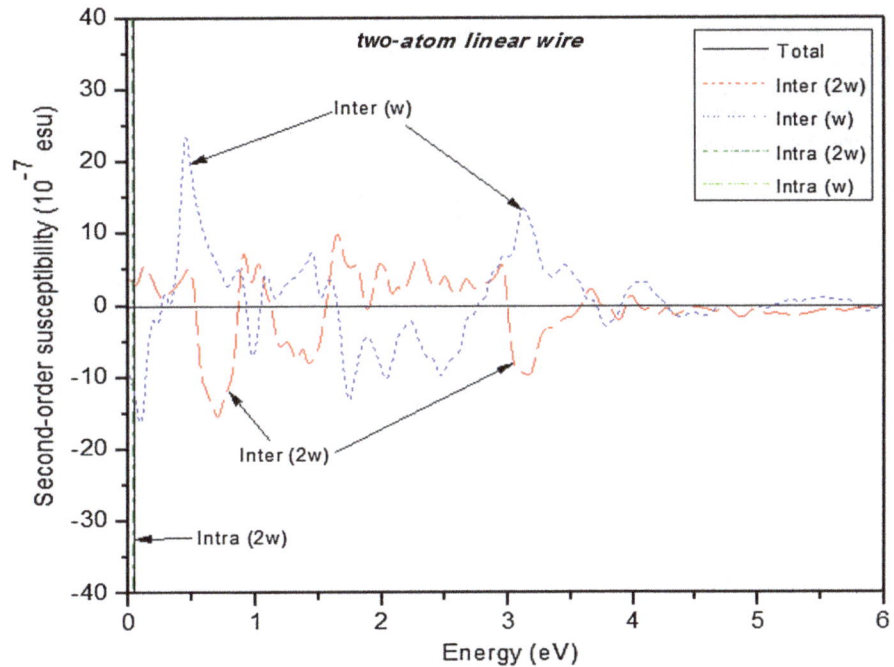

For the bulk spectra, the total SHG susceptibility is zero up to 3.5 eV as depicted in Fig. 3, in the high-energy region, the SHG optical spectra are dominated by inter (2ω) contribution towards the positive susceptibility and intra (2ω) towards the negative region in the graph. For the two-atom linear wire (Fig. 4), the total SHG peaks are dominant near lower energy values, having exceptionally high susceptibility with intense peaks magnitude; here, the major contributions come from inter (ω) transition towards positive

side and inter (2ω) transitions towards the negative side. In case of two-atom zigzag wire (Fig. 5), the complex peaks are found intermingled with various contributions. It reveals that major SHG spectrum occurs between 0.5 and 6 eV regions. We find peaks from inter (2ω) transition in upper part and inter (2ω) transition in lower part are obtained only two peaks near 0.5 and 1.5 eV possesses inter (2ω) transitions.

For four-atom square wire cross-section (Fig. 6), the total SHG as well as intra (ω) and intra (2ω) transitions

Fig. 5 Calculated second-order susceptibility $I_m [\chi_{zzz} (2\omega, \omega, \omega)]$ and different contributions for the two-atom GaAs zigzag wire

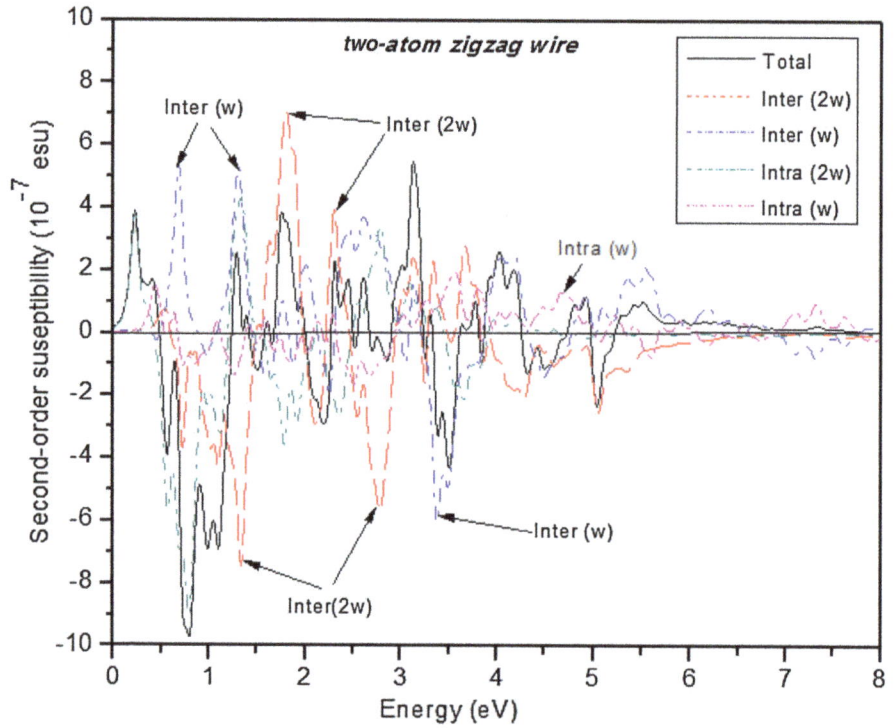

Fig. 6 Calculated second-order susceptibility $I_m [\chi_{zzz} (2\omega, \omega, \omega)]$ and different contributions for the four-atom GaAs square wire

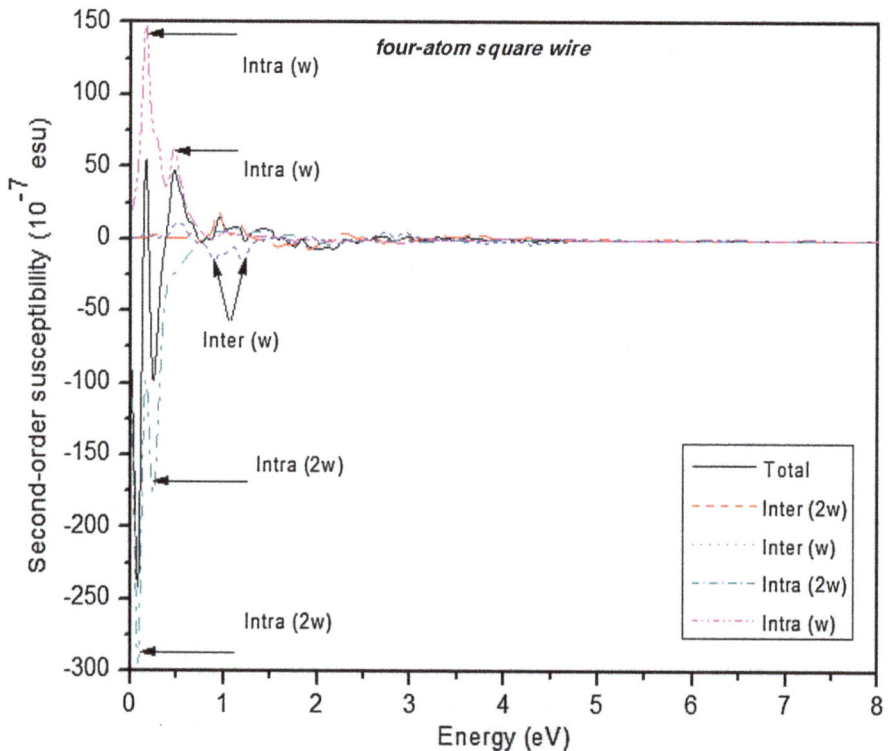

have greater contribution near lower energy region in contrast to higher energy side. The remarkable features account that the magnitude of SHG spectra is high in this type of configuration. In the positive SHG spectra, highest absorption spectra come from intra (ω) in the lower energy side, whereas in the negative SHG spectra the highest contribution comes from intra (2ω) plus total transition.

Figure 7 shows spectra of six-atom hexagonal wire, the spectra include all the transitions, i.e., inter- and intra-contribution. Many peaks of various magnitudes in

Fig. 7 Calculated second-order susceptibility I_m [χ_{zzz} (2ω, ω, ω)] and different contributions for the six-atom GaAs hexagonal wire

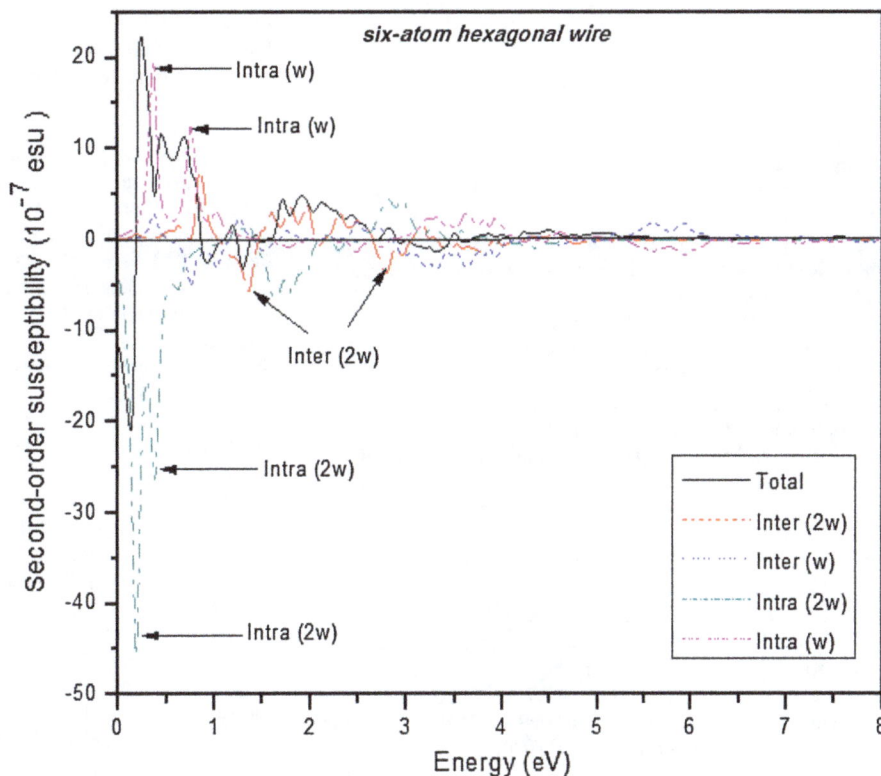

between 0.0 and 3.0 eV region are observed, but the magnitudes are of low value in comparison to other configurations; thus, the highest absorption in the positive as well as negative spectra occurs due to intra (ω) and intra (2ω) transitions, respectively.

Conclusions

We have calculated and analyzed the linear and nonlinear optical response of four different structures of GaAs nanowires. The aim of our analysis was to explore the best possible configuration of GaAs nanowire to be applied in photonic and optoelectronic devices. First, we have analyzed the linear response of the considered structures. The analysis of linear spectra reveals that strongest absorption occurs for four-atom square wire. In the stability analysis and electonic properties of GaAs nanowire, we found that the four atom square wire has the highest stability, thus it can be predicted that the structure that has the highest stability, must also have the strongest absorption. Second, we have investigated the SHG susceptibility spectra of these structures. In SHG spectra, we observe remarkable features for all structures. We predict that strongest absorption occurs for four-atom square nanowire configuration. We also

revealed that the calculated peaks not only get sharper, but also show pronounced energy shift. This is mainly due to interband contribution to the imaginary part of the dielectric function. All SHG spectra comprise total, inter- and intra-band contributions. The SHG spectra are rather complicated due to various microscopic features observed. We have not come across any of the experimental or theoretical results to compare such type of linear and SHG optical spectra for various structures of GaAs nanowires. The present investigation is important, because size, shape and structure are the important criteria in nano-regime which one must not ignore so far as nanowires are concerned. Thus these types of semiconductor GaAs nanowires offer many opportunities for the assembly of nanoscale electronic and optoelectronic devices and applicable as interconnects.

Acknowledgments The authors are thankful to Department of Physics of Shri Ram College of Engineering and Management, Banmore, Morena (M.P.), and Computational Nanoscience and Technology Laboratory (CNTL) of ABV-Indian Institute of Information Technology and Management (ABV-IIITM), Gwalior (M.P.), for providing the infrastructural facilities for computational work.

References

Collins PG, Avouris P (2000) Nanotubes for electronics. Sci Am 283:62

Dekker C (1999) Carbon nanotubes as molecular quantum wires. Phys Today 52:22

Duan X, Wangand J, Lieber CM (2000) Synthesis and optical properties of gallium arsenide nanowires. Appl Phys Lett 76:1116

Duan X, Huang H, Cui Y, Wang J, Lieber CM (2001) Indium phosphide nanowires as building blocks for nanoscale electronic and optoelectronic devices. Nature 409:66

Gonze X, Beuken JM, Caracas R, Detraux F, Fuchs M, Rignanese GM, Sindic L, Verstraete M, Zerah G, Jollet F, Torrent M, Roy A, Mikami M, Ghosez P, Raty JY, Allan DC (2002) First-principles computation of material properties: the ABINIT software project. Comput Math Sci 25:478

Hoang TB, Zhou H, Moses A, Dheeraj A, Helvoort A, Fimland BO, Weman H (2009) Observation of free exciton photoluminescence emission from single wurtzite GaAs nanowires. Appl Phys Lett 94:133105

Hohonberg P, Kohn W (1964) Inhomogeneous electron gas. Phys Rev 136:B864

Ihn SG, Song JI (2007) Morphology- and orientation-controlled gallium arsenide nanowires on silicon substrates. Nano Lett 7:39

Joyce HJ, Kim Y, Gao Q, Tan HH, Jagdish C (2006) Growth, structural and optical properties of GaAs/AlGaAs core/shell nanowires with and without quantum well shells, The Australian National University, Canberra, pp ACT0200

Joyce HJ, Gao Q, Tan HH, Jagdish C, Kim Y, Zhang X, Guo Y, Zou J (2007) Twin-free uniform epitaxial GaAs nanowires grown by a two-temperature process. Nano Lett 7:921

Khare SV, Gade V, Shi N, Ram Prasad R (2009) Effect of structure, surface passivation, and doping on the electronic and optical properties of GaAs nanowire: a first principles study. In: Bulletin of the American Physical Society 2009 APS March Meeting, vol 54:D26.0001

Kim Y, Gao Q, Joyce HJ, Tan HH, Jagdish C, Paladugu M, Zou J (2006) III–V nanowires for optoelectronics. Asia Pac Opt Commun 13:635226

Kohn W, Sham LJ (1965) Self-consistent equations including exchange and correlation effects. Phys Rev 140:A1133

Law M, Goldberger J, Yang Y (2004) Semiconductor nanowires and nanotubes. Ann Rev Mater Res 34:83

Lieber CM (2001) The incredible shrinking circuit. Sci Am 285:58

Lieber CM (2003) Nanoscale science and technology: building a big future from small things. MRS Bull 28:486

Martin RM (2009) Electronic structure. Cambridge University Press, Cambridge

Niquet YM (2006) Electronic and optical properties of InAs/GaAs nanowire superlattices. Phys Rev B 74:155304

Perdew JP, Burke K, Ernzerhof M (1996) Generalized gradient approximation made simple. Phys Rev Lett 77:3865

Redliski P, Peters FM (2008) Optical properties of free-standing GaAs semiconductor nanowires and their dependence on the growth direction. Phys Rev B 77:753329

Singh S, Srivastava P (2013) Optical properties of gallium phosphide (GaP) nanowires. Appl Nanosci 3:89

Singh S, Srivastava P, Mishra A (2009a) Ab-initio study of gallium arsenide nanowires. J Comput Theor Nanosci 6:1556

Singh S, Srivastava P, Mishra A (2009b) Phys E 42:46

Srivastava P, Singh S (2008) Linear and second-order optical response of different GaN nanowires. Phys E 40:2742

Srivastava P, Singh S (2011) Stability analysis of AlN nanowire. J Comput Theor Nanosci 8:1764

Srivastava P, Singh S, Mishra A (2008) Stability and electronic properties of GaN nanowires: an ab initio approach. J Comput Theor Nanosci 5:635

Srivastava P, Singh S, Mishra A (2011) Electronic properties of GaP nanowires of different shapes. J Nanosci Nanotechnol 11:10464

Titova LV, Hoang TB, Jackson HE, Smith LM, Yarrison JM, Kim Y, Joyce HJ, Tan HH, Jagdish C (2006) Temperature dependence of photoluminescence from single core–shell GaAs–AlGaAs nanowires. Appl Phys Lett 89:173126

Troullier N, Martins JL (1991) Efficient pseudopotentials for plane-wave calculations. Phys Rev B 43:1993

Xia X, Yang P, Sun Y, Wu Y, Mayers B, Gates B, Yin Y, Kim F, Yan H (2003) One-dimensional nanostructures: synthesis, characterization, and applications. Adv Mater 5:353

Zhang G, Tateno K, Sanada H, Tawara T, Gotoh H, Nakano H (2009) Synthesis of GaAs nanowires with very small diameters and their optical properties with radial quantum-confinement effect. Appl Phys Lett 95:123104

Zou J, Paladugu M, Wang H, Auchterlonie GJ, Gao Y, Kim Y, Gao QH, Joyce HJ, Tan HH, Jagdish C (2007) Growth mechanism of truncated triangular III–V nanowires. Small 3:389

8

Influence of carbon nanotubes support on the morphology of Fe_3O_4 nanoparticles

F. Zabihi · F. Taleshi · A. Salmani ·
A. Pahlavan · N. Dehghan-niarostami ·
M. M. Vadadi

Abstract In this paper, the effects of carbon nanotubes as a support to the morphology and size of Fe_3O_4 magnetic nanoparticles have been investigated. The synthesis of Fe_3O_4/CNTs nanocomposite powder was performed by the direct precipitation method through ferric chloride (II) and (III) at room temperature. The prepared samples were analyzed by X-ray diffraction spectra, Fourier transform infrared spectroscopy, scanning electron microscopy and transmission electron microscopy. The results demonstrated considerable changes in the Fe_3O_4 nanoparticle size, also the morphology of Fe_3O_4/CNTs nanocomposite powder from agglomerative into rode shape.

Keywords Carbon nanotube · Magnetic nanoparticles · Morphology · Nanocomposite · Direct precipitation

Introduction

Extensive investigations' have been conducted on improving the synthesis and characterization of magnetic nanoparticle due to their possessing unique properties and their wide application in the various industries (Tjong and Chen 2004; Laurent et al. 2008; Zhang et al. 2006; He et al. 2008). Among the various magnetic nanoparticles, Fe_3O_4 the nanoparticle has leading properties such as chemical stability, high dispersion in liquid circumferences, good bio-adaptability, stability in various physiological conditions and low dissolubility (Yan et al. 2009; Tartaj et al. 2003; Qiao et al. 2009; Ghandoor et al. 2012).

These kinds of nanoparticles have been successfully used and applied in magnetic recorders (Laurent et al. 2008; Ghandoor et al. 2012), cancer therapy (He et al. 2008; Figuerola et al. 2010) storing information processes (Yan et al. 2009), magnetic resonance imaging (MRI) (Qiao et al. 2009), and drug targeting (Figuerola et al. 2010).

Moreover, since the discovery of carbon nanotubes (CNTs), they are being used as a support for controlling the size and morphology of nanoparticles (Cheng et al. 2008). Therefore, every effort is being made to allow the placement of metal nanoparticles, metal oxides and semiconductors nanoparticles onto the nanotubes (Cheng et al. 2008; Ramin and Taleshi 2013). Due to the high aspect ratio of the carbon nanotubes, they prevent the agglomerating of the particles and increase their applicability considerably (Cheng et al. 2008; Ramin and Taleshi 2013; Liu et al. 2009). However, there are two basic problems in the preparation of such a composite. The first problem is spinning and the varied diffusion rate of the nanotube in the composite. The second is the suitable non-adhesion of the carbon nanotube surface with synthesized nanoparticles, which is due to a strong covalent bond between the considered nanoparticle and the white carbon nanotubes (Taleshi and Hosseini 2012). As a result, the nanotubes surface activation makes a suitable reaction between their surfaces, and with the nanoparticle, which is necessity (Taleshi and Hosseini 2012; Wang et al. 2010). The CNTs

F. Zabihi · A. Salmani · A. Pahlavan · N. Dehghan-niarostami
Department of Physics, Sari Branch, Islamic Azad University, Sari, Iran

F. Taleshi (✉)
Department of Applied Science, Qaemshahr Branch, Islamic Azad University, PO Box 163, Qaemshahr, Iran
e-mail: far.taleshi@gmail.com

M. M. Vadadi
Department of Chemistry, Payam Noor University, PO Box 19395-697, Tehran, Iran

activation of the surface is typically performed by a chemical oxidation process and produces the formation of various functional groups such as COOH, C–O, C=O and OH (Taleshi and Hosseini 2012; Wang et al. 2010). These chemical bonds could act as sites for nucleation of nanoparticle.

The effects of CNTs as a support for size and morphology of Fe_3O_4 nanoparticles prepared by direct precipitating were further explored in this research paper. Fourier transform infrared spectroscopy (FT-IR, Shimadzu—8400 s) has been used for the study of functional groups on the CNTs surface, as well as X-ray diffraction (XRD, Pw, 1800, Philips) was applied for identifying the crystalline structure of the nanoparticles. The morphology of Fe_3O_4 nanoparticle powder and Fe_3O_4/CNTs nanocomposite powder has been analyzed by the scanning electron microscope (SEM, Philips, SE, 3000K, and 15X) and transmission electron microscope (TEM, Philips, CM10, and HT100 kV). The results showed that carbon nanotubes had decreased considerably and the size of the Fe_3O_4 magnetic nanoparticles changed in morphology in the powder forming an agglomerative into a fibrous shape.

Experimental process

Iron (II) chloride tetrahydrate ($FeCl_2 \cdot 4H_2O$, 99 %+), Iron (III) chloride hexahydrate ($FeCl_3 \cdot 6H_2O$, >99 %), ammonium hydroxide solution (NH_4OH), carbon nanotube (CNTs, Neutrino china, pure percentage 99 %, 20 nm < d < 30 nm), sulfuric acid H_2SO_4 and nitric acid HNO_3 were used for preparing the samples for this research.

The synthesis of Fe_3O_4 nanoparticles was conducted by direct precipitating method for this reason, we initially poured 4 mmol (1.08 g) $FeCl_3 \cdot 6H_2O$ and 2 mmol (0.4 g) $FeCl_2 \cdot 4H_2O$ into the rotating beaker containing 50 ml distilled water and prepared a brown solution. Then 2.5 ml of ammonium hydroxide was quickly added to the solution, during rotation black sediment resulted. We then poured the obtained sediment through filter paper and washed it with distilled water. The resulting sample was dried in the oven at 110 °C for 3 h.

For application of the reagent to the CNTs surface, a required amount of CNTs was mixed with a 6 M composite solution of nitric and sulfuric acid and placed under the ultrasonic waves for 1 h, the resulting product was rotated over the hotplate at 80 °C for 2 h. We poured the sample through filter paper to neutralize the acidic phase form, then washed the sample with distilled water several times to reach to a neutralized state (pH = 7).

To synthesize the nanocomposite powder Fe_3O_4/CNTs to various composition ratios of 2/1, 1/1, 1/2, 1/4 and 1/8,

we poured 0.25 g of the CNTs with the applied regent into 50 ml of distilled water and placed it under ultrasonic waves for 10 min; we then added a required amount of ferric chloride salts to the solution which contained the carbon nanotubes, for each composition ratio, the obtained solution was rotated by a magnetic mixer at room temperature. Subsequently, after 5 min of rotating the solution, 2.5 ml of ammonium hydroxide was added to the obtained solution and was separated by pouring the solution through a filter paper, the resulting samples were dried in the oven at 110 °C for 2 h.

Results and discussion

Figure 1 represents the XRD patterns of pure Fe_3O_4 nanoparticles and Fe_3O_4/CNTs nanocomposite with various composition ratios. In Fig. 1a, the peaks placed on 2θ of 18.4°, 30.2°, 35.6°, 43.2°, 53.6°, 57.1°, 62.6°, 73.9° and 74.2° represent the synthesis of magnetite phase (Fe_3O_4) with reversed spin cubic structure that is, respectively, the reflex of crystalline planes (111), (220), (311), (400), (422), (511), (440) and (622) (Wang et al. 2010; Hoa et al. 2009). In this chemical structure, oxygen ions are in a cubic crystalline arrangement and the iron ions have occupied the octahedral and tetrahedral (Han et al. 1994). Moreover, the peaks being sharp like represents the crystalline order and high crystalline structure of Fe_3O_4 nanoparticles, there were no additional peaks, due to the existing impurity in the prepared powder.

In the spectra obtained from Fe_3O_4/CNTs nanocomposite powders with various ratios (Fig. 1b–f), the $2\theta = 25.9°$, 42.9° as related to reflex of geraphen planes of carbon nanotubes in (002) and (100) direction, respectively (Taleshi and Hosseini 2012). Investigation of the sample spectra of Fig. 1c–f shows that the heights of peaks related to this nanotubes have reduced by reduction of Fe_3O_4 concentration and its width has increased. These changes indicated the considerable reduction of Fe_3O_4 nanoparticles size. On the other hand, the height of peaks related to carbon nanotubes at $2\theta = 25.9°$ and 42.9° did increased. This increase could be due to the reduction of Fe_3O_4 concentration on the CNTs surface and more exposure to the geraphen planes in the beam of X-ray. Studying the spectra of samples (Fig. 1c–f) shows that the width of basic peak related to Fe_3O_4 nanoparticles has increased by increasing the Fe_3O_4 concentration and the separation and visibility of peaks improved. The mentioned results represented the nanoparticles size that increased, but the width of peaks was associated to the carbon nanotubes decreasing.

According to maximum peak of XRD at $2\theta = 35.8°$, the sizes of nanoparticles calculated by Scherrer formula following as (Venkateswarlu et al. 2010):

Fig. 1 XRD spectra from various synthesized samples; pure Fe$_3$O$_4$ nanoparticles (*a*) and Fe$_3$O$_4$/CNTs nanocomposite powder with different weight ratios of 2/1 (*b*), 1/1 (*c*), 1/2 (*d*), 1/4 (*e*) and 1/8 (*f*)

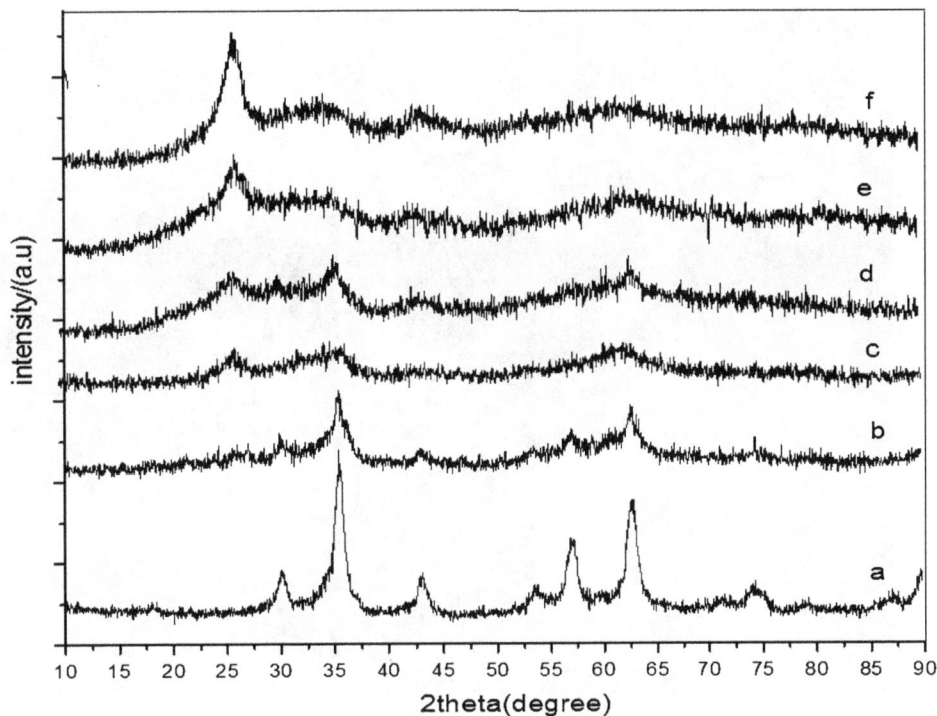

Fig. 2 FT-IR spectra from initial (*a*) and purified CNTs (*b*)

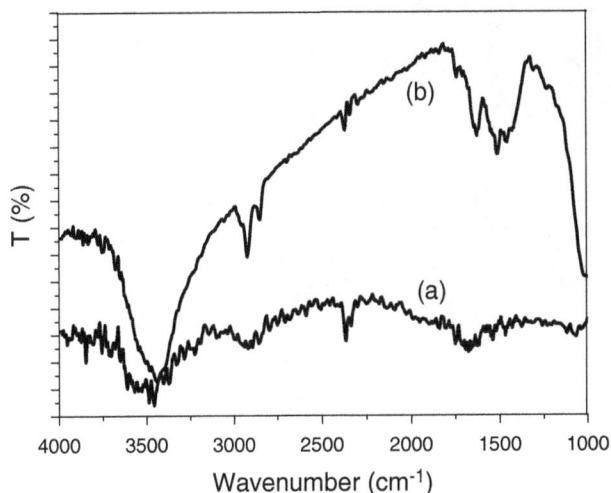

$$D = \frac{K \cdot \lambda}{\beta \cdot \cos \theta}$$

where K is Scherrer constant (about 0.9), λ is the X-ray wave length equal to 1.54 Å, θ is position of the maximum peak in term of the degree and β is semi-width of spectra in the half of maximum peak (FWHM) in term of radian.

Results of calculations show that average of nanoparticles sizes for weight ratio of 1/0, 2/1, 1/1, 1/2, 1/4 and 1/8 is, respectively, 5.1, 2.7, 1.7, 1.5, 1.3 and 1.2 nm.

To investigate the influence of the acidification process on the nanotubes surface and to study the reagent groups prepared from carbon nanotubes before and after purification process, the FT-IR spectra were prepared (Fig. 2). The FT-IR spectra from CNTs purified (Fig. 2a), over to initial CNTs (Fig. 2b), and the presence of some absorptive peaks reflect the formation of binding between the reagent groups placed on the carbon nanotubes through ultrasonic and acidification processes in H$_2$SO$_4$ and HNO$_3$ acids. These reagent groups include O–C=O 'C–O–O 'C–O and C=O, which respectively related to 1,210, 1,415, 1,475 and 1,720 cm^{-1} wavelengths (Taleshi and Hosseini 2012). The above reagent groups have an important role in nucleation and formation of Fe$_3$O$_4$ nanoparticles. Due to a strong chemical binding and the binding between ions of iron atom with reagent groups of O–C=O 'C–O–O 'C–O and C=O placed on the carbon nanotubes surface, Fe$_3$O$_4$/CNTs will form.

In Fig. 3, SEM images of prepared pure Fe$_3$O$_4$ nanoparticles (Fig. 3a) and Fe$_3$O$_4$/CNTs nanocomposite powder (Fig. 3b–f) with various weight ratios of 2/1, 1/1, 1/2, 1/4 and 1/8 are showed. SEM images of pure Fe$_3$O$_4$ nanoparticles (Fig. 3a), represent the high assembly of synthesized nanoparticles. This high agglomeration could be due to synthesis of nanoparticles in the liquid medium and because they were small the nanoparticles, due to the surface effects, had a high tendency for surface adhesion with the adjacent nanoparticles. SEM images of Fe$_3$O$_4$/CNTs nanocomposite powder (Fig. 3b–f) demonstrated that nanoparticles have different morphologies in agglomerating amount, irrespective to reaction condition. The size of nanoparticles was reduced with the increase of the concentration of carbon nanotube support in composite,

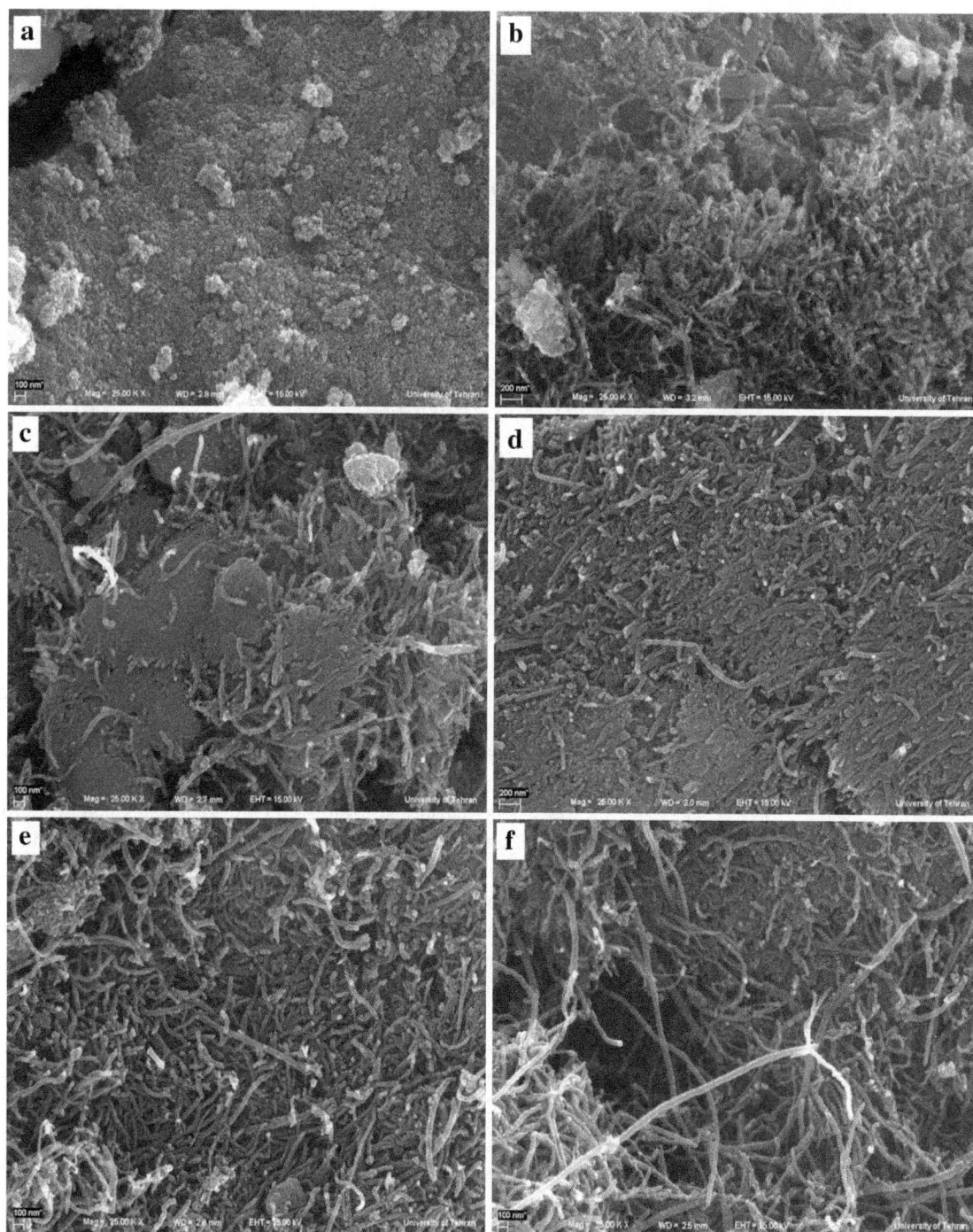

Fig. 3 SEM images of pure Fe_3O_4 nanoparticles (**a**); Fe_3O_4/CNTs nanocomposite powder with weight ratio of; 2/1 (**b**), 1/1 (**c**), 1/2 (**d**), 1/4 (**e**), and 1/8 (**f**). (The magnification of all images is $\times25k$)

and the agglomerations of the particles were reduced considerably. Thus, the change of Fe_3O_4 nanoparticles concentration ratio to carbon nanotubes has a bad fundamental influence on the morphology and sizes of nanocomposite, consequently the prepared powders were converted from agglomerative state of nanoparticles (at

weight ratio of 2/1, 1/1) into a nanorode state (at 1/8 ratio). Synthesis of Fe_3O_4/CNTs nanocomposite into a fibrous state and their homogenous distribution on the nanotubes surface could change their physical and chemical properties considerably, and show better performance in the industrial application.

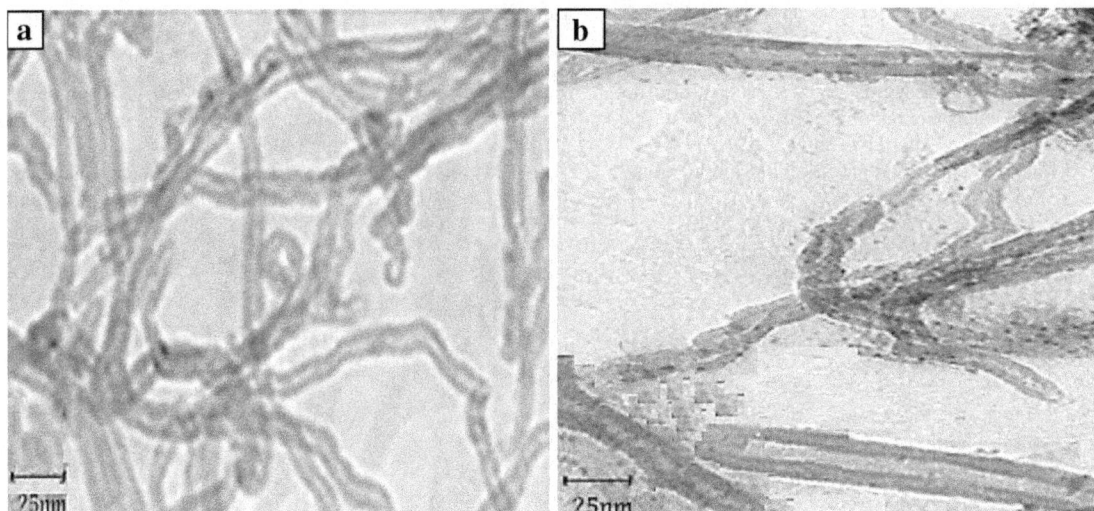

Fig. 4 TEM images of initial pure nanotubes (**a**), and synthesized Fe_3O_4/CNTs nanocomposite powder (**b**) with 1/8 weight ratio

Figure 4 represents the images of TEM prepared from pure carbon nanotubes and prepared Fe_3O_4/CNTs nanocomposite with 1/8 ratio. According to Fig. 4a, image of the carbon nanotube was shown after the reagent processing, in which no covering on the surface or in nanotubes was observed. However, the image of Fe_3O_4/CNTs nanocomposite (Fig. 4b) showed Fe_3O_4 nanoparticles with round and separated geometrical shapes, and with very small improper dimensions, adhering to the nanotubes surface.

Conclusion

Results of X-ray diffraction spectra show that the application of carbon nanotubes caused a considerable reduction of peaks height of the Fe_3O_4 nanoparticles and an increase of their widths, consequently reducing the nanoparticles sizes. Also, we reduced the high agglomeration of synthesized Fe_3O_4 nanoparticles in the liquid medium, by applying the carbon nanotubes. Thus, the presence of the carbon nanotubes as a support for increasing Fe_3O_4 nanoparticles, in addition to reduction of nanoparticles size, has produced a change in the morphology of the prepared powder from agglomerative to a fibrous state. TEM images revealed the formation of nanoparticles on the purified carbon nanotubes surface, and this adhesion could be a document suitable for the reagent processing of the nanotubes surface under the acidification process.

Acknowledgments The authors would like to acknowledge the Islamic Azad University of Qaemshahr, Islamic Azad University of Sari and Iranian National Nanotechnology Initiation Council (INNIC) for their financial support of this project.

References

Cheng JP, Zhang XB, Yi GF, Ye Y, Xia MS (2008) Preparation and magnetic properties of iron oxide and carbide nanoparticles in carbon nanotube matrix. J Alloy Compd 445:5–9

Figuerola A, Corato RD, Manna L (2010) From iron oxide nanoparticles towards advanced iron-based inorganic materials designed for biomedical applications. Pharmacol Res 62:126–143

Ghandoor HE, Zidan HM, Khalil MMH, Ismail MIM (2012) Synthesis and some physical properties of magnetite Fe_3O_4 nanoparticles. Int J Electrochem Sci 7:57734–57745

Han DH, Wang JP, Luo HL (1994) Crystallite size effect on saturation magnetization of fine ferromagnetic particles. J Magn Magn Mater 136:176–182

He Q, Zhang Z, Xiongm J (2008) A novel biomaterial-Fe_3O_4:TiO_2 core-shell nano particle with magnetic performance and high visible light photocatalytic activity. Optic Mater 31:380–384

Hoa TM, Dung TT, Danh TM, Duc NH, Chien DM (2009) Preparation and characterization of magnetic nanoparticles coated with polyethylene glycol. J Phys Conf Ser 187:012048

Laurent S, Forge D, Port M, Roch A, Robic C, Elst LV, Muller RN (2008) Magnetic iron oxide nanoparticles: synthesis, stabilization, vectorization, physicochemical characterizations, and biological applications. Chem Rev 108:2064–2110

Liu Y, Jiang W, Li S, Li F (2009) Electrostatic self-assembly of Fe_3O_4 nanoparticles carbon nanotubes. Appl Surf Sci 255:7999–8002

Qiao RR, Yang CH, Gao MY (2009) Superparamagnetic iron oxide nanoparticles: from preparations to in vivo MRI applications. J Mater Chem 19:6274–6293

Ramin M, Taleshi F (2013) The effect of carbon nanotubes as a support on morphology and size of silver nanoparticles. Int Nano Lett.

Taleshi F, Hosseini AA (2012) Synthesis of uniform MgO/CNT nanorods by precipitation method. J Nanostruct Chem 3:1–5.

Tartaj P, Morales MP, Veintemillas-Verdaguer S, Gnźalez-Carre˜no T, Serna CJ (2003) The preparation of magnetic nanocrystals for applications in biomedicine. Appl Phys 36:182–197

Tjong SC, Chen H (2004) Nano crystalline materials and coatings. Mat Sci Eng R 45:1–88

Venkateswarlu K, Chandra Bose A, Rameshbabu N (2010) X-ray peakbroadening studies of nanocrystalline hydroxyapatite by Williumson-Hall analysis. Phys B 405:4256–4261

Wang X, Zhao Z, Qu J, Wang Z, Qiu J (2010) Fabrication and characterization of magnetic Fe_3O_4–CNT composites. J Phys Chem Solids 71:673–676

Yan H, Zhang J, You C, Song Z, Yu B, Shen Y (2009) Influences of different synthesis conditions on properties of Fe_3O_4 nanoparticles. Mater Chem Phys 113:46–52

Zhang L, He R, Gu HC (2006) Synthesis and kinetic shape and size evolution of magnetite nanoparticles. Mater Res Bull 41:260–267

9

First-principles study of the interaction of hydrogen molecular on Na-adsorbed graphene

Nurapati Pantha · Kamal Belbase ·
Narayan Prasad Adhikari

Abstract We have performed density functional theory-based first-principles calculations to study the stability, geometrical structures, and electronic/magnetic properties of pure graphene, sodium (Na)-adsorbed graphene and also the adsorption properties of H_2-molecular ranging from one to five molecules on their preferred structures. Using the information of binding energy of Na at different adsorption sites of varying sized graphene supercell, it has been observed that hollow position is the most preferred site for Na adsorption, and the same in 3×3 supercell has been used for further calculations. The band structure and density of states calculations have been performed to study the electronic/magnetic properties of Na-atom graphene. On comparing adsorption energy per H_2-molecular in pure and Na-adsorbed graphene, we find that presence of Na atom, in general, enhances binding strength to H_2-moleculars.

Keywords Atom-adsorbed graphene · Hydrogen storage · Renewable energy · Density functional theory (DFT)

Introduction

Graphene is a two-dimensional layer of carbon atoms in honeycomb lattice structure, arranged with sp^2-hybridization (Geim and Novoselov 2007). With a number of wonderful properties, graphene has a long history (more than 60 years) of being studied first theoretically in 1947 (Wallace 1947) and later on experimentally in 2004 (Novoselov et al. 2004). Besides the attractive physical properties as a thin, strong and stretchable material (Geim 2009), graphene has interesting electronic/magnetic properties which enhances its potential use in academic and industrial world.

Graphene, its derivatives and other similar two-dimensional compounds have become frontier area of research activities, and also the materials of commercial interest in recent days (Medeiros et al. 2010; Johll et al. 2009; Thapa et al. 2011; Chan et al. 2008; Karki and Adhikari 2014; Ci et al. 2010; Kaloni et al. 2011; Park and Louie 2010; Kaloni et al. 2013; Esquinazi et al. 2003; Mukherjee and Kaloni 2012; Ugeda et al. 2010; Singh et al. 2013; Kaloni et al. 2014; Saha et al. 2009; Kaloni et al. 2012; Wu and Yang 2012; Dai et al. 2009; Choi and Jhi 2009). Graphene, hexagonal boron nitride (h-BN) and their doped structures have been studied by a number of research groups to see their potential applications in advanced electronics and optics (Ci et al. 2010; Mukherjee and Kaloni 2012; Singh et al. 2013). Also the multilayer graphene and other graphitic arrangements with electron donor/acceptor atoms, at varying concentrations, are covered to see their modified magnetism, and band structures (Kaloni et al. 2011, 2014, 2008; Park and Louie 2010; Kaloni et al. 2012; Ataca et al. 2008). The varying band gaps of these new materials and other carbon nanostructures include extensively a wide range of conductivity with all the potential applications in semiconductor (Mukherjee and Kaloni 2012; Kaloni et al. 2014), conductor (Ataca et al. 2008) to superconductor (Ugeda et al. 2010). The easily tunable structural stability, band structure and other electronic/magnetic properties, and also photonic and vibrational properties of graphene

N. Pantha · K. Belbase · N. P. Adhikari (✉)
Central Department of Physics, Tribhuvan University, Kirtipur,
Kathmandu, Nepal
e-mail: npadhikari@gmail.com

have been achieved through a number of techniques like: via vacancies (Kaloni et al. 2013; Ugeda et al. 2010; Singh et al. 2013; Kaloni et al. 2012), atomic/molecular doping (Medeiros et al. 2010; Johll et al. 2009; Thapa et al. 2011; Chan et al. 2008; Karki and Adhikari 2014; Saha et al. 2009; Wu and Yang 2012), functionalization (Wood et al. 2012; Ulman et al. 2014), irradiation (Esquinazi et al. 2003) and external fields (Park and Louie 2010). The modified compounds carry potential applications over many dimensions like in electronics (Geim 2009; Novoselov et al. 2005, 2007), spintronics (Palacios et al. 2008; Ding et al. 2011), chemical sensors (Saha et al. 2009), and energy storage (Pumera 2011). Graphene with the adsorption of relevant metal atoms, clusters and other functional groups as impurities have been seriously considered and studied as substrates for gaseous adsorption (Wu and Yang 2012), including energy carrying gases methane and hydrogen (Wood et al. 2012; Ulman et al. 2014).

Hydrogen, either in gaseous form or as a component of abundant compounds like water and hydrocarbons, is a widely available energy resource in the Earth and extraterrestrial planets. As hydrogen releases only the water vapour when it burns, it is considered as one of the clean and green sources of energy. In spite of having many advantages to be a potential gas for future energy carrier, user friendly storage and transportation have become one of the major challenges. There are a number of traditional techniques which have been practiced to store natural gases and hydrogen, like in tanks under high pressure (Compressed natural gases, CNG) (Burchell and Rogers 2000; Duren et al. 2004) and in liquid form at low temperature (Liquefied natural Gases, LNGs) (Nakanishi and Reid 1971) or in chemical hydrides (Orimo et al. 2007). The techniques, however, are not user friendly due to weight, space and economy concerned perspectives and also they display the risk of leakage and explosion (Chen et al. 1999). Adsorption of hydrogen in different porous materials (Ma et al. 2008) and carbon nanomaterials including activated carbons is some of the highly searched techniques in recent time (Pan et al. 2008). It has been reported that an appreciable amount of hydrogen could be stored at low temperature and the continuous development is going on to develop the devices which are suitable to store hydrogen at operating conditions. Among the carbon nanotubes and graphene allotropes as substrates for hydrogen adsorption, adatoms including alkali elements have performed catalytic effect in dissociating the H_2 molecule, promoting atomic adsorption and enhancing binding strength at moderate temperature and ambient pressure (Kwon 2010; Chen et al. 2008; Chandrakumar and Ghosh 2008). The United State's department of energy (US, DOE) has set targets of about 6 wt% (Durgun et al. 2008) and adsorption energy range (0.2–0.7 eV) (Zhou and Willians 2011) for hydrogen

storage in practical applications like vehicular transportation. Approaching towards these goals by designing the proper substrate will explore interesting scientific and commercial implications.

The adsorption of alkali metal atoms on graphitic materials has been widely studied to see the interaction between them (Medeiros et al. 2010; Liu et al. 2011). Being as electron donor atoms, electronic charge transfers from adsorbed alkali atoms (Li, Na and K) to graphene, which causes change in band structure and neutrality of the new materials (Mukherjee and Kaloni 2012). The properties of atom-adsorbed graphene can also be tuned using mechanical strain (Zhou et al. 2010). Among the alkali metal atoms, Li with its light weight has been previously tested to metalize graphene, which enhances hydrogen storage capacity up to 12.8 wt% (Ataca et al. 2008). Having similar electronic properties and slightly higher molecular weight, Na could be a good option for catalyzing graphene to store H_2-moleculars. However, in our best knowledge, this element neither as an isolated atom nor as a cluster in carbon nanostructures has been checked for enhancing gaseous adsorption. In the present work, we first see the nature of interaction between the Na atom with monolayer graphene and then Na-graphene system is used to study its binding strength for H_2-moleculars.

Computational details

Density functional theory (DFT)-based first-principles calculations (Hohenberg and Kohn 1964; Kohn and Sahm 1965) are carried out to investigate the structural stability and electronic/magnetic properties of graphene and single sodium atom added graphene. The optimized geometry of 3×3 Na-added graphene, based on compromise between the size effect and the computational cost, is then used as substrate to study the adsorption behavior of a number of H_2-molecular. The calculations are performed by incorporating van der Waals (vdW) interactions (Klimes and Michaelides 2012) via London dispersion effects in DFT-D2 approach, implemented with the quantum ESPRESSO package (Giannozzi et al. 2009). The algorithm has used Rappe–Rabe–Kaxiras–Joannopoulos (RRKJ) model of ultrasoft pseudopotential to account the interaction between the ion cores and valence electrons and generalized gradient approximation (GGA) formalism to treat the electronic exchange and correlation effects, as described by Perdew–Burke–Ernzerhof (PBE) (Perdew et al. 1996), for all the species (C, Na, H). The semi-core states: $2p3s$ of sodium, $1s$ of hydrogen and $2s2p$ of carbon atoms are treated explicitly as valence in the pseudopotential description of the corresponding atoms. The inner cores for carbon and sodium, on the other hands, are included as ion cores by the corresponding pseudo potentials Quantum (2014).

A hexagonal unit cell with the basis of two atoms in honeycomb lattice structure was initially constructed using experimental value (Castro Neto et al. 2006, 2009). The structure was then optimized with respect to lattice parameter (a), cut off energy of the plane waves and the number of k-points along x and y axes, respectively. Based on these convergence tests, a plane wave basis set with the kinetic energy cutoff 35 Ry is used for the expansion of the ground state electronic wave function. The plane waves are chosen to have a periodicity compatible with the periodic boundary conditions of the simulating cell. In case of lattice constant, we have used the calculated value (2.46 Å) which comes to agree with the experimental value for the planner sides of the unit cell, whereas the dimension along z-direction is kept large enough (20 Å) to avoid the interaction between the graphene layers and also to provide the space for upcoming hydrogen adsorption calculations.

Similarly, we have used $15 \times 15 \times 1$ mesh of k-points from convergence test for the unit cell calculations (Fig. 1).

The Na-adsorbed graphene system is modeled using single Na atom in the 2×2, 3×3 and 4×4 supercell of graphene containing 8, 18, and 32 number of carbon atoms. In this work, the adsorption of sodium on graphene is performed at three different occupation sites: the Top (T) site directly above the carbon atom, the Hollow (H) site at the center of hexagon and the bridge (B) site at the midpoint of the C–C bond (Fig. 2). For each adsorption site of the adatom-graphene system, the foreign atom (Na) and C atoms on graphene are relaxed in all x, y and z directions. To estimate the binding energy of Na, the calculations for the isolated Na, isolated graphene and Na-adsorbed graphene system are performed in same-sized graphene supercell.

During the process of calculations, we have first performed the relaxation of initial geometries to obtain the optimized structures. The systems thus calculated are

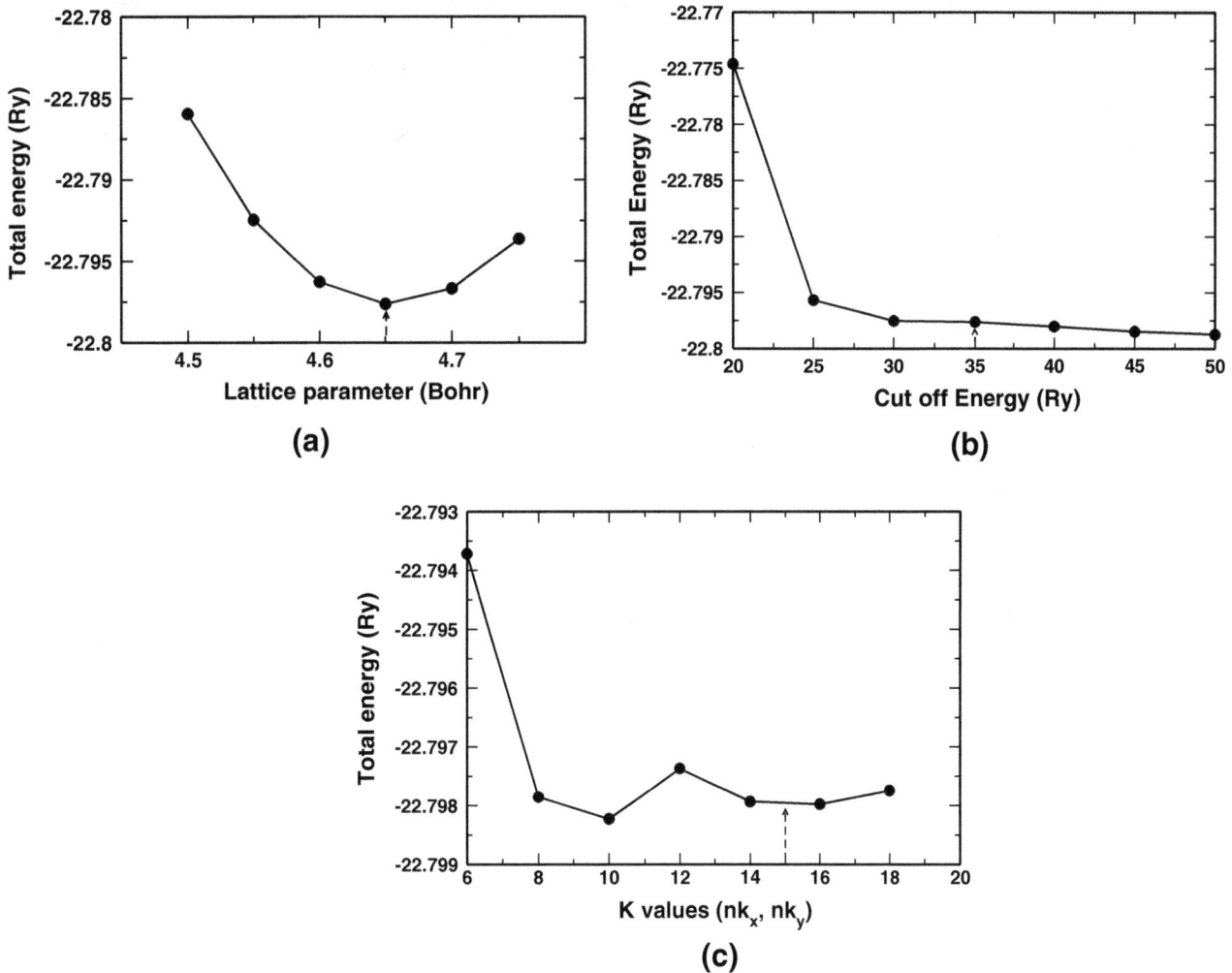

Fig. 1 The convergence of total energy with respect to the lattice parameter (**a**), cut off energy (**b**) and k-points (**c**)

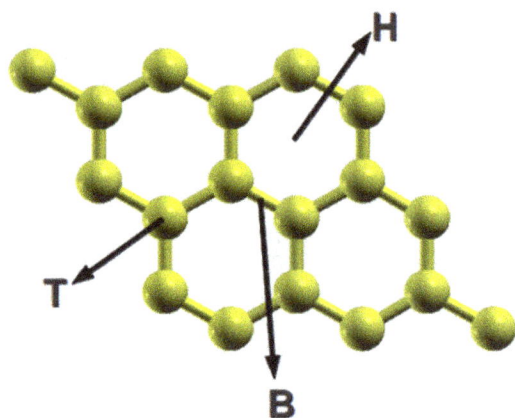

Fig. 2 Three occupation sites top (T), bridge (B) and hollow (H) of the graphene sheet are shown

allowed fully to relax using BFGS (Broyden–Fletcher–Goldfarb–Shanno) scheme until the total energy changes between two consecutive self-consistent field steps are less than 10^{-4} Ry and force acting is less than 10^{-3} Ry/Bohr. The first Brillouin zone of graphene is sampled in the reciprocal space using the Monkhorst–Pack scheme with an appropriate number of k-points, as described by the convergence test. Smearing was incorporated to aid the convergence. We have used 'Marzarri–vanderbilt' method (Marzari et al. 1999) or cold smearing with a small Gaussian spread of 0.001 Ry. Furthermore, we have chosen 'david' diagonalization method with the mixing factor 0.6 for self consistency. Spin polarized calculations are accommodated to study the magnetic properties of the systems. For the density of states (DOS) calculations of pure graphene and sodium-adsorbed graphene systems, we have used 3×3 supercell of graphene with a denser mesh of $(15 \times 15 \times 1)$ k-points.

We have also investigated the stability of hydrogen adsorption on sodium added graphene system. As a first step, H_2-molecular was relaxed on similar box of 3×3 graphene supercell and then defined its optimized geometry accordingly. The bond length of H_2 molecule calculated in this way has been found as 0.75 Å which agrees with the previously reported value (Arellano et al. 2012). Due to the periodicity of the crystal, the distance between two H_2-moleculars of nearest supercells becomes equal to the cell dimension (i.e., 9.299 Bohr) which is large enough compared to the molecular bond length of H_2 (i.e., 1.418 Bohr). This situation ensures that there is no interaction between two H_2-moleculars located at adjacent supercells.

Results and discussion

In the present work, we study the adsorption of single sodium atom on three different occupation sites of 2×2,

3×3 and 4×4 supercells of pure graphene. Furthermore, we extend our work to study the adsorption of H_2-moleculars on pure graphene and Na-adsorbed graphene.

Adsorption of sodium atom on graphene

We discuss structural, electronic and magnetic properties of pure and sodium added graphene in this section.

The binding energy of sodium atom on graphene sheet is calculated using relation,

$$\Delta E = E_{Na} + E_G - E_{G+Na} \tag{1}$$

where E_{Na}, E_G and E_{G+Na} are the ground state energies of sodium atom, pure graphene and sodium-adsorbed graphene systems in the fixed supercells, respectively. The positive binding energy from Eq. (1) indicates stable configurations. We have defined the most preferred sites/geometries of the systems for 2×2, 3×3, and 4×4 supercells separately by taking information of the largest binding energy values from these calculations.

The adatom height (h) is defined as the difference in z coordinate of the Na and an average of z coordinates of the C atoms in graphene layer. We have also calculated the distance (d_{Ac}) between the adsorbed atom and its nearest carbon atom. The adsorption of sodium on graphene produces a small distortion which is quantified by computing the maximum deviation in z direction of C atoms in graphene layer from their average initial position. This distortion of the graphene layer upon the adsorption of a foreign atom (Na in this case) is also calculated in terms of change in dihedral angles.

The binding energy of sodium atom on hexagonal 2×2 graphene supercell, containing 8 carbon atoms, has been calculated as −0.29, −0.38, −0.39 eV on H, B, and T sites, respectively. This signifies that Na-adsorbed graphene in 2×2 supercell is less stable than its constituent systems (the pristine graphene and isolated Na atom), which agrees with the previously reported results for two-atom unit cell (Medeiros et al. 2010).

The supercells of larger size are more favorable for the adsorption of Na atom. Table 1 presents the binding energy values, the equilibrium distances from the average graphene sheet (h), the distance of adatom (Na) from the nearest carbon atom (d_{AC}) of graphene and also the distortion of the graphene sheet d_{GC} in 3×3 supercell. The table quantifies binding energies of sodium atom at the hollow (H), bridge (B), and top (T) sites as 0.45, 0.31 and 0.33 eV, respectively. The values imply that sodium atom is bound to all the tested sites of the graphene sheet and the H site is the most favored one among them. This conclusion agrees with the previously reported results (Chan et al. 2008; Oli et al. 2013). The equilibrium distance of sodium atom from the surface of graphene sheet is found to be 2.32 Å for H

Table 1 The adsorption energy of sodium atom ΔE, its height from graphene sheet (h), the distance of the nearest carbon atom of the sheet (d_{AC}), and the distortion observed in 3×3 graphene sheet due to adsorption of sodium atom (d_{GC}) are listed in this table

Atom	site	ΔE (eV)	h (Å)	d_{AC} (Å)	d_{GC} (10^{-2} Å)
Na	H	0.45	2.32	2.72	0.94
	B	0.31	2.42	2.57	−0.08
	T	0.33	2.41	2.48	−0.53

site, 2.42 Å for B site, and 2.41 Å for T site. The smallest equilibrium distance at H site signifies that there exists the strongest interaction between the sodium atom and carbon atoms of this site compared to those at B and T sites. The distortion produced by sodium is found only about 0.009 Å for hollow (H) site. The magnitude, however, is not significant (noticeable distortion is defined for ≥ 0.07 Å (Chan et al. 2008)) and this is true for bridge and top sites as well. Since the distortion in the graphene sheet (3×3 supercell) is not significant, C−C bonds near the Na atom retain their sp^2 character and do not rehybridize significantly with any adsorbed atom orbitals.

In the case of 4×4 supercell, the binding energy of sodium atom at hollow site is found to be 0.78 eV. This value is significantly higher than that for 3×3 supercell at the same (H) site (0.45 eV). It has been found that the binding energy of adsorbed metal atoms is size dependent and gets almost constant value beyond certain number of C atoms in graphene. Oli et al. (2013) have reported rapid change in binding energy of alkali metal atoms (Li, Na and K) on graphene before reaching its saturation size, i.e., 48 number of carbon atoms. The quantities revealed by the authors for Na in passivated graphene structure of 16 carbon atoms of $C_{16}H_{10}$ and 30 carbon atoms of $C_{30}H_{14}$ are 0.32 eV and 0.72 eV, which are reasonably close to the present work of 18 (3×3 supercell) and 32 (4×4 supercell) atoms of graphene sheets, respectively (Oli et al. 2013). Ding et al. (2011), on the other hand, have reported different trends of size dependency of binding energy for different atoms, while varying graphene supercell from 3×3 to 4×4. The authors find binding energy difference in between 3×3 and 4×4 supercells for the elements of transition metals moving from Sc to Cu, in general, in decreasing trend. In separate calculations, we have observed very slight change in binding energy of Pt in 4×4 graphene supercell with respect to that in 3×3 graphene, while it is noticeable in between 2×2 and 3×3 (Lamichhane et al. 2014). It can, therefore, be concluded that size dependency of binding energy and also the saturation size of graphene depend on the chemistry of elements and usually higher for monovalent atoms.

The equilibrium distance between the sodium atom and graphene sheet in 4×4 supercell, $h = 2.25$ Å, is lower

than that in smaller supercells. Also the distance of the nearest carbon atom at hexagonal site from the sodium, d_{AC}, is found to be 2.69 Å. Both the geometrical parameters, h and d_{AC}, from the present results come in good agreement with the previous work performed by Chan et al. (2008) where the values are reported as 2.28 and 2.70 Å respectively, in the same-size (4×4) supercell. Sodium atom produces a small distortion, 0.008 Å, to the graphene sheet of 4×4 supercell as well. The distortion falls in the similar range to that of the 3×3 supercell of graphene system with slightly higher value in the smaller size. The small positive distortion implies that graphene sheet gently moves towards the sodium atom (upwards), due to interaction between them.

The density of states (DOS) and the band structure calculations of 3×3 pure and Na-adsorbed-graphene systems disclose the changes in electronic/magnetic properties of the systems, due to presence of Na, over the pure graphene.

The DOS for spin up and spin down states with reference to Fermi level (represented by the vertical dotted line) are shown in Figs. 3, and 4. In case of pure graphene (Fig. 3), the Fermi level lies at −2.35 eV and the DOS for spin up and spin down are seen symmetrical. Dirac point, where DOS is zero, lies exactly at the Fermi level. This symmetry of DOS about the Fermi level implies that pure graphene is non-magnetic. However, Dirac point shifts below the Fermi level and also the symmetry of DOS is broken with the extra peaks at/near the Fermi level when Na is adsorbed in graphene (Fig. 4). The shifting of Dirac point below the Fermi-level has also been discussed in Mukherjee and Kaloni (2012) for nitrogen-doped (N-doped) graphene system, where N contributes as an electron donor as similar to Na in the present work. Figure 4 also indicates its modified conductivity with reference to zero-band gap pristine graphene, which is consistent with the Medeiros et al. (2010). The breaking of symmetry, on the other hand, introduces the magnetic behavior. In case of

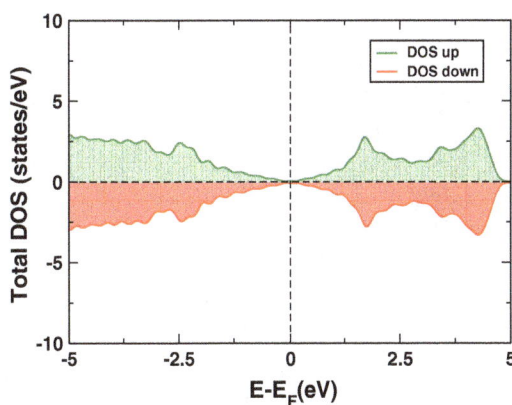

Fig. 3 Density of states (DOS) of pure graphene

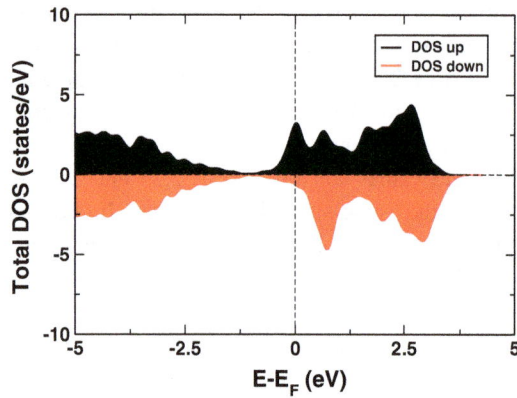

Fig. 4 Density of states (DOS) of Na-adsorbed graphene

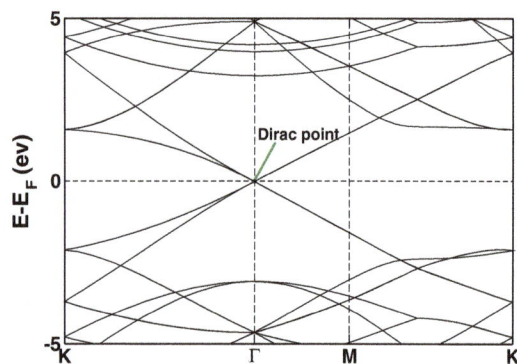

Fig. 5 The band structure of the pure graphene

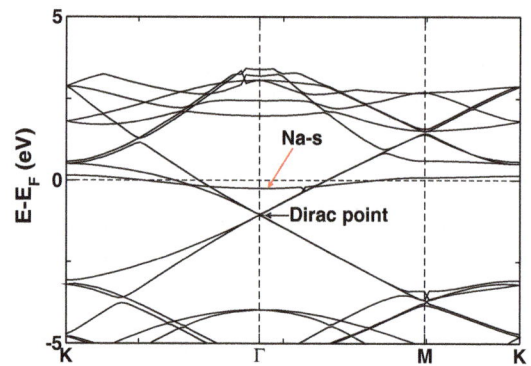

Fig. 6 The bands structure of Na-graphene system. The *red arrow* shows the s-orbital contribution of Na atom

localized and form σ bonds. These electrons do not contribute to the transport phenomenon. The fourth electron, perpendicular to the plane of graphene sheet, occupies the $2p_z$ orbital and forms the π-bond with $2p_z$ electron of neighboring carbon atom (Novoselov et al. 2005). These electrons forming π-bond, however, are delocalized over the entire lattice and have higher energy than the electrons that form the σ-bonds. Most of the interesting properties of graphene arise due to these delocalized π-electrons (Novoselov et al. 2007). The bands above and below the Fermi level are referred to π* and π bands, respectively. In case of pure graphene, these conical π bands meet at the Fermi level and the band gap formed by them becomes zero (Fig. 5). The pure graphene thus is also known as zero band gap semiconductor. The meeting point of the conical bands (also known as Dirac point) and the conical region around the Dirac point (also known as Dirac cone) are the special properties of graphene and they are successfully produced in our calculations.

The band structure of Na-graphene system, on the other hands, shows that Dirac point shifts below the Fermi level by 1.084 eV (Fig. 6), similar to the results described by Liu et al. (2011). When the electron donor Na atom is adsorbed on graphene, a fraction of electronic charge transfers from the sodium atom towards the carbons of graphene. Since the π band of the graphene is already occupied, the transferred charge from sodium moves towards the π* band (Medeiros et al. 2010). If we compare the Fermi level of pure graphene and Na-adatom graphene, it lies at −2.346 eV in the former case and rises to −0.695 eV after the adsorption of Na atom. It means that the Fermi level shifts by 1.651 eV towards the conduction band (upwards) due to Na contribution. Furthermore, an extra band nearby Fermi level is clearly seen and marked as s orbital contribution of Na in Fig. 6, which is similar to Li contribution on graphene reported in Kaloni et al. (2013). The conduction band and valence band overlap in sodium-added graphene, and this suggests that the new material is conductor.

3×3 supercell of graphene, the value of magnetization is 0.08 μ_B where as that for 4×4 supercell is 0.24 μ_B. The total magnetization of 4×4 supercell from this work agrees well with the previously reported value 0.27 μ_B (Chan et al. 2008). However, these values of magnetic moments for Na-graphene systems are lower than that for isolated sodium atom estimated theoretically (using electronic configurations) and computationally (present work) (1.0 μ_B). We believe that this reduction in magnetization could be due to partial transfer of electronic charge from Na to the graphene.

We have performed the band structure calculations by defining high symmetry points (i.e., K, Γ and M) on the edges of irreducible Brillouin zone in reciprocal space. In this case, we have taken 100 k-points along the specific direction of irreducible Brillouin zone to obtain the fine band structure. Figures 5 and 6 represent the band structure of pure graphene and Na-graphene systems, respectively.

Graphene is a single atomic layer of graphite where the carbon atoms condense in a honeycomb lattice due to their sp² hybridization of 2s, $2p_x$, and $2p_y$ orbitals of three valence electrons out of four valence electrons. The mixing of the 2s and two 2p orbitals gives the planar sp² hybridization. The electrons involved in sp² hybridization are

Adsorption of H_2-molecular/s

In this section, we discuss the geometry and the stability of adsorption of H_2-molecular/s on pure and sodium-adsorbed graphene systems.

Adsorption of H_2-molecular on the pure graphene

We have considered a number of samples to study the interaction between pure graphene sheet and single H_2-molecular in perpendicular and parallel orientations at different positions of 3×3 supercell of graphene. The perpendicular configurations, H_2 molecule perpendicular to the graphene plane at top, bridge and hollow sites, are shown in Figs. 7, 8, and 9, respectively. Figure 10, on the other hand, defines the parallel configuration of H_2 at hollow position of the graphene.

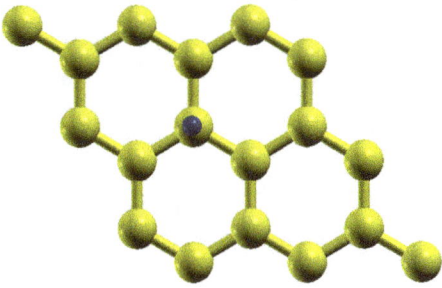

Fig. 7 H_2 molecule perpendicular to the carbon atom (top site)

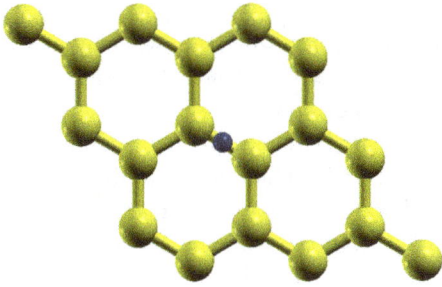

Fig. 8 H_2 molecule perpendicular to the carbon–carbon bond (bridge site)

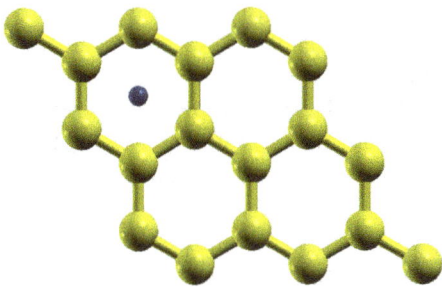

Fig. 9 H_2 molecule perpendicular to the center of hexagonal ring of carbon atoms (hollow site)

The adsorption/binding energy (ΔE) of H_2 molecule, will be referred as adsorption energy hereby, is calculated from the optimized geometry of H_2-graphene system using equation

$$\Delta E = E_G + E_{H_2} - E_{G+H_2} \tag{2}$$

with appropriate replacements of combined and individual components of the system. In this case, E_G, E_{H_2}, and E_{G+H_2} are the ground state energy of pure graphene, H_2-molecular, and H_2-graphene (combined) systems, respectively. The values are calculated using same-sized supercell of graphene. The magnitudes of adsorption energy of H_2 molecule and its distance from the graphene plane are presented in Table 2.

Table 2 shows that the adsorption energy is maximum and the equilibrium distance is minimum for H_2 molecular of its perpendicular configuration at hollow (H) site of graphene. This confirms that H_2 molecular perpendicular to the hexagonal ring is most stable than any other configurations. The calculated values of adsorption energy of H_2 molecular from the present work (61–69 meV) in the pure graphene sheet lie within the range of previously reported values (55–80 meV) (Kwon 2010). Small adsorption energy for single hydrogen adsorption could be because of very strong intra-molecular H–H bond, extremely small induced dipole moment, and the chemically inert nature of the our system (i.e., graphene). The small adsorption causes the weak binding of H_2 mainly due to van der Waals

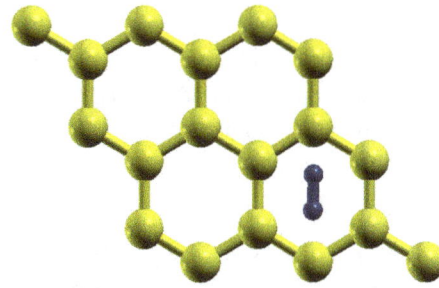

Fig. 10 H_2 molecule parallel to the hexagonal ring of carbon atoms

Table 2 The table quantifies the adsorption energy of single H_2 molecule when adsorbed in perpendicular and parallel configurations at bridge (B), top (T) and hollow (H) positions of the graphene sheets (ΔE). The symbol 'h' represents the equilibrium distance of H_2 molecular, measured from the lower atom of H_2 in perpendicular case, from the graphene sheet

Position of H_2	Binding energy ΔE (meV)	h (Å)
Perpendicular at B	62.9	2.67
Perpendicular at T	61.9	2.72
Perpendicular at H	68.6	2.48
Parallel at H	66.2	2.82

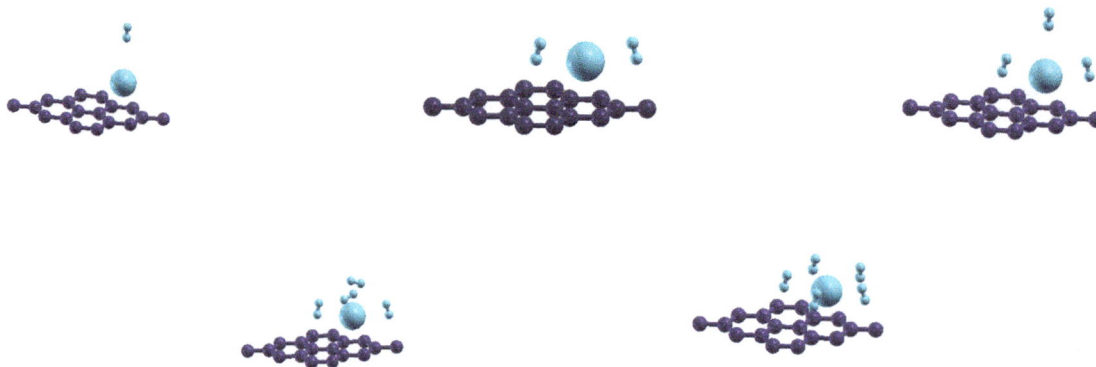

Fig. 11 Adsorption of a varying number of H_2-molecular on Na-adsorbed graphene system

Table 3 The adsorption energy of H_2-moleculars (E_{NH_2}) and adsorption energy per H_2-molecular (E_{H_2}) are listed

Number of H_2	E_{NH_2} (eV)	E_{H_2} (eV)
1	0.023	0.023
2	0.174	0.087
3	0.405	0.135
4	0.628	0.157
5	0.960	0.192

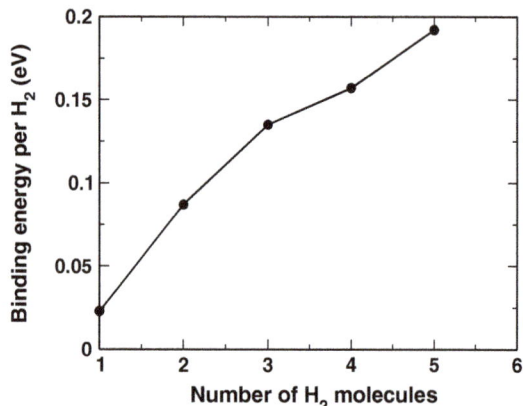

Fig. 12 The binding energy per H_2-molecular is in increasing order up to 5 H_2 molecular

interaction, and can be easily taken out from system, even at low temperature.

Adsorption of H_2-molecular/s on the sodium-adsorbed graphene

The adsorption of varying number of H_2-moleculars (one to five) has been tested on sodium-adsorbed 3×3 graphene supercell, and the adsorption energy of the same is computed by

$$\Delta E = E_{G+Na} + NE_{H_2} - E_{G+Na+H_2} \qquad (3)$$

where E_{G+Na+H_2} is the ground state energy of the graphene supercell after adsorbing sodium atom and H_2-moleculars. Similarly, E_{G+Na} and E_{H_2} are the ground state energies of sodium-adsorbed graphene and molecular hydrogen, respectively. The number of H_2-moleculars is represented by N in the equation. The positive adsorption energy signifies that H_2-molecular bounds to the adatom-graphene system. The binding energy per H_2-molecular has been further calculated by,

B.E. per H_2-molecular $= \frac{\Delta E}{N}$

Figure 11 shows the optimized structures of hydrogen–sodium–graphene systems with a varying number of H_2-molecular.

Table 3 shows that the adsorption energy of H_2 molecular as well as adsorption energy per H_2-molecular increases continuously as we increase the number of

molecules in the system. If we compare the adsorption energy per H_2 molecular with and without adatom in graphene supercell, it can be found that presence of Na enhances the binding strength except for $N = 1$. The adsorption energy per H_2 molecular ranges from 0.023 to 0.192 eV/H_2 for the adsorption of one to five numbers of H_2 molecular (Fig. 12). This range of adsorption energy comes in good agreement with the previously reported average binding energy of hydrogen, 0.09 to 0.22 eV for one to four molecules, in the Lithium-adsorbed graphene system (Ataca et al. 2008).

Figure 12 shows that the adsorption energy per H_2-molecular increases and attains the maximum value (0.192 eV) as we increase the number of H_2-molecular from one to five. The maximum value (0.192 eV/H_2) approaches nearby one of the goals set by DOE, 0.2–0.7 eV for the practical applications (Zhou and Willians 2011) . However, this increasing pattern is seen dissimilar with our calculations of hydrogen adsorption in other metallic atom adsorbed-graphene (Pd/Pt-graphene) systems. The reason could be because of different structural/electronic structure of sodium compared to other atoms. We believe that further analysis of charge distribution before and after the adsorption of gaseous molecules will explore the causes beyond these results.

Graphene and carbon nanomaterials have been considered as porous materials/substrates for the storage of energy carrying gases via adsorption at operating conditions. Similar to the previously accepted results, our calculations have found low adsorption energy of H_2 in pure graphene. However, the binding strength is enhanced significantly for 2–5 numbers of H_2-molecular in Na-adsorbed graphene. The content of hydrogen in the proposed system can also be analyzed on the basis of its presence by weight. For 5 molecular of hydrogen, the maximum number of molecules adsorbed in our calculations reaches 4.02 wt % (weight percentage) which is progressive towards the DOE (Department of Energy) target (around 6 wt %) for the practical applications (Durgun et al. 2008).

Conclusions

The structural and electronic properties of pure graphene and Na-graphene systems are studied. The binding energy values for Na at different sites suggest that the hollow site is the most favorable position for the adsorption of sodium atom on graphene sheet. The band structure and DOS calculations of the pure graphene find zero band gap and non-magnetic nature, which are consistent with the previously established facts. In case of sodium-adsorbed graphene system, the conduction band and valence band are overlapped. DOS are not symmetrical, which indicates that the system is magnetic, and the calculations find magnetic moment 0.24 μ_B for Na in 4 × 4 graphene supercell.

We have also performed the adsorption of H_2-molecular on pure graphene sheet and sodium-adsorbed graphene system. The present study reveals that the adsorption energy per H_2-molecular is enhanced due to the presence of Na atom and reaches 0.192 eV in case of Na-adsorbed graphene compared to its value 0.068 eV in pure graphene. The hydrogen storage capacity of Na-adsorbed graphene for the adsorption of five H_2-molecular has been calculated as 4.02 wt % (weight percentage), which is progressive towards the DOE (Department of Energy) target (around 6 wt %) for the practical applications.

Acknowledgments We acknowledge the partial support from The Abdus Salam International Center for Theoretical Physics (ICTP) through office of external activities within NET-56 project. We also extend our gratitude to S. Narasimhan and K. Ulman for their valuable suggestions and inspiration to work in this area. N. Pantha acknowledges to Nepal Academy of science and Technology (NAST), Nepal, for its partial financial support. We also acknowledge re-viewer's feedback for the improvement of this manuscript.

References

Arellano JS, Molina LM, Rubio A, Lopez MJ, Alonso JA (2012) Interaction of molecular and atomic hydrogen with (5,5) and (6,6) single-wall carbon nanotubes. J Chem Phys 117:2281

Ataca C, Aktürk E, Ciraci S, Ustunel H (2008) High-capacity hydrogen storage by metallized graphene. Appl Phys Lett 93:043123

Burchell T, Rogers M (2000) Low pressure storage of natural gas for vehicular applications. SAE Tech. Pap. Ser. 2000–01-2205

Castro Neto AH, Guinea F, Peres NMR (2006) Drawing conclusions from graphene. Phys World 19:33

Castro Neto AH, Guinea F, Peres NMR, Novoselov KS, Geim AK (2009) The electronic properties of graphene. Rev Mod Phys 81:109

Chan KT, Neaton JB, Cohen ML (2008) First-principles study of metal adatom adsorption on graphene. Phys Rev B 77:235430

Chandrakumar KRS, Ghosh SK (2008) Alkali-metal-induced enhancement of hydrogen adsorption in C60 fullerene: an ab initio study. Nano Lett 8:13

Chen P, Wu X, Lin J, Tan KL (1999) High H_2 uptake by alkali-doped carbon nanotubes under ambient pressure and moderate temperatures. Science 285:91

Chen B, Li B, Chen L (2008) Prompted hydrogenation of carbon nanotubes by doping light metals. Appl Phys Lett 93:043104

Choi S-M, Jhi SH (2009) Electronic property of Na-doped epitaxial graphenes on SiC. Appl Phys Lett 94:153108

Ci L, Song L, Jin C, Jariwala D, Wu D, Li Y, Srivastava A, Wang ZF, Storr K, Balicas L, Liu F, Ajayan PM (2010) Atomic layers of hybridized boron nitride and graphene domains. Nat Mater 9:430

Dai J, Yuan J, Giannozzi P (2009) Gas adsorption graphene doped with B, N, Al, and S: a theoretical study. Appl Phys Lett 95:232105

Ding J, Qiao Z, Feng W, Yao Y, Niu Q (2011) Engineering quantum anomalous/valley Hall states in graphene via metal-atom adsorption: an ab-initio study. Phys Rev B 84:195444

Duren T, Sarkisov L, Yaghi OM, Snurr RQ (2004) Design of new materials for methane storage. Langmuir 20:2683

Durgun E, Ciraci S, Yildirim T (2008) Functionalization of carbon-based nanostructures with light transition-metal atoms for hydrogen storage. Phys Rev B 77:085405

Esquinazi P, Spemann D, Höhne R, Setzer A, Han K-H, Butz T (2003) Induced Magnetic Ordering by Proton Irradiation in Graphite Phys. Rev Lett 91:227201

Geim AK, Novoselov KS (2007) The rise of Graphene. Nat Mater 6:183

Geim AK (2009) Graphene: status and prospects. Science 324:1530

Giannozzi P et al (2009) QUANTUM ESPRESSO: a modular and open-source software project for quantum simulations of materials. J Phys Condens Matter 21:395502

Hohenberg P, Kohn W (1964) Inhomogeneous electron gas. Phys Rev 136:B864

Johll H, Kang HC, Tok ES (2009) Density functional theory study of Fe Co, and Ni adatoms and dimers adsorbed on graphene. Phys Rev B 79:245416

Kaloni TP, Upadhyay Kahaly M, Cheng YC, Schwingenschlögl U (2008) K-intercalated carbon systems: effects of dimensionality and substrate. EPL 98:67003

Kaloni TP, Upadhyay Kahaly M (2011) Induced magnetism in transition metal intercalated graphitic systems. J Mater Chem 21:18681

Kaloni TP, Cheng YC, Schwingenschlögl U (2012) Fluorinated monovacancies in graphene: even–odd effect. EPL 100:37003

Kaloni TP, Upadhyay Kahaly M, Faccio R, Schwingenschlögl U (2013) Modelling magnetism of C at O and B monovacancies in graphene. Carbon 64:281

Kaloni TP, Balatsky AV, Schwingenschlögl U (2013) Substrate-enhanced superconductivity in Li-decorated graphene. EPL 104:47013

Kaloni TP, Joshi RP, Adhikari NP, Schwingenschlögl U (2014) Band gap tunning in BN-doped graphene systems with high carrier mobility. Appl Phys Lett 104:073116

Karki DB, Adhikari NP (2014) First-principles study of the stability of graphene and adsorption of halogen atoms (F, Cl and Br) on hydrogen passivated graphene Int. J Mod Phys B 28:1450141

Klimes J, Michaelides A (2012) Perspective: advances and challenges in treating van der Waals dispersion forces in density functional theory. J Chem Phys 137: 120901

Kohn W, Sahm LJ (1965) Self-consistent equations including exchange and correlation effects. Phys Rev 140:1133

Kwon YK (2010) Hydrogen adsorption on sp2-bonded carbon structures: Ab-initio study. J Korean Phys Soc 57:778

Lamichhane S, Pantha N, Adhikari NP (2014) Hydrogen storage on platinum decorated graphene: a first-principles study. Bibechana 11(1):107

Liu X, Wang CZ, Yao YX, Lu WC, Tringides MC, Ho KH (2011) Bonding and charge transfer by metal adatom adsorption on graphene. Phys Rev B 83:235411

Ma S, Sun D, Simmons JM, Collier CD, Yuan D, Zhou HC (2008) Metal-organic framework from an anthracene derivative containing nanoscopic cages exhibiting high methane uptake. J Am Chem Soc 130:1012

Marzari N, Vanderbilt D, de Vita A, Payne MC (1999) Thermal contraction and disordering of the Al(110) surface. Phys Rev lett 82:3296

Medeiros VCP, de Mota FN, Mascarenhas JSA, de Castilho CMC (2010) Adsorption of monovalent metal atoms on graphene: a theoretical approach. Nanotechnology 21:115701

Mukherjee S, Kaloni TPJ (2012) Electronic properties of boron-and nitrogen-doped graphene: a first principles study. Nanopart Res 14:1059

Nakanishi E, Reid RC (1971) Liquid natural gas–water reactions. Chem Eng Prog 67:36

Novoselov KS, Jiang Z, Zhang Y, Morozov SV, Stormer HL, Zeitler U, Maan JC, Boebinger GS, Kim P, Geim AK (2007) Room-temperature quantum hall effect in graphene. Science 315:1379

Novoselov KS, Geim AK, Morozov SV, Jiang D, Katsnelson MI, Dubonos SV, Grigorieva IV, Firsov AA (2004) Electric field effect in atomically thin carbon films. Science 306:666

Novoselov KS, Geim AK, Morozov SV, Jiang D, Katsnelson MI, Zhang Y, Grigorieva IV, Dubonos SV, Firsov AA (2005) Two-dimensional gas of massless Dirac fermions in graphene. Nature 438:197

Oli BD, Bhattarai C, Nepal B, Adhikari NP (2013) First-Principles study of adsorption of alkali metals (Li, Na, K) on graphene. Adv Nanomater Nanotechnol 143:515

Orimo S, Nakamori Y, Eliseo JR, Zttel A, Jensen CM (2007) Complex hydrides for hydrogen storage. Chem Rev 107:4111

Palacios JJ, Rossier JF, Brey L (2008) Vacancy-induced magnetism in graphene and graphene ribbons. Phys Rev B 77:195428

Pan L, Sander MB, Huang X, Li J, Smith M, Bittner E, Bockrath B, Johnson JK (2008) Microporous metal organic materials: promising candidates as sorbents for hydrogen storage. J Am Chem Soc 126:1308

Park CH, Louie SG (2010) Tunable excitons in biased bilayer graphene. Nano Lett 10:426

Perdew JP, Burke K, Ernzerhof M (1996) Generalized gradient approximation made simple. Phys Rev 77:3865

Pumera M (2011) Graphene-based nanomaterials for energy storage. Energy Environ Sci 4:668

Quantum ESPRESSO user manual, version 4.3.1

Saha SK, Chandrakanth RC, Krishnamurthy HR, Waghmare UV (2009) Mechanisms of molecular doping of graphene: a first-principles study. Phys Rev B 80:155414

Singh N, Kaloni TP, Schwingenschlögl U (2013) Additional information on Appl. Phys Lett Appl Phys Lett 102:023101

Thapa R, Sen D, Mitra MK, Chattopadhyay KK (2011) Palladium atoms and its dimers adsorbed on graphene: first-principles study. Phys B Condens Matt 406:368

Ugeda MM, Brihuega I, Guinea F, Gómez-Rodríguez JM (2010) Missing atom as a source of carbon magnetism. Phys Rev Lett 104:096804

Ulman K, Bhaumik D, Wood BC, Narasimhan S (2014) Physical origins of weak H2 binding on carbon nanostructures: insight from ab initio studies of chemically functionalized graphene nanoribbons. J Chem Phys 140:174708

Valencia F, Romero A, Ancilotto F, Silvestrelli P (2006) Lithium adsorption on graphite from density functional theory calculations. J Phys Chem B 110:14832

Wallace PR (1947) The band theory of graphite. Phys Rev 71:622

Wood BC, Bhide SY, Dutta D, Kandagal VS, Pathak AD, Punnathanam SN, Aayappa KG, Narasimhan S (2012) Methane and carbon dioxide adsorption on edge-functionalized graphene: a comparative DFT study. J Chem Phys 137:054702

Wu BR, Yang CK (2012) Electronic structures of graphane with vacancies and graphene adsorbed with fluorine atoms. AIP Adv 2:012173

Zhou JG, Willians QL (2011) Hydrogen storage on platinum decorated carbon nanotubes with boron, nitrogen dopants or sidewall vacancies. J Nano Res 15, 29.

Zhou M, Lu Y, Zhang C, Feng YP (2010) Strain effects on hydrogen storage capability of metal-decorated graphene: a first-principles study. Appl Phys Lett 97:103109

10

Optimization of process parameters for ruthenium nanoparticles synthesis by (w/o) reverse microemulsion

S. U. Nandanwar · J. Barad · S. Nandwani ·
M. Chakraborty

Abstract Taguchi OA factorial design method was used to identify the several factors that might affect the particle size of ruthenium nanoparticles prepared by the mixing of two reactive microemulsions. In the present work, the objective of evaluating the factors influencing the particle size had been improvised by studying two qualitative factors viz., effect of different reducing agents and effect of different co-surfactants. Using orthogonal experimental design and analysis technique, the system performance could be analyzed with more objective conclusion through only a small number of simulation experiments. Analysis of variance was carried out to identify the significant factors affecting the response and the best possible factor level combination was determined through. It was found that the formation of ruthenium nanoparticles, microemulsions were greatly influenced by the type of reducing agent used in the technique followed by water-to-surfactant molar ratio.

Keywords Taguchi method · ANOVA · Ru nanoparticles · Microemulsions technique

Introduction

Nowadays, there is a great interest in transition metal nanotechnologies and the development of simple and reproducible methods to synthesize the nanoparticles by the various research groups as they have shown maximum application in magnetism, semiconductor, optoelectronics, and especially in the field of catalysis (Kim et al. 2007). Ruthenium, transition metal has shown very unique and interesting catalytic activities for many reactions (Lu et al. 2008). Owing to huge application in field of catalysis many research groups were focused to synthesize the Ru nanoparticles. Generally, various methods were available for synthesis of Ru nanoparticles, such as refluxing a polyol solution (Yan et al. 2001), thermal decomposition of $Ru_3(CO)_{12}$ (Motoyama et al. 2006), sonochemical reduction (He et al. 2006), microwave-assisted reduction (Zhang et al. 2007a, b; Zawadzki and Okal 2008), organometallic synthesis (Deboutiere et al. 2009), solvothermal (Nandanwar et al. 2011a, b), chemical reduction method (Patharkar et al. 2013), and also by microemulsion technique (Kim et al. 2007; Xiong and Manthiram 2005; Rojas et al. 2005; Zhang et al. 2007a, b; Nandanwar et al. 2011a, b).

Kim et al. (2007), synthesized the bimetallic, platinum–ruthenium transition metal nanoparticles by chemical reduction using sodium borohydride in reverse microemulsion of water/isooctane/Igepal CA-630/2 propanol for fuel cell catalyst. Pt–Ru/C catalysts were synthesized using a reverse microemulsion using sodium *bis*(2-ethylhexyl) sulfosuccinate (AOT) as the surfactant and heptane as the oil phase (Xiong and Manthiram 2005). Carbon supported Pt and Pt–Ru electrocatalyst was prepared by the microemulsion technique (Rojas et al. 2005). Zhang et al. (2007a, b), prepared the ternary platinum–ruthenium-nickel nanoparticles by w/o reverse microemulsion. The composition and size of ternary Pt–Ru–Ni nanoparticles were controlled by adjusting the initial metal salt solution and preparation conditions. Recently, Nandanwar et al. (2011a, b), synthesized ruthenium nanoparticles using two reactants ruthenium chloride and sodium borohydride by microemulsion method.

S. U. Nandanwar · J. Barad · S. Nandwani ·
M. Chakraborty (✉)
Department of Chemical Engineering, S. V. National Institute of Technology, Surat 395 007, Gujarat, India
e-mail: mch@ched.svnit.ac.in

The microemulsion is one of the best techniques to synthesize ruthenium nanopaticles at ambient condition with narrow size distribution. Microemulsions are amphiphile stabilized transparent, optically isotropic and thermodynamically stable dispersions of water, oil and surfactant. Water-in-oil (w/o) microemulsion is one of the most recognized methods due to its several advantages, such as demanding no extreme pressure or temperature control, easy to handle, soft chemistry, and requiring no special or expensive equipment (Charinpanitkul et al. 2005). This technique has already 25 years of history, but the mechanisms to control the final size and the size distribution are still not known.

In the present article, highly monodisperse ruthenium nanoparticles were synthesized by w/o microemulsion system at room temperature. The various parameters that influenced on the size of particles like surfactant concentration, water-to-surfactant molar ratio (ω), concentration of the precursor, reducing agent-to-ruthenium trichloride molar ratio (R), effect of cosurfactant and effect of different reducing agent etc. were studied. Taguchi OA fractional factorial design method is used to optimize all the six parameters to synthesize the smallest ruthenium nanoparticles.

Taguchi OA fractional factorial design

The experimental work had been designed in a sequence of steps to insure that data are obtained in a way that its analysis will lead immediately to valid statistical inferences. This research methodology is termed as design of experiment (DOE) methodology. DOE using Taguchi approach attempts to extract maximum important information with minimum number of experiments (Lazic 2004). Taguchi techniques are experimental design optimization techniques that use standard orthogonal arrays (OA) for forming a matrix of experiments. An orthogonal array is a fractional factorial experimental matrix that is orthogonal and balanced. The ASQC (1983) Glossary & Tables for Statistical Quality Control defines fractional factorial design in the following way: a factorial

experiment is one in which only an adequately chosen fraction of the treatment combinations required for the complete factorial experiment is selected to be run.

In the present work, six parameters each at three levels are selected to evaluate the size of ruthenium nanoparticles obtained after mixing two microemulsions during each run. The factors to be studied are mentioned in Table 1. Based on Taguchi orthogonal array factorial designs method, the L_{27}-OA is constructed. A L_{27}-OA is chosen to evaluate some of the two-way interactions.

Experimental data

Materials

Transition metal, ruthenium trichloride (RuCl$_3$·nH$_2$O, Ru Content $\geq$37 %), cyclohexane, n-butanol, hydrazine hydrate (80 %) and methanol all were of analytical grades and purchased from Finar chemicals, India. The hydroxylammonium sulphate and n-pentanol were purchased from National chemicals, India. Sodium borohydride (NaBH$_4$, 95 %) and non-ionic surfactant, polyoxyethylene octyl phenyl ether (Triton X-100) were purchased from S. D. fine chemicals, India. All the chemicals were used without further purification. Distilled water was used throughout the experiments for preparing all the aqueous solutions.

Synthesis of ruthenium nanoparticle

Synthesis of monodisperse ruthenium nanoparticles by reverse microemulsion was reported in our previous paper (Nandanwar et al. 2011a, b). To prepare (0.2 mol/L) concentration of surfactant (Triton X-100), known amount of Triton X-100 was dissolved into cyclohexane and vigorously stirring by high-speed blender. The concentration of surfactant was varied by increasing amount of Triton X-100 in cyclohexane. RuCl$_3$-aqueous/cyclohexane/Triton X-100 microemulsion was prepared by drop wise addition of RuCl$_3$ aqueous solution (0.1 M) into the prepare mixture of cyclohexane-Triton X-100. The known quantity of

Table 1 Factors and their levels in the experimental design

Level	Water-to-surfactant molar ratio	Effect of surfactant concentration	Effect of RuCl$_3$ concentration	Effect of molar ratio of reducing agent	Effect of co-surfactant	Different reducing agent
1	3	0.2	0.1	3	No co-surfactant	NaBH$_4$
2	5	0.3	0.2	5	Pentanol	Hydrazine
3	7	0.4	0.3	10	Butanol	Hydroxylammonium

Table 2 L_{27}-OA and response values

Run no.	Water-to-surfactant molar ratio	Effect of surfactant concentration	Effect of RuCl$_3$ concentration	Effect of molar ratio of reducing agent	Effect of co-surfactant	Different reducing agent	(Number %) particle size (nm)
1	3	0.2	0.1	3	No co-surfactant	NaBH$_4$	19.01
2	3	0.2	0.1	3	n-Pentanol	Hydrazine	8.29
3	3	0.2	0.1	3	n-Butanol	Hydroxylammonium	9.47
4	3	0.3	0.2	5	No surfactant	NaBH$_4$	17.21
5	3	0.3	0.2	5	n-Pentanol	Hydrazine	8.01
6	3	0.3	0.2	5	n-Butanol	Hydroxylammonium	9.35
7	3	0.4	0.3	10	No co-surfactant	Hydrazine	15.07
8	3	0.4	0.3	10	n-Pentanol	NaBH$_4$	7.83
9	3	0.4	0.3	10	n-Butanol	Hydroxylammonium	9.08
10	5	0.2	0.2	10	No co-surfactant	Hydrazine	22.07
11	5	0.2	0.2	10	n-Pentanol	Hydroxylammonium	8.34
12	5	0.2	0.2	10	n-Butanol	NaBH$_4$	10.17
13	5	0.3	0.3	3	No co-surfactant	Hydrazine	23.09
14	5	0.3	0.3	3	n-Pentanol	Hydroxylammonium	10.74
15	5	0.3	0.3	3	n-Butanol	NaBH$_4$	11.52
16	5	0.4	0.1	5	No co-surfactant	hydrazine	23.12
17	5	0.4	0.1	5	n-Pentanol	Hydroxylammonium	10.75
18	5	0.4	0.1	5	n-Butanol	NaBH$_4$	11.72
19	7	0.2	0.3	5	No co-surfactant	Hydroxylammonium	24.91
20	7	0.2	0.3	5	n-Pentanol	NaBH$_4$	10.23
21	7	0.2	0.3	5	n-Butanol	Hydrazine	11.27
22	7	0.3	0.1	10	No co-surfactant	Hydroxylammonium	24.39
23	7	0.3	0.1	10	n-Pentanol	NaBH$_4$	10.19
24	7	0.3	0.1	10	n-Butanol	Hydrazine	11.37
25	7	0.4	0.2	3	No co-surfactant	Hydroxylammonium	25.07
26	7	0.4	0.2	3	n-Pentanol	NaBH$_4$	10.31
27	7	0.4	0.2	3	n-Butanol	Hydrazine	11.48

aqueous solution was added into Triton X-100 in cyclohexane mixture, to get desired water-to-surfactant molar ratio (ω) of 5. Uniform stirring was maintained with ultraturax T25 high-speed mechanical stirrer (Ultraturax® IKA WERKE, GmBH & Co. KG) at 6,500 rpm for 5 min at room temperature for proper mixing. Secondly, NaBH$_4$-aqueous/cyclohexane/Triton X-100 microemulsions with same ω value was prepared by using aqueous solution of NaBH$_4$ (0.5 M). Two microemulsions were mixed directly under adequate stirring for 15 min at room temperature. The mixture was destabilized by methanol. Finally, metal

nanoparticles were separated by high-speed centrifugation to accumulate concentrated Ru colloidal nanoparticles for further characterization.

Design of experiments

The experiments were carried out according to the L_{27}-OA. The size of ruthenium nanoparticles represented as (number %) particle size (nm) was considered as Taguchi array response. After an experimental design is created, the

Fig. 1 Main effect plot

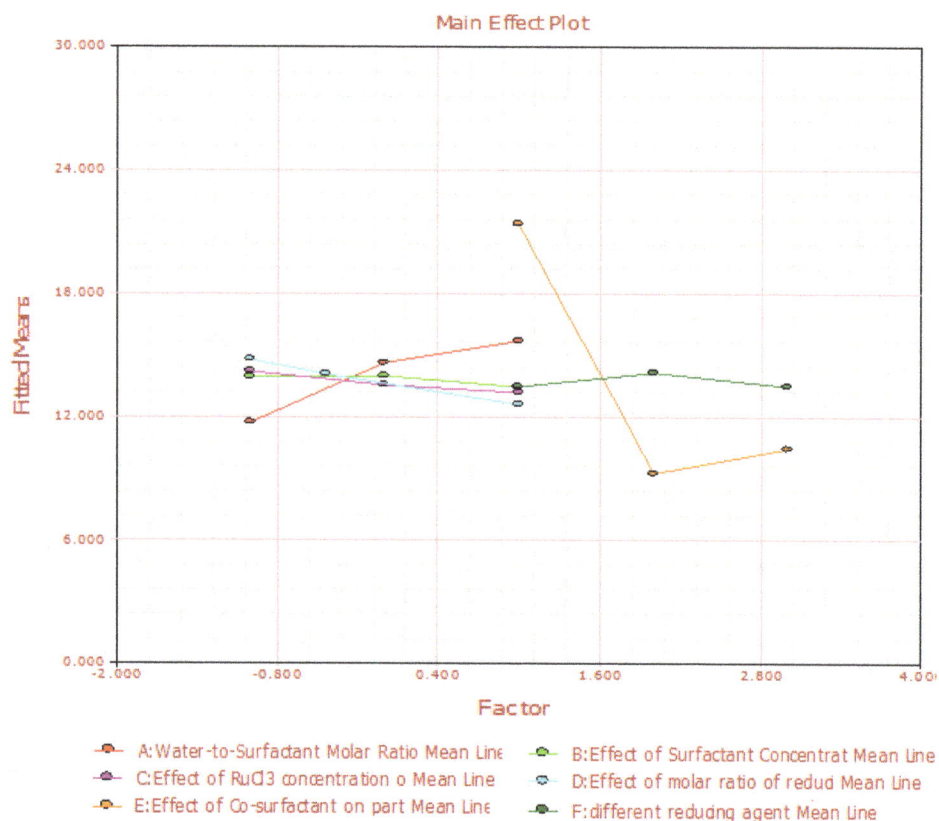

combinations of factor levels are randomly run and the data is grouped with the corresponding experimental run. The L_{27}-OA and response values (Number %) in terms of particle size (nm) are shown in Table 2.

Results and discussions

Main effect plot and interaction plot

Main effects plot for the main effect terms viz. factors A, B, C, D, E and F are shown in Fig. 1. However, the main effect plot does not tell, which of the main effect factors are statistically significant (Mendenhall and Sincich 1989). From the main effect plot, it has been observed that the size of ruthenium nanoparticles increases with increasing water-to-surfactant molar ratio (A). Since the interest of the present study lies in achieving smaller ruthenium nanoparticles, a water-to-surfactant molar ratio of 3 is found to be desirable. Because of low water content, the water solubilized in the polar core is bound by the surfactant molecules, which increases the boundary strength and decreases the intermicellar exchange rate among the reverse micelles which controls micellar sizes as well as sizes of the nanoparticles (Nandanwar et al. 2011a, b).

From the plot, it is clear that the size of ruthenium nanoparticles is almost independent of the two factors viz., surfactant concentration (B) and RuCl$_3$ concentration (C).

It has also been found that the size of ruthenium nanoparticles decreases with an increase in molar ratio of reducing agent (D). At higher molar ratio, higher intramicelles nucleation and growth will be promoted. As reduction takes place at a faster rate so bigger size particles will be generated.

Effect of different types of co-surfactants (E) on the size of ruthenium nanoparticles has also been observed. From the plot, it has been found that ruthenium nanoparticles obtained without co-surfactant is larger than when co-surfactants are used. However, among the two types of co-surfactants used, n-pentanol and n-butanol, the former gives smaller sized ruthenium nanoparticles. The co-surfactant increases the fluidity of the interface and thus the kinetics of the intermicellar exchange, which in turns ensures a more homogeneous repartition of reactants among droplets (Lopez-Quintela et al. 2004).

Different types of reducing agents (F) also affect the size of the ruthenium nanoparticles. Sodium borohydride is a strong reducing agent in comparison with hydrazine hydrate and hydroxylammonium sulphate. This leads to rate of reduction begin fast and thus rate of nucleation and the particles growth being controlled by the collision,

Fig. 2 Interaction plot for $A \times B$ where (B effect of surfactant concentration)

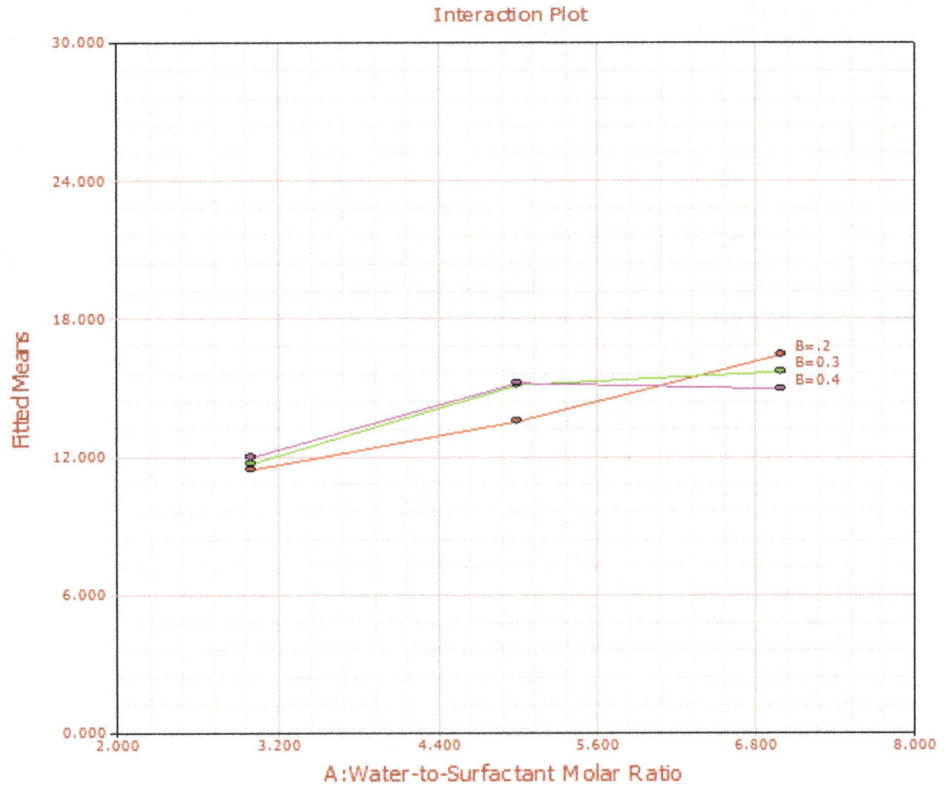

Fig. 3 Interaction plot for $A \times C$ where (C effect of $RuCl_3$ concentration)

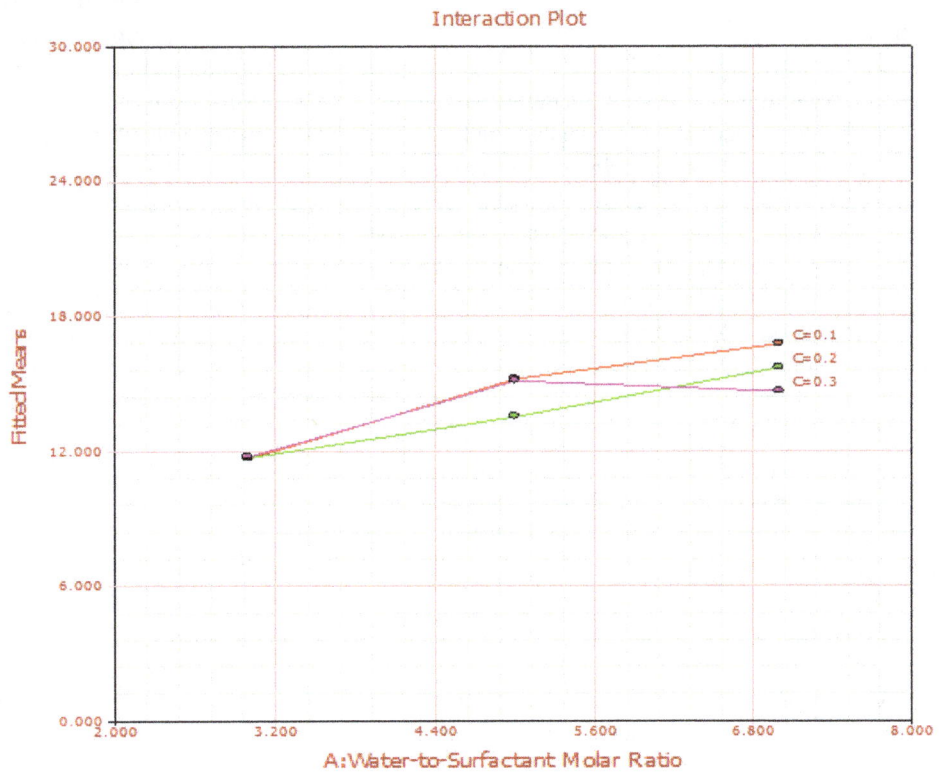

Fig. 4 Interaction plot for
$A \times E$ where (E effect of
cosurfactant)

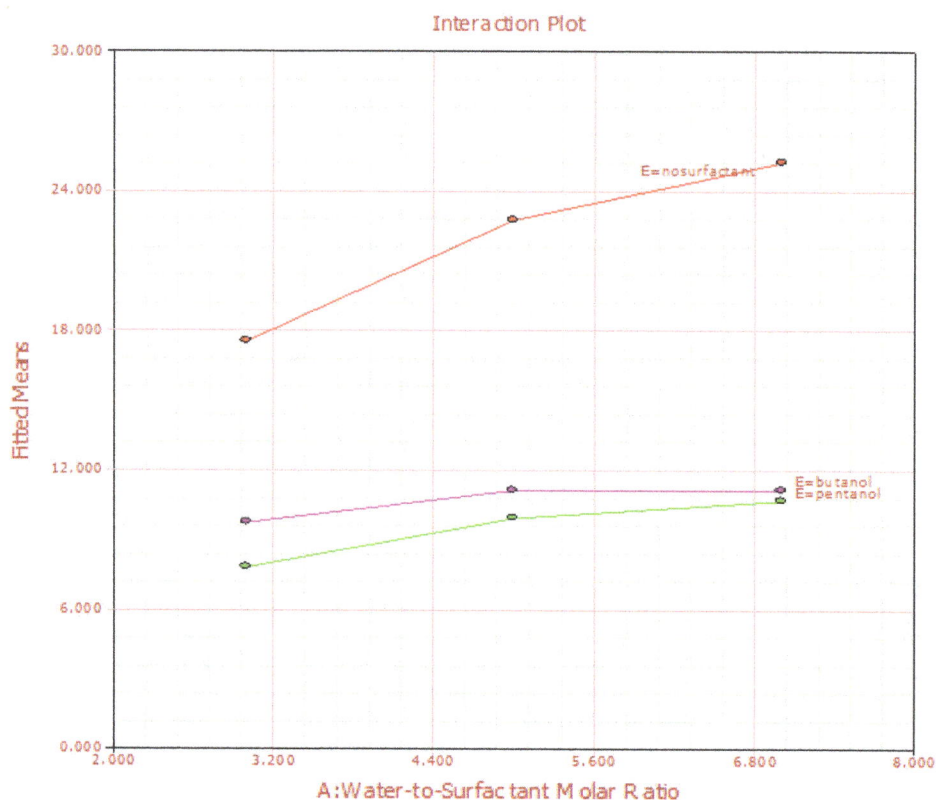

fusion and split of the droplets, which results in bigger size but highly monodispersed nanoparticles.

Interaction effects

Interaction means influence of one operating variable on the other operating variable (Mendenhall and Sincich 1989). In the present work three interaction terms have been considered viz., $(A \times B)$, $(A \times C)$ and $(A \times E)$. Whether interactions between factors exist or not can be shown by plotting an interaction plot. Parallel lines in an interaction plot indicate no interaction. However, the interaction plot doesn't tell if the interaction is statistically significant (Clements 1991). Interaction plots are most often used to visualize interactions during DOE. *Reliasoft DOE++1* displays interaction plots for each of the three interaction terms viz., $(A \times B)$, $(A \times C)$ and $(A \times E)$ in Figs. 2, 3 and 4 respectively. Since all the interaction plots have non-parallel lines, interaction exists between all the three interacting factors.

Optimum settings via term effect plot

In the present study, a term effect plot has been used to predict the optimum settings for obtaining smaller ruthenium nanoparticles. The level of each factor that gives smaller sized ruthenium nanoparticles (smaller response 'Y') is considered. From the term effect plot shown in Fig. 5, it is obvious that a water-to-surfactant molar ratio of '3' gives smaller ruthenium particles. Similarly for the effect of surfactant concentration, effect of $RuCl_3$ concentration and effect of molar ratio of reducing agent, the levels at which we obtain smallest ruthenium nanoparticles are '0.4', '0.3' and '10' respectively. When using different reducing agents, $NaBH_4$ gives the smallest nanoparticles and for different types of co-surfactants, *n*-pentanol gives result at optimum level. Thus from the term effect plot, it is observed that the optimum settings for the experimentation of producing ruthenium nanoparticles might be the one shown in Table 3. Using the diagnostic tool from the *Reliasoft* software, it has been observed that the smallest nanoparticles were obtained for the same settings, i.e., a smallest size of '7.83 nm'.

ANOVA table

The analysis of variance (ANOVA) is one of the most commonly used methods of analyzing experiments. In any experiment where several factors are allowed to vary, a situation called experimental error exists. This experimental error creates a background "noise" in the data. ANOVA measures this background noise and also the amount of signal each factor under study creates. If a factor is creating a signal that has more magnitude than the background noise then that factor has a significant effect (3).

Fig. 5 Term effect plot

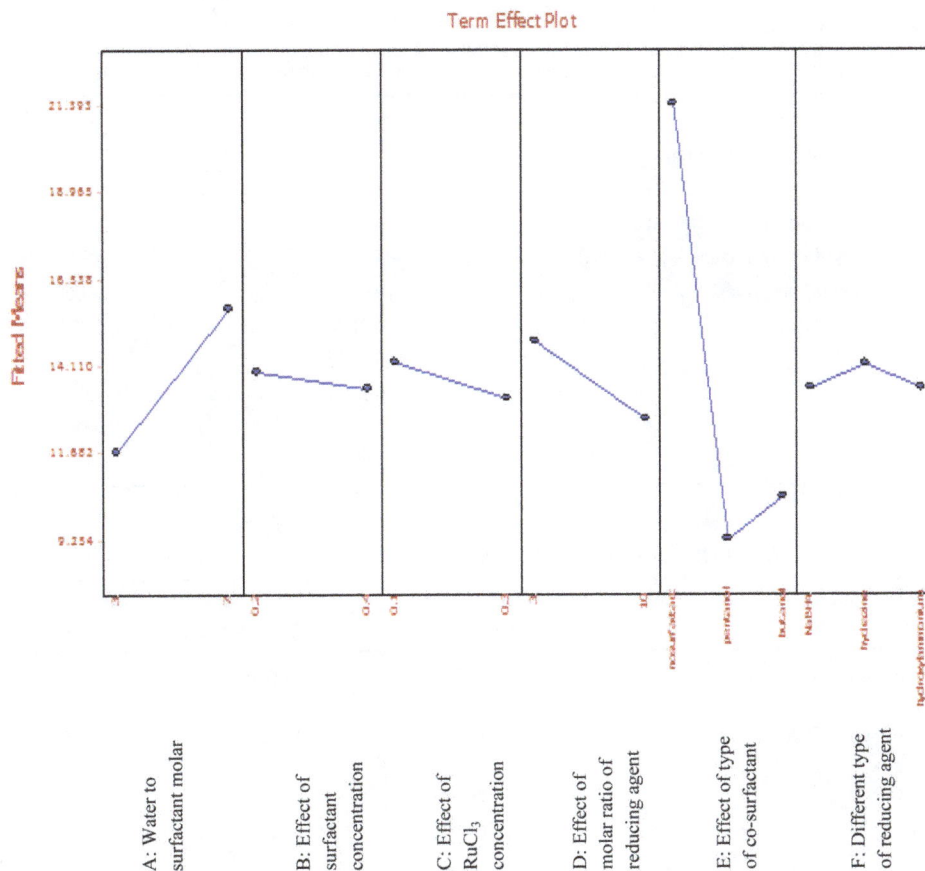

Term Effect Plot

Fitted Means

A: Water to surfactant molar

B: Effect of surfactant concentration

C: Effect of RuCl$_3$ concentration

D: Effect of molar ratio of reducing agent

E: Effect of type of co-surfactant

F: Different type of reducing agent

Table 3 Optimum settings for the experimentation of producing ruthenium nanoparticles

Factors	Optimum level
Water-to surfactant molar ratio	3
Effect of surfactant concentration	0.4
Effect of RuCl$_3$ concentration	0.3
Effect of molar ratio of reducing agent	10
Effect of type of co-surfactant	Pentanol
Different type of reducing agent	NaBH$_4$

Statistical tests are applied to the data to test for significance or association. The analysis of experiment is done by using *Reliasoft DOE++1* software. The present work has been based on 95 % confidence level ($\alpha = 0.05$) which means that there is a probability of at least 95 % that the result is reliable. In the present study, since orthogonal arrays do not test all variable combinations, the interaction effect of all the parameters could not be taken into optimization process. As a result, necessary two factor interactions, i.e., the combined effect of the factors ($A \times B$), ($A \times C$) and ($A \times E$) where [A Water-to-surfactant molar ratio, B effect of surfactant concentration, C effect of

RuCl$_3$ concentration and E effect of different co-surfactants] are only considered.

An ANOVA table for the above experiment is given in Table 4. The table reveals the significant effects (in red) and the non-significant effects. It also shows the standard deviation for the statistical analysis to be $s = 0.9674$. Regression coefficients R^2 and R^2 (adj) are found to be 98.59 and 97.39 respectively which indicates that model can explain the variation in size of ruthenium nanoparticles to the extent of 98.59 % which makes the model adequate to represent the process.

The final ANOVA table for formation of ruthenium nanoparticles via microemulsion technique, showing the significant factor (main effects and interactions) and their percentage contribution, is shown in Table 5. The significant main effect terms and interaction terms in the ANOVA Table 5 are computed using the F ratio as a test statistic (2).

Pareto chart

A PARETO chart is used to visually describe the significant and non-significant terms in synthesis of ruthenium nanoparticles by microemulsion method (Fig. 6).

Table 4 ANOVA Table

Source of variation	Degrees of freedom	Sum of squares (sequential)	Mean squares (sequential)	F ratio	P value
A water-to-surfactant molar ratio	1	71.6006	71.6006	76.5091	4.77E−07
B effect of surfactant concentration	1	0.0249	0.0249	0.0266	0.8727
C effect of RuCl$_3$ concentration	1	1.1603	1.1603	1.2398	0.2843
D effect of molar ratio of reducing agent	1	6.6576	6.6576	7.114	0.0184
E effect of type of co-surfactant	2	805.7456	402.8728	430.4913	2.68E−13
F different reducing agent	2	16.0046	8.0023	8.5509	0.0037
$A \times B$	1	0.5833	0.5833	0.6233	0.443
$A \times C$	1	0.4975	0.4975	0.5316	0.478
$A \times E$	2	16.2621	8.1311	8.6885	0.0035
Residual (error)	14	13.1018	0.9358		
Total	26	931.6383			

Table 5 Final ANOVA table showing only significant factors

Source of variation	Degrees of freedom	Sum of squares (sequential)	Mean squares (sequential)	F ratio	P value	Percentage contribution (%)
A water-to-surfactant molar ratio	1	71.6006	71.6006	39.5375	4.77E−07	7.6855
E effect of type of co-surfactant	2	805.7456	402.8728	222.4588	2.68E−13	86.487
$A \times E$	2	16.2621	8.1311	4.4898	0.0035	1.7455
Residual (pooled error)	21	38.0300	1.8110			4.082
Total	26	931.6383				

Fig. 6 PARETO chart for Ru nanoparticles by microemulsion method

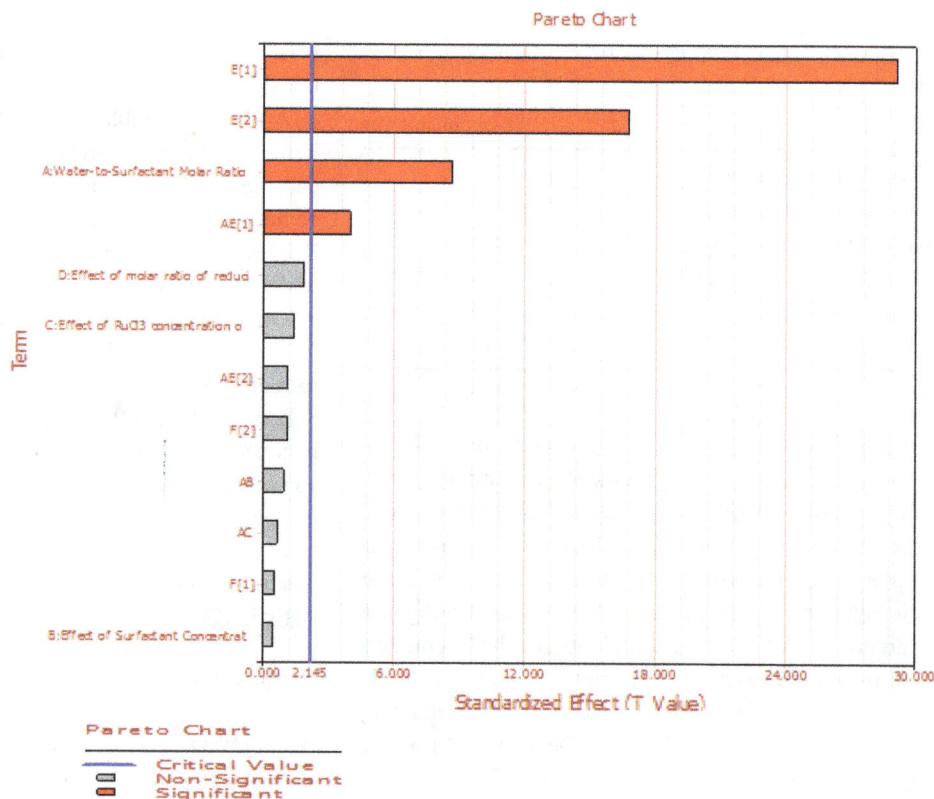

A PARETO chart is used to determine the importance of an effect. The standardized PARETO chart shown contains a bar for each effect, sorted from most significant to least significant. The length of each bar is proportional to the standardized effect, which is equal to the magnitude of the statistic that would be used to test the statistical significance of that effect. A vertical line is drawn at the location of the 0.05 critical values for Student's. Any bars that extend to the right of that line indicate effects that are statistically significant at the 5 % significance level.

In the present study, from PARETO chart it is seen that the effect of different co-surfactants and water-to-surfactant molar ratio are the significant factors. In addition, interaction between the above two factors is significant.

Conclusion

This study had shown the application of Taguchi OA Factorial Design method to identify the performance evaluation of microemulsion technique for synthesis of ruthenium nanopaticles. Authors tried to find out optimum parameters to synthesize the smallest ruthenium nanoparticles with narrow size distribution by this method. By using orthogonal experimental design and analysis technique, the system performance was analyzed in better way through only a small number of simulation experiments. ANOVA was carried out to identify the significant factors affecting particle size as well as size distribution and the best possible factor level combination was determined through.

References

Charinpanitkul T, Chanagul A, Dutta J, Rungsardthong U, Tanthapanichakoon W (2005) Effects of cosurfactant on ZnS nanoparticles synthesis in microemulsion. Sci Technol Adv Mater 6:266–271

Clements RB (1991) Handbook of statistical methods in manufacturing. Prentice Hall, Englewood Cliffs, NJ

Deboutiere P-J, Martinez V, Philippot K, Chaudret B (2009) An organometallic approach for the synthesis of water-soluble ruthenium and platinum nanoparticles. Dalton Trans 38(46): 10172–10174

He Y, Vinodgopal K, Muthupandian A, Grieser F (2006) Sonochemical synthesis of ruthenium nanoparticles. Res Chem Intermed 32:709–715

Kim T, Kobayashi K, Nagai M (2007) Preparation and characterization of platinum–ruthenium bimetallic nanoparticles using reverse microemulsions for fuel cell catalyst. J Oleo Sci 56(10): 553–562

Lazic ZR (2004) Design of experiments in chemical engineering. Wiley, New Jersey

Lopez-Quintela MA, Tojo C, Blanco MC, Garcia Rio L, Leis JR (2004) Microemulsion dynamics and reactions in microemulsion. Curr Opin Colloid Interface Sci 9:264–278

Lu F, Liu J, Xu J (2008) Synthesis of chain-like Ru nanoparticle arrays and its catalytic activity for hydrogenation of phenol in aqueous media. Mater Chem Phys 108:369–374

Mendenhall W, Sincich T (1989) Statistics for the engineering and computer sciences, 2nd edn. Maxwell Macmillian International Editions, San Francisco

Motoyama Y, Takasaki M, Higashi K, Yoon SH, Mochida I, Nagashima H (2006) Highly-dispersed and size-controlled ruthenium nanoparticles on carbon nanofibers: preparation, characterization, and catalysis. Chem Lett 35:876–877

Nandanwar SU, Chakraborty M, Mukhopadhyay S, Shenoy KT (2011a) Stability of ruthenium nanoparticles synthesized by solvothermal method. Cryst Res Technol 46:393–399

Nandanwar SU, Chakraborty M, Murthy ZVP (2011b) Formation of ruthenium nanoparticles by the mixing of two reactive microemulsions. Ind Eng Chem Res 50:11445–11451

Patharkar RG, Nandanwar SU, Chakraborty M (2013) Synthesis of colloidal ruthenium nanocatalyst by chemical reduction method. J Chem 2013:1–5

Rojas S, Garcia-Garcia FJ, Jaras S, Martinez-Huerta MV, Garcia Fierro JL, Boutonnet M (2005) Preparation of carbon supported Pt and PtRu nanoparticles from microemulsion electrocatalysts for fuel cell applications. Appl Catal A 285(1–2):24–35

Xiong L, Manthiram A (2005) Catalytic activity of Pt–Ru alloys synthesized by a microemulsion method in direct methanol fuel cells. Solid State Ion 176:385–392

Yan X, Liu H, Liew KY (2001) Size control of polymer-stabilized ruthenium nanoparticles by polyol reduction. J Mater Chem 11:3387–3391

Zawadzki M, Okal J (2008) Synthesis and structure characterization of Ru nanoparticles stabilized by PVP or γ-Al$_2$O$_3$. Mater Res Bull 43:3111–3121

Zhang X, Zhang F, Guan RF, Chan KY (2007a) Preparation of Pt–Ru–Ni ternary nanoparticles by microemulsion and electrocatalytic activity for methanol oxidation. Mater Res Bull 42(2): 327–333

Zhang Y, Yu J, Niu H, Liu H (2007b) Synthesis of PVP-stabilized ruthenium colloids with low boiling point alcohols. J Colloid Interface Sci 313:503–510

11

Watermelon rind-mediated green synthesis of noble palladium nanoparticles: catalytic application

R. Lakshmipathy · B. Palakshi Reddy ·
N. C. Sarada · K. Chidambaram · Sk. Khadeer Pasha

Abstract The present study reports the feasibility of synthesis of palladium nanoparticles (Pd NPs) by watermelon rind. The aqueous extract prepared from watermelon rind, an agro waste, was evaluated as capping and reducing agent for biosynthesis of palladium nanoparticles. The formation of Pd NPs was visually monitored with change in color from pale yellow to dark brown and later monitored with UV–Vis spectroscopy. The synthesized Pd NPs were further characterized by XRD, FTIR, DLS, AFM and TEM techniques. The synthesized Pd NPs were employed in Suzuki coupling reaction as catalyst. The results reveal that watermelon rind, an agro waste, is capable of synthesizing spherical-shaped Pd NPs with catalytic activity.

Keywords Watermelon rind · Pd nanoparticles · Green synthesis · Catalysis · Particle size

Introduction

Biosynthesis of nanoparticles is receiving increased attention due to the advancements in chemical and physical methods. Green synthesis of nanoparticles is cost effective and environmentally friendly as it has unique properties and enormous applications in biological tagging, pharmaceutical and optoelectronics (Jacob et al. 2012). Palladium nanoparticles (Pd NPs) are of interest because of their catalytic properties and unique application in sensors and catalysis (Bankar et al. 2010). Palladium nanoparticles are conventionally synthesized by chemical, electrochemical or sonochemical methods, but in recent years many researchers have reported green and eco-friendly way of synthesis of palladium nanoparticles using plant extracts, microbes and agricultural wastes (Sathishkumar et al. 2009; Saxena et al. 2012; Narayanan and Sakthivel 2011). Recently, agricultural wastes such as banana peel (Bankar et al. 2010), and custard apple peel (Roopan et al. 2011) which are rich in polyphenols, lignin and pectin were explored in the synthesis of palladium nanoparticles. The potential of other fruit peel extracts as surfactants/reductants for synthesis of Pd NPs needs to be fully explored.

Watermelon (*Citrullus lanatus*) being the largest and heaviest fruit, is one of the most abundant and cheapest available fruits in India with 3 lakh tones produced every year. Red flesh of watermelon present inside is sweet, edible and used for juices and salads but the outer rind is considered as waste and has no commercial value. Watermelon rind (WR) consists of pectin, citrulline, cellulose, proteins and carotenoids (Rimando and Perkins-Veazie 2005; Andrew et al. 2008; Quek et al. 2007) which are rich in functional groups such as hydroxyl (cellulose) and carboxylic (pectin). In this paper, for the first time, watermelon rind powder extract (WRPE) was used as capping or reducing agent for synthesis of Pd NPs.

R. Lakshmipathy · N. C. Sarada
Environmental and Analytical Chemistry Division, School of Advanced Sciences, VIT University, Vellore 632014, Tamilnadu, India

B. Palakshi Reddy
Department of GEBH, Sree Vidyanikethan Engineering College, Tirupati, AndhraPradesh, India

K. Chidambaram · Sk. Khadeer Pasha (✉)
Centre for Excellence in Nanomaterials, School of Advanced Sciences, VIT University, Vellore 632014, Tamilnadu, India
e-mail: khadheerau@yahoo.com; khadheerbasha@lycos.com

Materials and methods

Preparation of the extract

Watermelon rinds (WR) were obtained from local fruit market and washed with tap water followed by double distilled water. After thorough washing, WR was cut into small pieces and dried under sun light for 7 days. The dried WR pieces were washed repeatedly with hot water (70 °C) to remove any soluble matter present and then dried in an oven at 85 °C for 48 h. The oven-dried WR was powdered using a conventional grinder and sieved through 100 mesh sieve standard. To 50 mL of distilled water, 1 g of sieved WR was added and boiled in a water bath for 30 min to prepare aqueous extract. The extract was cooled to room temperature and filtered through 0.45 μm filters and freshly prepared extract was used for synthesis. The aqueous extract was termed as watermelon rind aqueous extract (WRAE).

Biosynthesis of palladium nanoparticles

The source of palladium was palladium chloride in distilled water. For synthesis of Pd NPs, 20 mL of 1 mM $PdCl_2$ solution was added with 10 mL of WRAE and incubated in a temperature controlled shaker at 150 rpm for 24 h at 30 °C. Parameters such as reactant ratio, pH and reaction time were optimized. The formation of Pd NPs was visually confirmed and later by UV–Vis spectrophotometer. At optimized conditions, the obtained colloidal solution was centrifuged at 10,000 rpm for 10 min and the solid was washed with water, methanol and acetone to remove biomolecules. The solid was dried at 85 °C and used for further characterization.

Characterization of palladium nanoparticles

The nanoparticles synthesized on reduction by WRPE were initially characterized using UV–Vis spectrometer (Hitachi, Model: U-2800) from 200 to 600 nm. Further characterization was done using FTIR (Thermo Nicolet, Avatar 330, USA) for identification of functional groups present in WRAE and Pd NPs between 4,000 and 400 cm^{-1}, crystalline nature of synthesized Pd NPs was recorded with XRD (Bruker Germany, D8 Advance diffractometer), surface morphology and surface roughness were examined with AFM (Nanosurf 2 Easyscan), TEM (Tecnai 10, Philips) and dynamic light spectroscopy (Delsa Nano S, Backman Coulter, USA) was equipped to know the size and shape of the as-synthesized Pd NPs.

Application in catalysis

The as-synthesized Pd NPs were employed as catalyst for Suzuki coupling. To a solution of Pd NPs (2 mol%, 0.01 g in water 10 mL), K_2CO_3 (2 mmol, 0.27 g) was added followed by iodobenzene (1 mmol, 0.11 g) and phenylboronic acid (1.5 mmol, 0.18 g). The reaction was stirred at room temperature until completion of the reaction. The progress of the reaction was monitored through TLC. After completion of the reaction, the reaction mixture was centrifuged and filtered. The filtrate was evaporated and the crude product was purified by column chromatography (petroleum ether).

Results and discussion

Polyhydroxylated molecules present interesting dynamic supramolecular associations facilitated by inter- and intramolecular hydrogen bonding resulting in molecular level capsules, which can act as templates for nanoparticles growth (Hebeish et al. 2010). Recently, watermelon rind was reported to bind cations and cationic dyes from aqueous solution (Lakshmipathy et al. 2013; Lakshmipathy and Sarada 2013a, b). Similarly, the negatively charged groups of extract facilitate the attraction of positively charged palladium cations towards the polymeric chain followed by reduction with the existing reducing groups. The mechanism predicted was to be a two-step process, i.e., atom formation and compounding of the atoms. In the first step, a portion of palladium ions in the solution is reduced by the available reducing groups of WRAE. The atoms thus produced act as nucleation centers and catalyze the reduction of remaining palladium ions present in the bulk solution.

Optimization of reaction conditions

Preliminary experiments were carried to standardize the reaction conditions. The reactant ratio, pH and contact time were monitored for the standardization of Pd NPs synthesis. Initially, the reaction time was optimized for synthesis of Pd NPs based on the absorbance. It was observed that with the increase in time, the number of Pd NPs was found to increase and attained saturation at 24 h. Further experiments were carried out at 24 h of contact time. The effect of pH on formation of Pd NPs was monitored at pH 2, 4, 6 and 8 and found that pH 6 showed the rapid formation of Pd NPs. The Pd source to extract ratio was varied from 1:0.2 to 1:2 and it was observed that 1:0.5 ratio of $PdCl_2$ to WRAE was found to yield maximum amount of Pd NPs beyond that found to be constant. Hence, bulk synthesis of Pd NPs was carried out at pH 6, 1:0.5 reactant ratio for 24 h of reaction time.

Characterization of Pd NPs

The formation of Pd NPs was visually monitored by change in color of solution from pale yellow to dark brown.

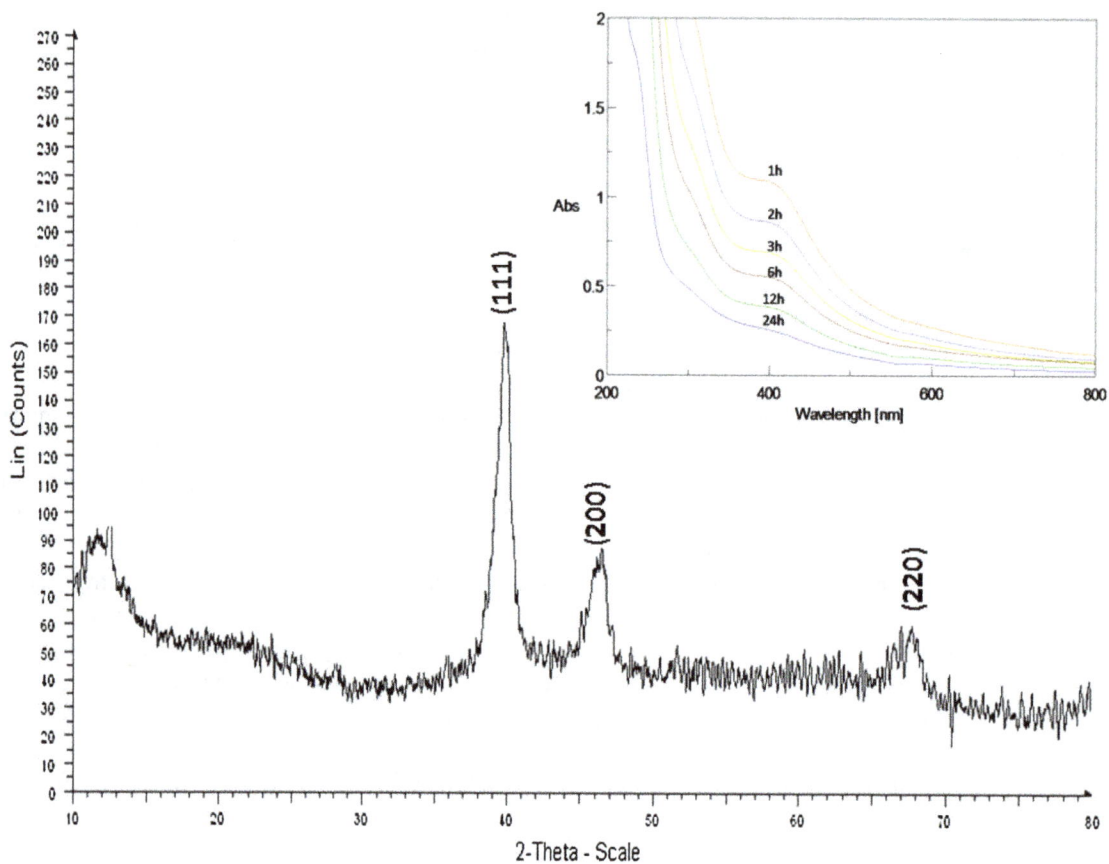

Fig. 1 XRD patterns of synthesized Pd NPs mediated by WRAE (*inset* UV–Vis spectra of reaction mixture with respect to time)

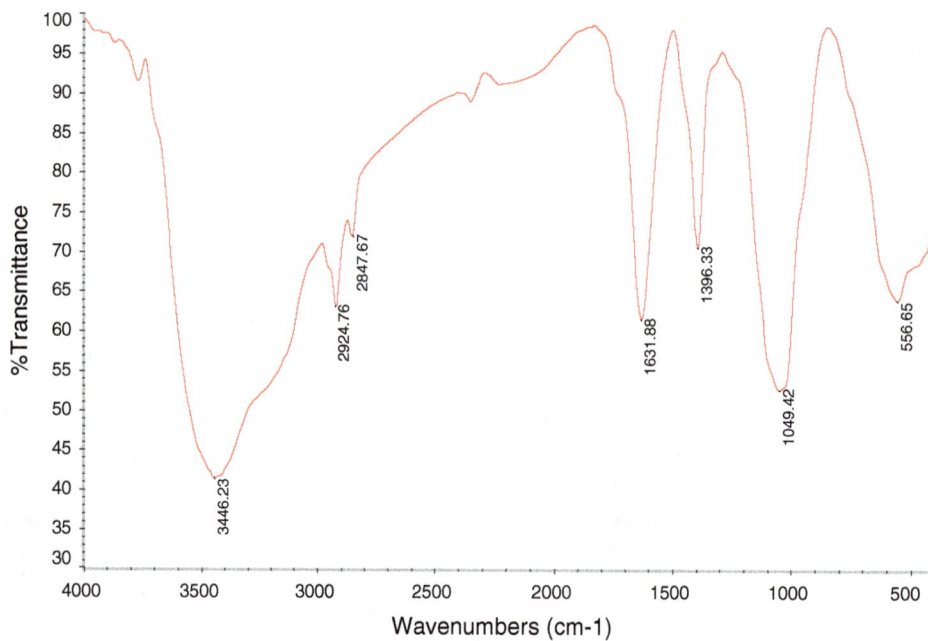

Fig. 2 FTIR spectra of Pd NPs mediated by WRAE

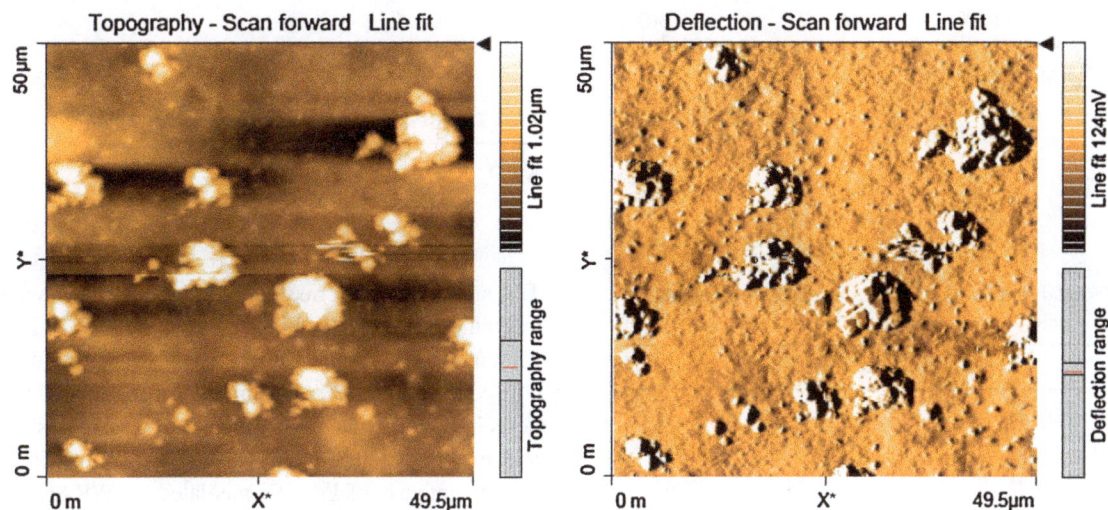

Fig. 3 AFM images of Pd NPs mediated by WRAE

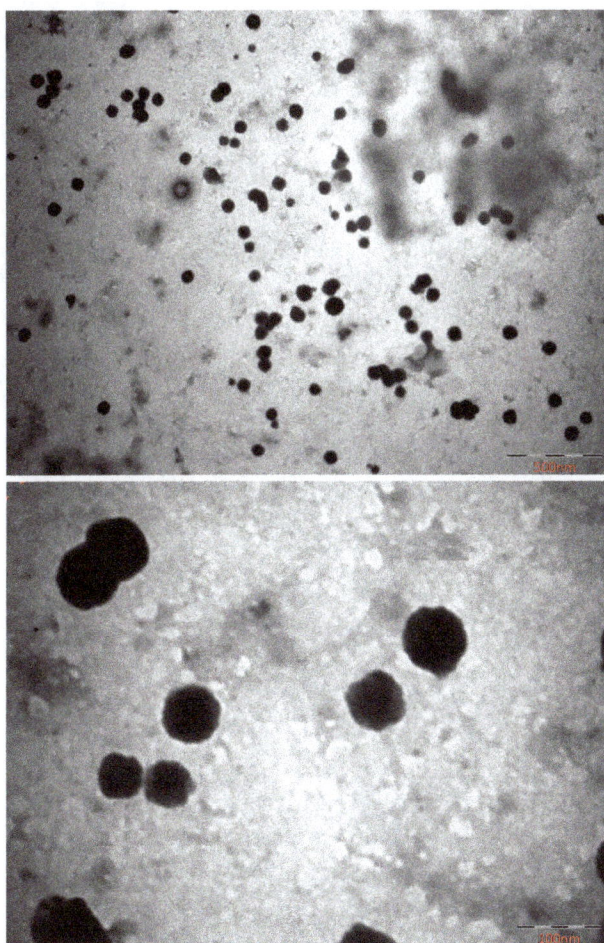

Fig. 4 TEM images of Pd NPs mediated by WRAE at different magnifications

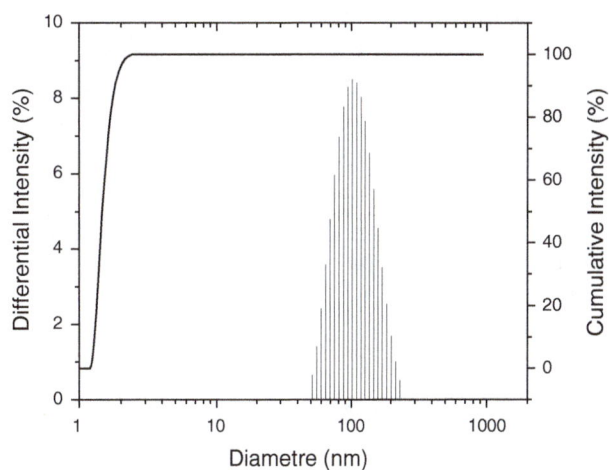

Fig. 5 DLS pattern of as-synthesized Pd NPs mediated by WRAE

Figure 1 (inset) shows the UV–Vis spectra of reaction mixture (PdCl$_2$ solution and WRAE) with respect to time. The observed peaks at 415 nm at 1 h indicate the presence of Pd^{2+} ions in reaction mixture. Over the time period, the peak at 415 starts to disappear and color of the solution turned intense dark brown due to excitation of surface plasmon vibrations in the Pd NPs. After 24 h, the complete disappearance of peak at 415 nm depicts the complete conversation of Pd^{2+} to Pd^0 state. Similar observations have been reported by Bankar et al. (2010) for banana peel and Yang et al. (2010) for *Cinnamomum camphora* leaf extract. The XRD pattern of WRAE-mediated synthesis of Pd NPs is shown in Fig. 1. The intense peaks at 40°, 46° and 68° can be assigned to (111), (200) and (220) Bragg's reflection. The lattice constant of Pd NPs synthesized by WRAE was in agreement with joint committee of powder diffraction standards (JCPDS).

To identify the capping/reducing agents present and predict their role in the synthesis of Pd NPs, FTIR measurement was done. The FTIR spectrum of Pd NPs shows peaks at various wavenumbers corresponding to different

Fig. 6 Scheme representing the Suzuki coupling reaction mediated by Pd NPs synthesized by WRAE

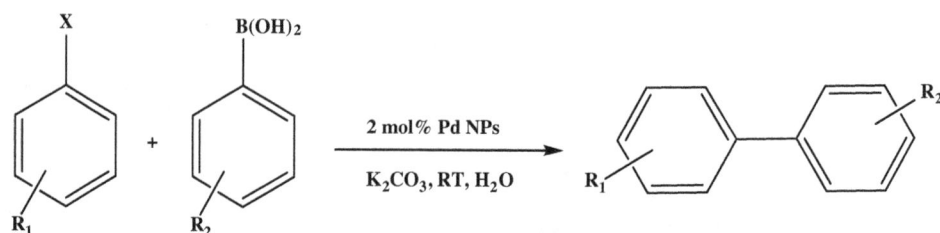

Table 1 Pd NPs catalyzed Suzuki coupling reaction with aryl halide and phenylboronic acid

Entry	X	R1	R2	Reac. time (h)	%Yield[a]	mp (°C)[b]
1	I	H	H	2	98	66–67
2	I	4-NO$_2$	H	2	93	113–115
3	Br	H	H	3	96	66–67
4	Br	3-NO$_2$	H	2.5	94	58
5	Br	4-NO$_2$	H	2	95	114–115
6	Br	4-CH$_3$	H	3	92	42–44
7	Br	3-OCH$_3$	H	3.5	90	liq
8	Br	4-OCH$_3$	H	3	91	liq
9	Br	4-COCH$_3$	H	2.5	93	110–112
10	Br	4-COCH$_3$	2-CH$_3$	3	83	94
11	Br	4-CH$_3$	2-CH$_3$	3	90	273
12	Br	4-NO$_2$	2-CH$_3$	2.5	92	102

Reaction conditions: aryl halide (1 mmol), boronic acid (1.5 mmol), K$_2$CO$_3$ (2 mmol), Pd NPs in water (2 mol%, 10 mL)

[a] Isolated yield

[b] (Li et al. 2012)

functional groups in Fig. 2. A strong peak at 3,447 cm^{-1} corresponds to symmetrical stretching vibrations of –OH groups of hydroxyl molecules. The peaks at 2,924 and 2,847 cm^{-1} correspond to stretching vibrations of methyl and methoxy groups. A peak at 1,636 corresponds to bending vibrations of –OH groups. A strong peak at 1,049 corresponds to asymmetrical bending vibrations of –CH$_2$ and –CH$_3$ groups. These results confirm that the polyhydroxyl groups from WRAE have been adsorbed as a layer over the Pd NPs and in turn stabilize the synthesis of nanoparticles.

Figure 3 shows the AFM images of Pd NPs mediated by WRAE. For the AFM studies, the synthesized Pd NPs were dispersed in distilled water and were air dried on glass plates. As a result of air drying on the surface, a coffee ring phenomenon was observed for Pd NPs. As observed in Fig. 3, at the center of the drop, nanoparticles and aggregates were formed. It is well known that when liquids containing fine particles are evaporated on a flat surface, the particles accumulate on the outer surface and form typical structures (Chen and Evans 2009). Similar type of observations has been reported for Pd NPs mediated by Banana peel extract for SEM studies (Bankar et al. 2010).

TEM analysis was carried out to study the shape and size of the synthesized Pd NPs. Figure 4 shows the TEM images at different magnifications. The Pd NPs were found to be well dispersed and spherical in shape. Shape-controlled synthesis of nanoparticles is of interest to researchers due to its unique applications in catalysis. The display of shape-controlled synthesis of Pd NPs mediated by WRAE makes the study unique for its industrial applications. Further, DLS measurements were carried out to obtain the size distribution of synthesized Pd NPs (Fig. 5). The average particle diameter of the synthesized Pd NPs was found to be ~96.4 nm with polydispersity index of 0.243.

Application in Suzuki coupling reaction

The Pd NPs synthesized using WRAE were employed as catalyst for Suzuki coupling reaction. The schematic reaction is represented in Fig. 6. The reaction was completed in the presence of 2 mol% Pd NPs at room temperature which resulted in good to excellent isolated yields of the products within short reaction times. A variety of substituted aryl bromides featuring either an electron-withdrawing or an electron-releasing group were then examined and results are summarized in Table 1. The reactions proceeded smoothly with nearly quantitative yields. These results reveal the potential of Pd NPs synthesized by WRAE as efficient catalyst for industrial applications.

Conclusion

An attempt was made to study the potential of watermelon rind aqueous extract in synthesizing palladium nanoparticles. Conditions such as reactant ratio, pH and reaction time were standardized prior to characterization of synthesized Pd NPs. The synthesized Pd NPs were found to be shape controlled with spherical in shape with average particle size of 96 nm. The FTIR peaks of Pd NPs revealed that polyhydroxyl groups of WRAE were involved in fabrication and stabilization of Pd NPs. Further, the synthesized Pd NPs were employed in Suzuki coupling reaction as catalyst and found to be efficient. These results suggest that watermelon rind, an agro waste, is capable of fabricating Pd to Pd NPs with catalytic applications.

Acknowledgments Authors gratefully acknowledge the Management of VIT University, Vellore for encouraging and providing necessary facilities for carrying out this work.

References

Andrew M, Zheng Y et al (2008) Structure of xylogalacturonan fragments from watermelon cell-wall pectin. Endopolygalacturonase can accommodate a xylosyl residue on the galacturonic acid just following the hydrolysis site. Carbohydr Res 343:1212–1221

Bankar A, Joshi B et al (2010) Banana peel extract mediated novel route for the synthesis of palladium nanoparticles. Mater Lett 64:1951–1953

Chen L, Evans JRG (2009) Arched structures created by colloidal droplets as they dry. Langmuir 25:11299–11301

Hebeish AA, El-Rafie MA et al (2010) Carboxymethyl cellulose for green synthesis and stabilization of silver nanoparticle. Carbohydr Polym 82:933–941

Jacob SJP, Finub JS et al (2012) Synthesis of silver nanoparticles using *Piper longum* leaf extracts and its cytotoxic activity against Hep-2 cell line. Colloids Surf B 91:212–214

Lakshmipathy R, Sarada NC (2013a) Adsorptive removal of basic cationic dyes from aqueous solution by chemically protonated watermelon (*Citrullus lanatus*) rind biomass. Des Water Treat.

Lakshmipathy R, Sarada NC (2013b) Application of watermelon rind as sorbent for removal of Nickel and Cobalt from aqueous solution. Int J Miner Process 122:63–65

Lakshmipathy R, Vinod AV et al (2013) Watermelon rind as biosorbent for removal of Cd^{2+} from aqueous solution: FTIR, EDX, and Kinetic studies. J Indian Chem Soc 90:1147–1154

Li X, Yan XY et al (2012) Suzuki–Miyaura cross-couplings of arenediazonium tetrafluoroborate salts with arylboronic acids catalyzed by aluminum hydroxide-supported palladium nanoparticles. Org Biomol Chem 10(3):495–497

Narayanan KB, Sakthivel N (2011) Extracellular synthesis of silver nanoparticles using the leaf extract of *Coleus amboinicus* Lour. Mater Res Bull 46:1708–1713

Quek YS, Chok KN et al (2007) The physicochemical properties of spray-dried watermelon powders. Chem Eng Process 46:386–392

Rimando MA, Perkins-Veazie MP (2005) Determination of citrulline in watermelon rind. J Chromatogr A 1078:196–200

Roopan SM, Bharathi A et al (2011) Acaricidal, insecticidal, and larvicidal efficacy of aqueous extract of *Annona squamosa* L peel as biomaterial for the reduction of palladium salts into nanoparticles. Colloids Surf B 92:209–212

Sathishkumar M, Sneha K et al (2009) Palladium nanocrystal synthesis using *Curcuma longa* tuber extract. Int J Mater Sci 4:11–17

Saxena A, Tripathi RM et al (2012) Green synthesis of silver nanoparticles using aqueous solution of *Ficus benghalensis* leaf extract and characterization of their antibacterial activity. Mater Lett 67:91–94

Yang X, Li Q et al (2010) Green synthesis of palladium nanoparticles using broth of *Cinnamomum camphora* leaf. J Nanopart Res 12:1589–1598

12

Synthesis and characterization of undoped and cobalt-doped TiO$_2$ nanoparticles via sol–gel technique

S. Mugundan · B. Rajamannan · G. Viruthagiri ·
N. Shanmugam · R. Gobi · P. Praveen

Abstract TiO$_2$ nanoparticles doped with different concentrations of cobalt (4, 8, 12 and 16 %) were synthesized by sol–gel method at room temperature with appropriate reactants. In general, TiO$_2$ can exist in anatase, rutile, and brookite phases. In this present study, we used titanium tetra iso propoxide and 2-propanol as a common starting materials and the obtained products were calcined at 500 °C and 800 °C to get anatase and rutile phases, respectively. The crystalline sizes of the doped and undoped TiO$_2$ nanoparticles were observed with X-ray diffraction (XRD) analysis. The functional groups of the samples were identified by Fourier transform infrared spectroscopy (FTIR). From UV–VIS diffuse reflectance spectra (DRS), the band gap energy and excitation wavelength of doped and undoped TiO$_2$ nanoparticles were identified. The defect oriented emissions were seen from photoluminescence (PL) study. The spherical uniform size distribution of particles and elements present in the samples was determined using two different techniques viz., scanning electron microscopy (SEM) with energy-dispersive spectrometer (EDX) and transmission electron microscope (TEM) with selected area electron diffraction (SAED) pattern. The second harmonic generation (SHG) efficiency was also found and the obtained result was compared with potassium di hydrogen phosphate (KDP).

B. Rajamannan (✉)
Department of Engineering Physics, (FEAT), Annamalai
University, Annamalainagar, Chidambaram, Tamilnadu 608002,
India
e-mail: mugugum@gmail.com

S. Mugundan · G. Viruthagiri · N. Shanmugam · R. Gobi ·
P. Praveen
Department of Physics, Annamalai University, Annamalainagar,
Chidambaram, Tamilnadu 608002, India

Keywords Cobalt · Doped TiO$_2$ · Nanoparticles ·
Crystalline size · FTIR · Optical properties

Introduction

Titanium dioxide or titania (TiO$_2$) was first produced commercially in 1923. It is obtained from a variety of ores. The bulk material of TiO$_2$ is widely nominated for three main phases of rutile, anatase and brookite. Among them, the TiO$_2$ exists mostly as rutile and anatase phases which both of them have the tetragonal structures. However, rutile is a high-temperature stable phase and has an optical energy band gap of 3.0 eV (415 nm), anatase is formed at a lower temperature with an optical energy band gap of 3.2 eV (380 nm) and refractive index of 2 (Thamaphat et al. 2008). Among these polymorphs, rutile and anatase have been widely studied. Brookite is rarely studied due to its complicated structure and difficulties in sample preparation (Hu et al. 2009). These three phases can be commonly described as constituted by arrangements of the same building block-Ti–O$_6$ octahedron in which Ti atom is surrounded by six oxygen atoms situated at the corners of a distorted octahedron. In spite of the similarities in building blocks of Ti–O$_6$ octahedra for these polymorphs, the electronic structures are significantly different (Guangshe et al. 2011). Photocatalysis using TiO$_2$ as a catalyst has been widely reported as a promising technology for the removal of various organic and inorganic pollutants from contaminated water and air because of its stability, low cost, and non-toxicity (Liu et al. 2008).

TiO$_2$ is the promising material as semiconductor having high photochemical stability and low cost. Well-dispersed titania nanoparticles with very fine sizes are promising in many applications such as pigments, adsorbents, and catalytic supports (Ramakrishna and Ghosh 2003).

Since Fujishima and Honda discovered the photocatalytic splitting of water on a TiO_2 electrode under ultraviolet (UV) light, many synthesis methods for preparing TiO_2 nanoparticles and their applications in the environmental (photo catalysis and sensors) and energy (photovoltaics, water splitting, photo/electrochromics, and hydrogen storage) fields have been investigated (Shan and Demopoulos 2010). Recently, fine particles of titania have attracted a great deal of attention, because of their specific properties as an advanced semiconductor material, such as a solar cell, luminescent material, and photocatalyst for photolysis of water or organic compounds and for bacteriocidal action (Sugimoto et al. 2003).

The Co-doped TiO_2 nanocrystals have consumed great attention due to its enhanced photocatalytic activity (Yang et al. 2007). In this paper, we report the preparation of different weight percentages of Co-doped TiO_2 nanoparticles by a sol–gel route.

Materials and methods

Sample preparation

Preparation of bare and cobalt-doped TiO_2 nanopowder

Sol–gel technique was used to prepare bare and cobalt-doped TiO_2 samples. 90 ml of 2-propanol was taken as a primary precursor and 10 ml titanium tetra isopropoxide was added to it drop wise with vigorous stirring during the process of TiO_2 formation. The solution was vigorously stirred for 45 min to form sols. Liquid solution cobalt nitrate of desired concentration (4, 8, 12, and 16 %) was poured slowly drop by drop to that mixture with continued stirring. To obtain nanoparticles, the obtained gels were dried at 80 °C for 5 h to evaporate water and organic material to the maximum extent. Finally, the powders were kept in muffle furnace and calcinated at 500 °C for 5 h for the harvest of anatase phase and 800 °C for rutile phase. The particle was pulverized to powder using an agate mortar at room temperature for further characterizations.

Sample characterization

The bare and Co-doped TiO_2 anatase are subjected into different characterizations, such as powder XRD, FTIR, UV-DRS, PL, SEM with EDX, TEM with SAED pattern, and SHG (NLO). The crystalline phase and particle size of TiO_2 nanoparticles were analyzed by X-ray diffraction (XRD) measurement, which was carried out at room temperature using XPERT-PRO diffractometer system (scan step of

0.05°, counting time of 10.16 s per data point) equipped with a Cu tube for generating Cu $K\alpha$ radiation ($\lambda = 1.5406$ Å). The incident beam in the 2-theta mode over the range of 20°–80°, operated at 40 kV and 30 mA. The chemical structure was investigated by AVATAR 330 Fourier transform infrared spectrometer (FTIR) in which the IR spectrum was recorded by diluting the mixed powder in KBr and in the wavelength between 4,000 and 400 cm^{-1}. The band gap energy and the particle size were measured at wavelength in the range of 200–2,500 nm by UV–VIS–NIR spectrophotometer (varian/carry 5000) equipped with an integrating sphere and the baseline correction was performed using a calibrated reference sample of powdered barium sulfate ($BaSO_4$). The photoluminescence spectra (PL) are recorded with Perkin Elmer LS fluorescence spectrophotometer. Scanning electron microscope (SEM) images were observed with a Hitachi S-4800 microscope, combined with energy-dispersive X-ray spectroscopy (EDX, Oxford 7021) for the determination of elemental composition. Transmission electron microscope (TEM) with selected area electron diffraction (SAED) images were taken using a technai t20 operated at a voltage of 200 kV. The NLO property of the materials was confirmed by the Kurtz powder second harmonic generation (SHG) test. A Q-switched Nd-YAG laser whose output was filtered through 1,064 nm narrow pass filter was used for this purpose. The input power of the laser beam was measured to be 4.5 m J/pulse.

Results and discussion

Powder X-ray diffraction study (XRD)

Figure 1 exhibits the XRD patterns of bare and cobalt-doped TiO_2 nanocrystals calcined at 500 °C for 5 h. From

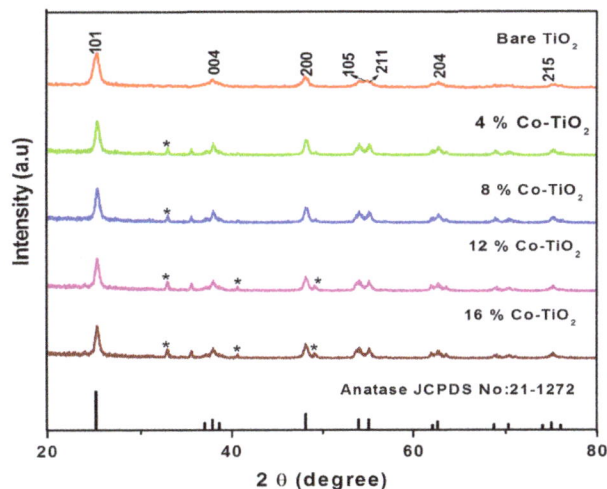

Fig. 1 X-ray diffraction pattern for bare TiO_2 and Co-doped TiO_2 nanoparticle

Table 1 Crystallite sizes for the bare and Co-doped TiO_2 nanoparticle

Samples	Crystallite size (nm)
Bare TiO_2	15.31
4 % Co-doped TiO_2	19.9092
8 % Co-doped TiO_2	20.9140
12 % Co-doped TiO_2	23.4380
16 % Co-doped TiO_2	25.9223

the diffraction patterns, it is noted that bare and cobalt-doped TiO_2 are in anatase phase rather than rutile or brookite (JCPDS Card no 21-1272). The XRD patterns obtained in the present study are identical with the earlier report (Karthik et al. 2010). At low level of cobalt incorporation, the XRD patterns do not show any cobalt phase, indicating that cobalt ions are uniformly dispersed on the host TiO_2. However, at high level of Co^{2+} incorporation (12 and 16 wt%), peaks related to cobalt are started to appear which are marked by a symbol star.

From the obtained peaks, the average nanocrystallite size was measured according to using Debye–Scherrer formula, the Eq. (1).

$$D = \frac{K\lambda}{\beta Cos\theta} \tag{1}$$

where D is the crystallite size, K is the shape factor, $\lambda = 0.154$ nm, β is the full width at half maximum, θ is the reflection angle and the results are presented in Table 1. In comparison with undoped TiO_2, all the doped products show increased particle size. The increased particle size may be explained by the fact that the ionic radius of Co^{2+} (0.74 Å) is greater than that of Ti^{4+} (0.60 Å) (Liu et al. 2008).

Fourier transform infrared spectroscopy (FTIR)

Figure 2 shows the FTIR spectra of the obtained bare and Co-doped TiO_2 nanoparticles after calcined at 500 °C for 5 h. The peak positioned at 3,415 cm^{-1} is attributed to O–H stretching vibration. The peaks appearing at 1,629 cm^{-1} were attributed to H–O–H bending vibration mode of physically observed water. Appearance of Ti–O–Ti frequency absorption is noted between 600 and 400 cm^{-1} (Choudhury and Choudhury 2012). The appearance of band at 2,860 cm^{-1} was due to the C–H bond of the organic compounds (Maensiri et al. 2006). The weak band at 2,833 cm^{-1} could be ascribed to the characteristic frequencies of residual organic species, which was not completely removed by ethanol and distilled water washing (Guo et al. 2007; Wang et al. 2007). The two prominent absorption bands at 3,400 and 1,633 cm^{-1} in the materials can be recognized as the stretching and bending vibrations

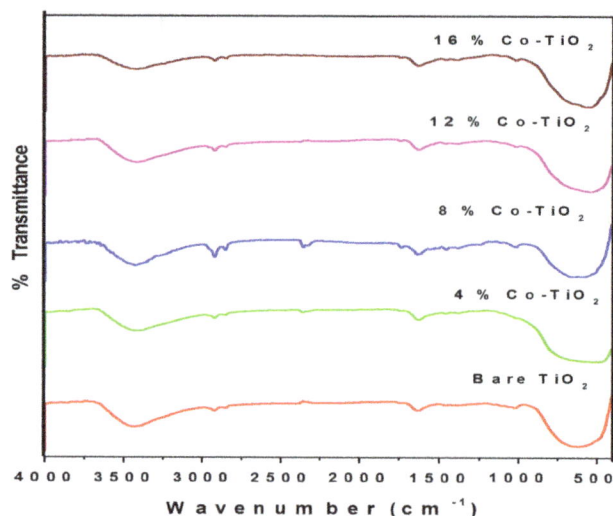

Fig. 2 FTIR spectra pattern for bare TiO_2 and Co-doped TiO_2 nanoparticle

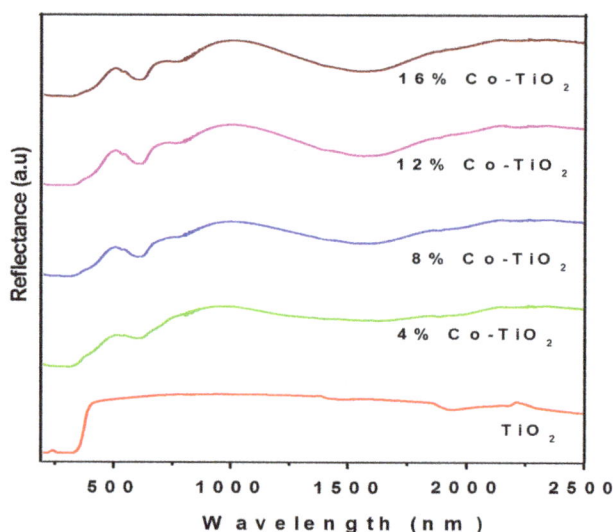

Fig. 3 UV-DRS spectra for bare TiO_2 and Co-doped TiO_2 nanoparticle-direct transition

of water molecules. The intensity of these two bands in Co is diminished compared with bare TiO_2. The peaks in between 2,924 and 2,843 cm^{-1} are assigned to C–H stretching vibrations of alkane groups. The alkane and carboxylic groups specified by different bands are arise from titanium tetra isopropoxide and 2-propanol (precursor material), when we used in the synthesis process. In addition, a broad absorption band between 500 and 1,000 cm^{-1} is ascribed to the vibration absorption of the Ti–O–Ti linkages in TiO_2 nanoparticles (Lu et al. 2008).

Ultra violet-diffuse reflectance spectra (UV-DRS)

The optical properties of samples were characterized by UV-DRS. The diffused reflectance spectra of undoped and different weight percentages (4, 8, 12 and 16 %) of Co-doped TiO_2 are shown in (Fig. 3).

In anatase phase of TiO_2, Ti^{4+} is surrounded by six oxygen atoms in an octahedral coordination, forming TiO_6 octahedra. When Co^{2+} substitutes Ti^{4+}, it forms bond with six oxygen atoms. Now, according to crystal field theory, the electrons in the d-orbital of Co^{2+} will undergo repulsion by the electrons of the six surrounding oxygen atoms. This results in the splitting of d-orbital of Co^{2+}, showing the aforementioned d–d transition. These types of electronic transition will be by Co^{2+} when it substitutes Ti^{4+} and remains in octahedral or pseudo octahedral coordination (Choudhury and Choudhury 2012). The DRS spectrum of TiO_2 showed a broad intense absorption at around 345 nm due to the charge transfer from the valence band formed by 2p orbitals of the oxide anions to the conduction band formed by 3d t_{2g} orbitals of the Ti $^{4+}$ cations (Sadanandam et al. 2013; Kirit and Dimple 2012).

From the UV–VIS DRS spectra (Fig. 3), bare TiO_2 particles exhibit strong absorption at $\lambda < 400$ nm. Compared to the spectrum of bare TiO_2, there is a new absorption band present between 400 and 800 nm of all samples, which means band gap absorption onset was extended into the visible light region for all Co-doped TiO_2 samples. The absorption bands at $200-400$ nm are due to the charge transfer from the valence band to the conduction band of TiO_2.

However, on 12 % Co^{2+} doping, the absorption spectrum is positioned at 748 nm. At 16 % of doping, the absorption peak is blue shifted to 742 nm. On increasing the concentration of doping, the value of blue shift increases with the enhancement in the intensity of visible band. The band gap energies were calculated from the cut-off wavelength of bare and Co-doped TiO_2 using the Eq. (2).

$$E_g = \frac{hc}{\lambda} \text{eV}; \; E_g = \frac{1,240}{\lambda} \text{eV} \qquad (2)$$

where E_g is the band gap energy (eV), h is the Planck's constant (6.626×10^{-34} Js), c is the light velocity (3×10^8 m/s), and λ is the wavelength (nm).

The calculated values are 3.58 eV (345 nm), 3.65 eV (339 nm), 3.75 eV (330 nm), 3.81 eV (325 nm), and 3.93 eV (315 nm) for bare TiO_2 and TiO_2 doped with 4, 8, 12, and 16 wt% of Co^{2+}, respectively, and are given in (Table 2).

Table 2 Band gap energy for bare and Co-doped TiO_2 nanoparticle

Samples	Band gap (eV)
Bare TiO_2	3.58
4 % Co-doped TiO_2	3.65
8 % Co-doped TiO_2	3.75
12 % Co-doped TiO_2	3.81
16 % Co-doped TiO_2	3.93

From the Table 2, it is noted that bare TiO_2 nanoparticle has a larger band gap (3.58 eV) than that of bulk TiO_2 nanomaterial (3.2 eV), which is quite expected due to quantum confinement effect. But, after various levels of cobalt doping, the band gap is running between 3.65 and 3.93 eV.

Photoluminescence (PL)

PL emission spectra of bare and 16 wt% of Co-doped TiO_2 with 300 nm excitation are shown in (Fig. 4). The PL spectrum of TiO_2 exhibits a UV emission (373 nm) and two visible emissions: one at 431 nm and the other at 587 nm. The UV emission is considered as the band edge emission of the host TiO_2, and the 431 nm peak can be ascribed to self-trapped excitations localized in TiO_6 octahedra (Wan et al. 2010). The existences of green emission peak at 587 nm explain the presence of oxygen vacancies (Abazovic et al. 2006). When compared with bare TiO_2, the PL emission bands of doped product are blue shifted with intensity reduction in the UV band and enhancement in the visible bands (Kirit and Dimple 2012).

Scanning electron microscope (SEM with EDX)

Figures 5 and 6 show typical scanning electron microscopic images of bare and 16 wt% Co-doped TiO_2 nanoparticle. From the figures, the most of the particles are almost spherical in shape with uniform size distribution (Reddy et al. 2010). It is evident that the growth of particle is restrained by Co on doping with TiO_2. A closer examination of these figures reveals a well-defined particle like morphology, having abundance of spherical-shaped particles, with the average agglomerated particle size in the range of $15.31-25.92$ nm. The spherical shape is

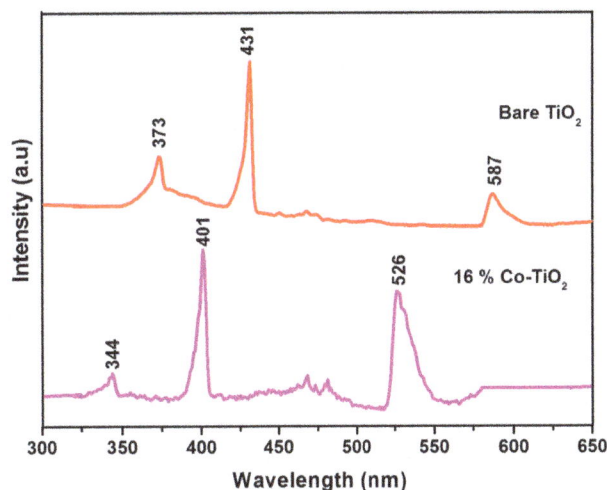

Fig. 4 PL spectra for bare TiO_2 and Co-doped TiO_2 nanoparticle

Fig. 5 SEM micrograph of TiO$_2$ nanoparticles (**a**, **b**) and corresponding to EDX diagram (**c**)

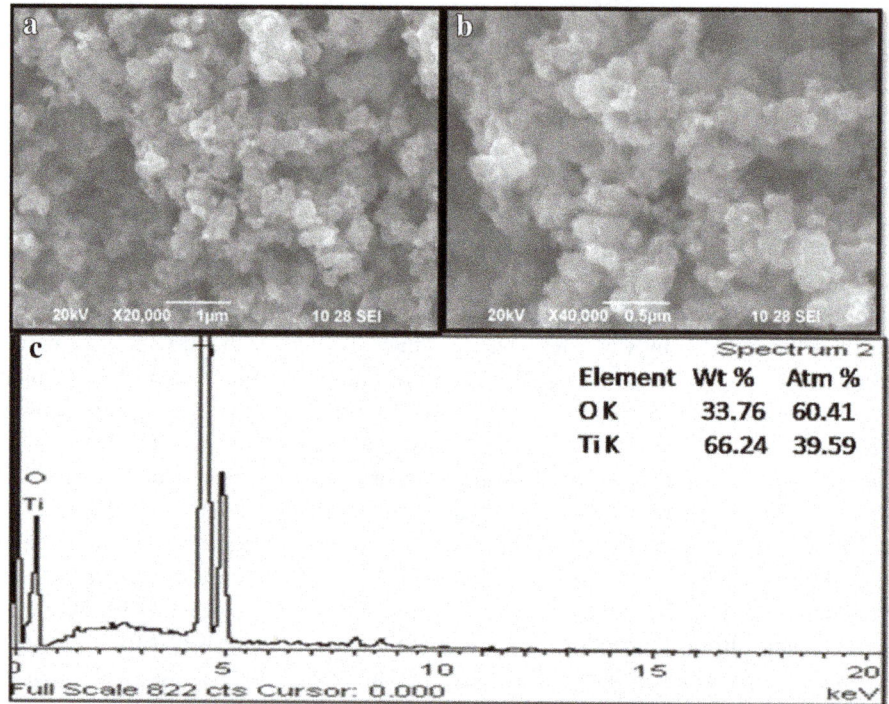

Element	Wt %	Atm %
O K	33.76	60.41
Ti K	66.24	39.59

Full Scale 822 cts Cursor: 0.000

Fig. 6 SEM micrograph of Co-doped TiO$_2$ nanoparticles (**a**, **b**) and corresponding to EDX diagram (**c**)

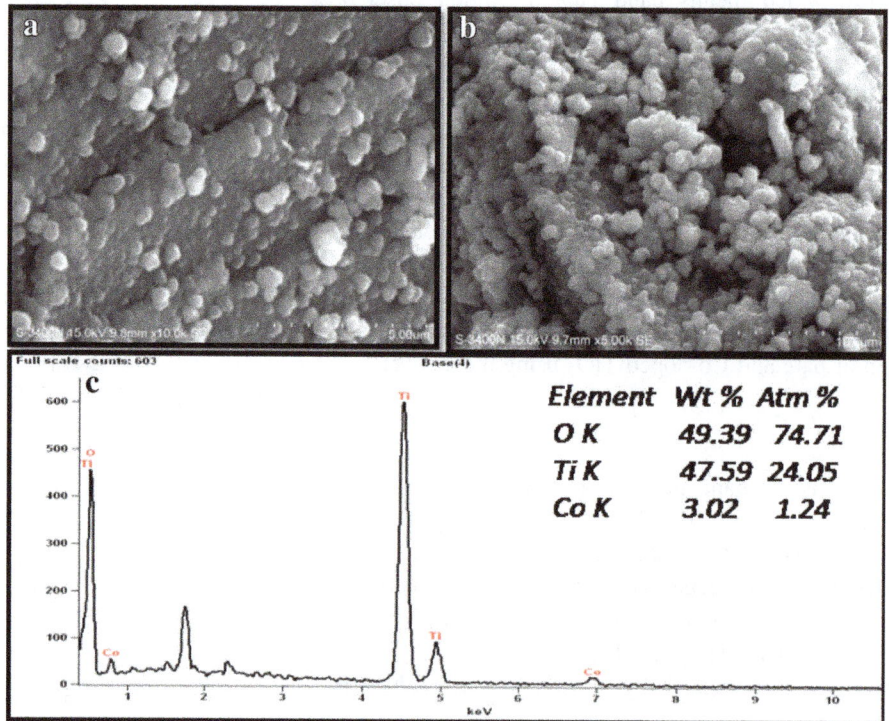

Element	Wt %	Atm %
O K	49.39	74.71
Ti K	47.59	24.05
Co K	3.02	1.24

significantly important not only for the design of surface properties and surface area, but also for tuning the electronic structure i.e., to make the visible light spectrum more active for the better photocatalytic activity (Narayana et al. 2011).

Energy-dispersive X-ray spectroscopy (EDAX) of 16 wt% of Co-doped TiO$_2$ (Figs. 5c, 6c) shows the presence of O, Ti and cobalt according to atomic weight stoichiometric of 74.71, 24.05 and 1.24 %, respectively (Kirit and Dimple 2012).

Fig. 7 TEM image of bare
TiO$_2$ nanoparticles (**a**, **b**) and
corresponding to SAED pattern
(**c**)

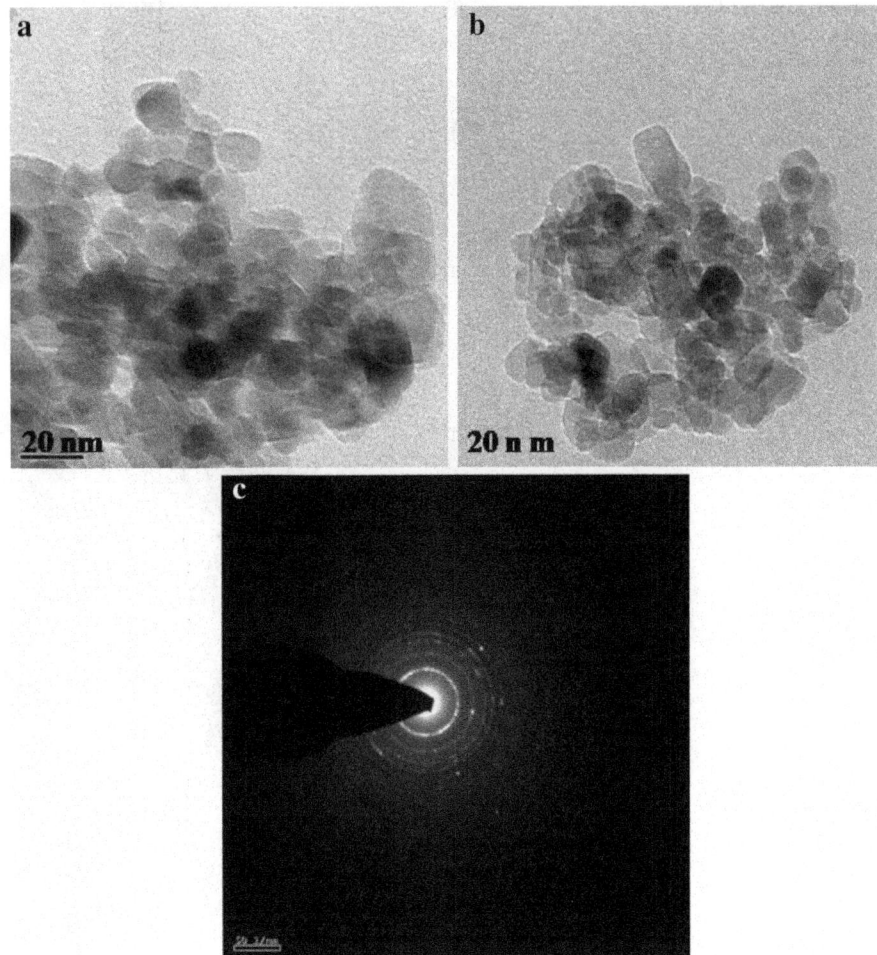

Transmission electron microscope (TEM)

Figures 7a, b and 8a, b illustrate the transmission electron microscope (TEM) images of the bare and 16 wt% Co-doped TiO$_2$ nanoparticles obtained after 500 °C of annealing. They showed almost spherical shape particles with uniform size distribution. As can be seen from the TEM image that the particle sizes are in the range of 15–25 nm, which is in good agreement with the crystallite size obtained from XRD results. In general, TiO$_2$ nanoparticles appear in transparent spherical shape. On incorporation of cobalt particles dark patches are seen on the surface of transparent TiO$_2$ nanoparticles. Electron diffraction patterns showed the brightness and intensity of polymorphic discrete ring of the highly crystalline nanoparticles are shown in (Figs. 7c and 8c).

Non-linear optical studies

For the preparation of NLO study, the oven-dried powders of bare TiO$_2$ and 16 wt% Co-doped TiO$_2$ were kept into muffle furnace and annealed at 800 °C for 5 h to obtain the rutile phase. The study of non-linear optical conversion efficiency was carried out using the experiment setup of Kurtz and Perry. A Q-switched Nd-YAG laser beam of wavelength 1,064 nm, with a repetition rate of 10 Hz was used. The bare TiO$_2$ and Co-doped TiO$_2$ were nanopowdered with a uniform particle size and then packed in a microcapillary of uniform bore and exposed to laser radiations. The SHG was confirmed by the emission of green radiation of 532 nm wavelengths and the parent ray was filtered using IR filter. A sample of potassium di hydrogen phosphate (KDP) also powdered to the same particle size as the experimental sample was used as a reference material in the present measurement. The relative SHG measurement gave conversion efficiency of bare TiO$_2$. Thus, the efficiency shows an output of 8.0 mV for bare TiO$_2$, 6.0 mV for Co-doped TiO$_2$ and 10.2 mV for KDP. When compared to 10.2 mV of KDP, for same input laser power, the bare and Co-doped TiO$_2$ show lesser second harmonic efficiencies. The calculated results are represented in Table 3.

Fig. 8 TEM image of Co-doped TiO_2 nanoparticles (**a**, **b**) and corresponding to SAED pattern (**c**)

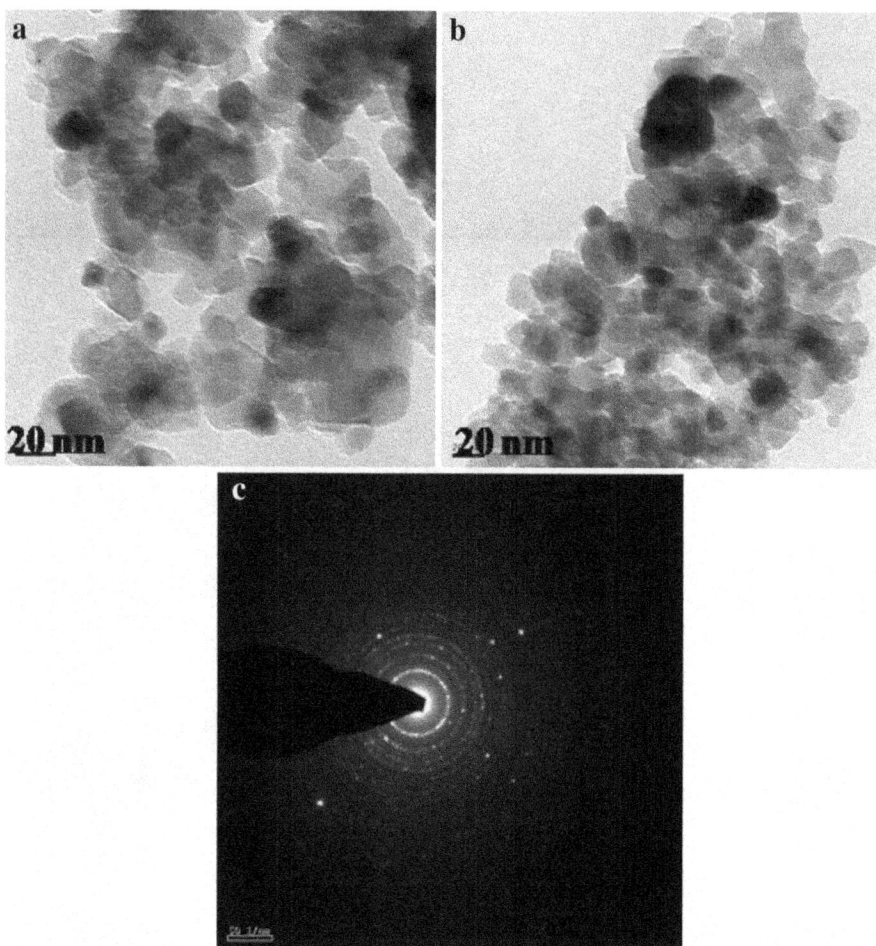

Table 3 Comparision of SHG efficiency of bare and 16 % Co-doped TiO_2 nanoparticles

Samples	Input (mJ/pulse)	Output (mV)	KDP (mV)	SHG efficiency
Bare TiO_2	4.5	8.0	10.2	0.78
Co-doped TiO_2	4.5	6.0	10.2	0.58

Conclusion

We have successfully synthesized the bare and different weight percentages (4, 8, 12 and 16 %) of Co-doped TiO_2 nanoparticles by means of sol–gel method at room temperature. The synthesized products were annealed at 500 and 800 °C for getting anatase and rutile phases, respectively. The annealed samples were characterized by the techniques like XRD, FTIR, and UV-DRS. The doping concentration of 16 wt% was chosen for further analysis like PL, SEM with EDX, TEM with SAED pattern, and NLO. From the results of XRD patterns, it is confirmed that the TiO_2 was in anatase phase with crystallite sizes in the range of 15.31–25.92 nm. In comparison with undoped TiO_2, all the doped products show increased particle size. The existences of functional groups were identified by FTIR analysis. The cut-off wavelengths were identified by UV–VIS DRS analysis and the band gap energies of the undoped and doped products are having values between 3.58 and 3.93 eV. The PL emission spectra show the creation of new luminescent centers. The SEM with EDX and TEM image with SAED pattern confirm the spherical morphology of the products. In SHG, bare TiO_2 shows greater NLO efficiency than the doped product.

References

Abazovic ND, Mirjana IC, Miroslav DD, Jovanovic DJ, Phillip AS, Jovan MN (2006) Photo luminescence of anatase and rutile TiO_2 particles. J Phys Chem B 110:25366–25370

Choudhury B, Choudhury A (2012) Luminescence characteristics of cobalt doped TiO_2 nanoparticles. J Lumin 132:178–184

Guangshe L, Liping L, Jing Z (2011) Understanding the defect chemistry of oxide nanoparticles for creating new functionalities. Inorg Solid State Chem Energy Mater 54:876–886

Guo GS, He CN, Wang ZH, Gu FB, Han DM (2007) Synthesis of titania and titanate nanomaterials and their application in environmental analytical chemistry. Talanta 72:1687–1692

Hu WB, Li LP, Li GS, Tang CL, Sun L (2009) High-quality brookite TiO_2 flowers: synthesis, characterization, and dielectric performance. Cryst Growth Des 9:3676–3682

Karthik K, Pandian SK, Kumar KS, Jaya NV (2010) Influence of dopant level on structural, optical and magnetic properties of Co-doped anatase TiO_2 nanoparticles. Appl Surf Sci 256:4757–4760

Kirit S, Dimple S (2012) Characterization of nanocrystalline cobalt doped TiO_2 sol–gel material. J Cryst Growth 352:224–228

Liu XH, He XB, Fu YB (2008) Effects of doping cobalt on the structures and performances of TiO_2 photocatalyst. Acta Chim Sinica 66:1725–1730

Lu X, Xiuqian L, Zhijie S, Zheng Y (2008) Nanocomposite of poly (L-lactide) and surface-grafted TiO_2 nanoparticles: synthesis and characterization. Eur Polym J 44:2476–2481

Maensiri S, Laokul P, Klinkaewnarong J (2006) A simple synthesis and room-temperature magnetic behavior of Co doped anatase TiO_2 nanoparticle. J Magn Magn Mater 302:448–453

Narayana RL, Matheswaran M, Aziz AA, Saravanan P (2011) Photocatalytic decolourization of basic green dye by pure and Fe, Co doped TiO_2 under daylight illumination. Desalination 269:249–253

Ramakrishna G, Ghosh HN (2003) Optical and photochemical properties of sodium dodecylbenzene sulfonate (DBS) capped TiO_2 nanoparticles dispersed in nonaqueous solvents. Langmuir 19:505–508

Reddy MV, Jose R, Teng TH, Chowdari BVR, Ramakrishna S (2010) Preparation and electrochemical studies of electrospun TiO_2 nanofibers and molten salt method nanoparticles. Electrochim Acta 55:3109–3117

Sadanandam G, Lalitha K, Kumari VD, Shankar MV, Subrahmanyam M (2013) Cobalt doped TiO_2: a stable and efficient photocatalyst for continuous hydrogen production from glycerol: water mixtures under solar light irradiation. Int J Hydrogen Energy 38:9655–9664

Shan GB, Demopoulos GP (2010) The synthesis of aqueous-dispersible anatase TiO_2 nanoplatelets. Nanotechnology 21:1–9

Sugimoto T, Zhou X, Muramatsu A (2003) Synthesis of uniform anatase TiO_2 nanoparticles by gel–sol method formation process and size control. J Colloid Interface Sci 259:43–52

Thamaphat K, Limsuwan P, Ngotaworchai B (2008) Phase characterization of TiO_2 powder by XRD and TEM. Nat Sci 42:357–361

Wan WY, Chang YM, Ting JM (2010) Room-temperature synthesis of single-crystalline anatase TiO_2 nanowires. Cryst Growth Des 10:1646–1651

Wang X, Song X, Lin M, Wang H, Hao Y, Zhong W, Du Q (2007) Surface initiated graft polymerization from carbon-doped TiO_2 nanoparticles under sunlight illumination. Polymer 48:5834–5838

Yang X, Cao C, Hohn K, Erickson L, Maghrang R, Hamal D, Klabunde K (2007) Highly visible-light active C- and V-doped TiO_2 for degradation of acetaldehyde. J Catal 252:296–302

13

Comparative study of carbon dioxide sensing by Sn-doped TiO$_2$ nanoparticles synthesized by microwave-assisted and solid-state diffusion route

K. R. Nemade · S. A. Waghuley

Abstract Gas sensor based on Sn-doped titanium dioxide (TiO$_2$) nanoparticles has been fabricated and evaluated for carbon dioxide (CO$_2$) sensing. The Sn-doped TiO$_2$ nanoparticles were synthesized by microwave-assisted and solid-state diffusion route. The structure and morphology of resulting samples were characterized by X-ray diffraction (XRD) and scanning electron microscopy (SEM). Ultraviolet–visible (UV–VIS) spectroscopy was employed to study the optical properties. The weight loss of samples was analyzed through thermo gravimetric analysis (TGA). The Sn-doped TiO$_2$ nanoparticles synthesized through microwave route exhibited good sensing characteristics.

Keywords Carbon dioxide · Microwave assisted · Solid-state diffusion

Introduction

Titanium dioxide (TiO$_2$) is a potent candidate material for the various modern applications due to its low cost, high chemical inertness, powerful oxidation potential and non-toxicity in water (Choi et al. 2012). TiO$_2$ is mainly attractive for their notable ability to change the electrical resistance in response to oxidizing and reducing gases (Radecka et al. 2010).

K. R. Nemade (✉)
Department of Applied Physics, J D College of Engineering and Management, Nagpur 441 501, India
e-mail: krnemade@gmail.com

S. A. Waghuley
Department of Physics, Sant Gadge Baba Amravati University, Amravati 444 602, India

In the last two decades, microwave synthesis has arisen as a novel synthesis strategy and method with many significant advances in practical aspects of chemistry (Surati et al. 2012). Lopez et al. (2013) reported the microwave-assisted synthesis of CdS nanoparticles and studied their size evolution. In this study, they reported that the microwave heating has uniform heating characteristics whereas in conventional heating thermal gradient is present.

Sui et al. (2010) reported the synthesis of Sn-doped TiO$_2$ by sol–gel synthesis with high aspect ratios. Duan et al. (2012) studied the photoanode performance of Sn-doped TiO$_2$ synthesized by the hydrothermal method. Sn-incorporated rutile TiO$_2$ nanorods synthesized by simple solvothermal route for photoelectrochemical water splitting have been reported by (Sun et al. 2013). Xiufeng et al. (2011) demonstrated the Sn-doped TiO$_2$ nanoparticles as a visible-light photocatalyst synthesized by vapor transport method. Radecka et al. (2010) demonstrated the flame spray synthesis route to grow TiO$_2$-based nanoparticles from which TiO$_2$:Cr nanosensors were obtained. Al-Homoudi et al. (2007) develop the TiO$_2$-based CO gas sensor. Resistive response of anatase TiO$_2$ films is discussed for different concentrations of CO gas.

Wu et al. (2012) studied the ethanol sensing characteristics of Sn-doped rutile TiO$_2$ nanowires synthesized by a thermal reactive evaporation route. The sensing response increased with an increase in the ethanol concentration. Benkara et al. reported the synthesis of Sn-doped ZnO/TiO$_2$ by two methods, first is anodic oxidation of Ti foil and another is hydrothermal process. These nanocomposites have been employed as hydrogen gas sensing materials. Chemisorption model was used to explain the H$_2$ sensing mechanism (Benkara et al. 2013). Nanostructured Sn-doped TiO$_2$ has been reported by (Raji et al. 2011), which is prepared by ball milling using SnO$_2$ and TiO$_2$ as raw materials. This work pointed out that Sn-doped TiO_2 possessed the highest humidity sensitivity, than pristine TiO_2 and SnO_2.

However, to the best of our knowledge, such kind of comparative study is still not reported in the literature of materials science. So, in the present work, we are studying the comparative sensing response of Sn-doped TiO_2 synthesized by microwave-assisted and solid-state diffusion route towards the carbon dioxide (CO_2). CO_2 is the primary greenhouse gas emitted through human activities, which is responsible for climate change. Therefore, its detection becomes crucial for human being.

The synthesized materials by both routes were characterized through X-ray diffraction (XRD), scanning electron microscope (SEM), ultraviolet–visible spectroscopy (UV–VIS) and thermo gravimetric analysis (TGA). The various gas sensing characteristics were also analyzed comparatively for both route of synthesis.

Experimental

Microwave-assisted synthesis of Sn-doped TiO_2 nanoparticles

The laboratory grade microwave synthesizer CEM Phoenix was used for synthesis of Sn-doped TiO_2 nanoparticles. The starting chemicals, $SnCl_4$ and TiO_2 of AR grade were used in this work. The 0.1 wt % $SnCl_4$ was added in TiO_2 in aqueous media. These solutions were mixed under constant magnetic stirring for 1 h. The mixture was heated in CEM supplied single use vessel. The mixture containing vessel was reacted in the CEM microwave at a temperature of 600 °C.

Synthesis of Sn-doped TiO_2 nanoparticles by solid-state diffusion routes

Sn-doped TiO_2 nanoparticles were prepared by the solid-state diffusion routes using the starting chemicals, $SnCl_4$ and TiO_2. The preparatory materials were mixed thoroughly by taking 0.1 wt % $SnCl_4$ in TiO_2 using the agate mortar pestle for 1 h. The crushed samples were placed in crucible and heated at 973 K in muffle furnace for 8 h. Then, allow to cool down, crushed in the mortar pestle again and heated at 1,173 K for 8 h. Lastly, the samples in the furnace were kept to cool down to room temperature.

Materials characterizations

X-ray diffraction patterns were recorded on Rigaku (Miniflex II) diffractometer with CuKα radiation in the range 10°–70°. The topography of samples was analyzed through SEM using JEOL JSM-7500F. The UV–VIS analysis was performed on Perkin Elmer UV spectrophotometer in the range 375–525 nm. The thermo gravimetric analysis (TGA) was obtained in a Shimadzu DTG-60 h thermal analyzer.

Sensors fabrication and gas sensing measurements

Fabrication of sensors was done through screen printing on chemically cleaned glass SiO_2 substrate of dimension 25×25 mm by screen printing technique using a temporary binder. In this way, sensing material deposited SiO_2 substrate kept for heating at 373 K for 1 h. During this stage, volatile organic compounds in binder were evaporated. After this step, highly conducting silver paint was deposited on both side of film to measure change in resistance during the exposure of gas. The voltage divider method was employed to measure the change in resistance. The temperature and pressure inside the sensing chamber were maintained precisely. To measure sensing response, air is used as background gas. The required concentration inside the chamber was maintained by injecting known volume of gas (Nemade and Waghuley 2013). The sensing response of chemiresistor was defined as (Nemade and Waghuley 2014):

$$S = \frac{\Delta R}{R_a} = \frac{|R_g - R_a|}{R_a} \tag{1}$$

where R_a and R_g is the resistance of chemiresistor in air and gas, respectively.

Results and discussion

Figure 1a–c depicts the XRD patterns of the pure TiO_2 and Sn-doped TiO_2 synthesized by microwave-assisted and solid-state diffusion route. The XRD data for both samples were acquired at room temperature (298 K). All diffraction peaks of pure TiO_2 (Fig. 1a) can be indexed to the anatase phase (JCPDS No. 01-070-7348). The average crystallite size was determined with Scherrer equation, $D = K\lambda/\beta Cos\theta$, where D is average crystallite size, K is shape factor ($K = 0.89$), λ is wavelength of X-ray used for analysis ($\lambda = 1.54$ Å). From the significant and characteristics peaks, the average crystallite size was found to be 18 nm for TiO_2 and 16.7 nm for Sn-doped TiO_2 samples. From XRD patterns, it is noticeable that particle size was not greatly influenced by synthesis route. The exact 2θ position and marginal difference in intensity reflect the good crystallinity of as-synthesized samples. The appearance of diffraction peaks in XRD pattern of Sn-doped TiO_2 at 27 and 34° was assigned to Sn incorporation. The larger effective radius of Sn^{4+} is 0.69 Å compared to that of Ti^{4+}

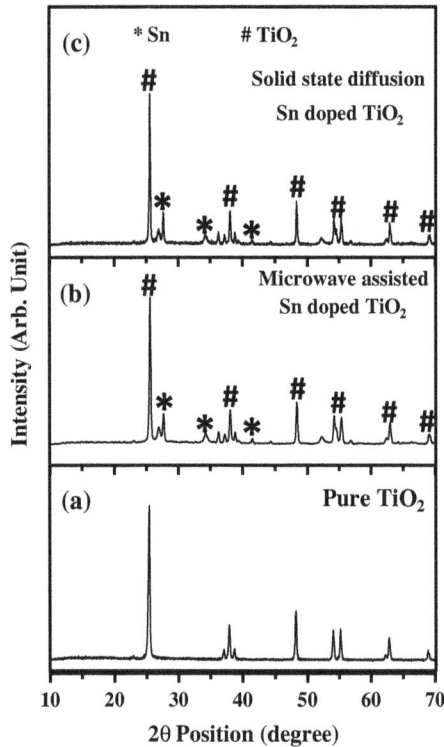

Fig. 1 XRD patterns for (a) pure TiO_2 and Sn-doped TiO_2 nanoparticles synthesized by (b) microwave-assisted and (c) solid-state diffusion route

Fig. 2 SEM image of Sn-doped TiO_2 nanoparticles synthesized by (a) microwave-assisted and (b) solid-state diffusion route

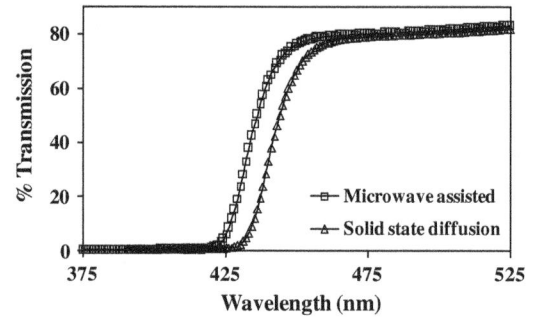

Fig. 3 UV–VIS spectrum of Sn-doped TiO_2 nanoparticles synthesized by microwave-assisted and solid-state diffusion route

that is 0.61 Å, which results in appearance of peaks (Duan et al. 2012).

To determine the morphology of Sn-doped TiO_2 samples, powder of the sample was suspended in acetone under sonication for 30 min. Figure 2a, b shows the SEM of Sn-doped TiO_2 synthesized by microwave-assisted and solid-state diffusion route. As can be seen from micrograph, nanoparticles take the shape like flakes of random size. From micrograph, it is observed that flakes of Sn-doped TiO_2 synthesized by microwave-assisted route are more separate than those synthesized by solid-state diffusion route.

Figure 3 shows the UV–VIS spectra of Sn-doped TiO_2 nanoparticles synthesized by microwave-assisted and solid-state diffusion route. The result shows that the absorption value of Sn-doped TiO_2 nanoparticle synthesized through microwave-assisted route shifts towards lower wavelength. According to effective mass approximation, particle size and band gap are inversely related to each other (Nemade and Waghuley 2014). Similarly, absorption of wavelength by particle is a function of particle size. Therefore, shift towards the lower wavelength results in the increase of band gap. This also shows the reduced number of defects on particle surface. Therefore, microwave-assisted synthesized nanoparticles have large band gap than nanoparticles synthesized through solid-state diffusion route. This change in band gap was related with various optical properties, such as optical conductivity, extinction coefficient and optical dielectric constant.

Figure 4 shows the TGA plot for the as-synthesized Sn-doped TiO_2 nanoparticles. TGA curve clearly shows that synthesized sample continuously lost its weight over the temperature range from room temperature to 725 K. TGA curves for both the samples do not show endothermic or exothermic peaks. This shows that during heating treatment nanoparticles do not undergo phase change.

Figure 5 shows the gas sensing response curve for Sn-doped TiO_2 nanoparticles synthesized by microwave-assisted and solid-state diffusion route towards the CO_2

gas. Response curves clearly show the good dependence on concentration of CO_2 gas. As the concentration of CO_2 gas increases, resistance of the sensor also increases. This may be due to the oxidizing nature of CO_2 gas. The sensing curves for both samples are almost linear up to 300 ppm, but beyond that diverted from linearity. This may be a CO_2 detection limit for as-synthesized samples. With view of comparison, Sn-doped TiO_2 nanoparticles synthesized by microwave-assisted route show higher response than solid-state diffusion route. This may be because microwave heating is more uniform and avoids thermal gradients present in conventional heating. In SEM analysis, it is observed that Sn-doped TiO_2 flakes are more separate than solid-state diffusion route. Therefore, sample provides larger area to test gas for adsorption. This may increase the interaction probability of gas–solid.

The mechanism for CO_2 gas detection for these materials is based on bridging oxygen mechanism. Reactions that occur at the sensor surface are mainly assisted by adsorbed oxygen. The change in concentration of adsorbed oxygen may alter the sensing response. Therefore, modification of pristine TiO_2 by Sn results in creating additional adsorption site and catalyzes the surface redox reaction.

We suggest that the oxygen adsorbed on the sensor surface may be involved in the sensing process of CO_2. The interface between adsorbed atmospheric oxygen and sensing surface is shown below (Fan et al. 2013).

$$O_2 \, (gas) \; \rightarrow \; O_2 \, (ads)$$
$$O_2 \, (ads) \; + \; e^- \; \rightarrow \; O_2^- \, (ads)$$
$$O_2^- \, (ads) \; + \; e^- \rightarrow 2O^- \, (ads)$$

Therefore, CO_2 may initially adsorb on pre-adsorbed oxygen and form surface carbonate. The plausible sensing

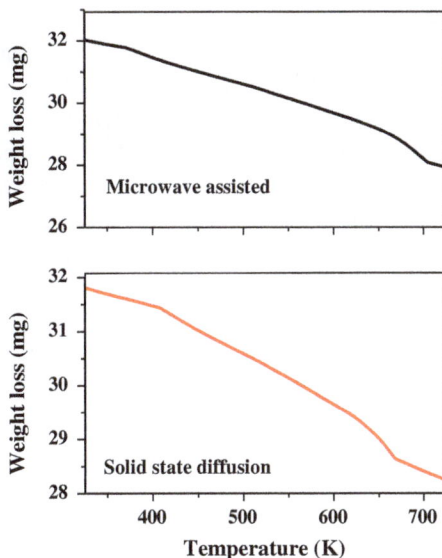

Fig. 4 TGA curves of Sn-doped TiO_2 nanoparticles synthesized by microwave-assisted and solid-state diffusion route

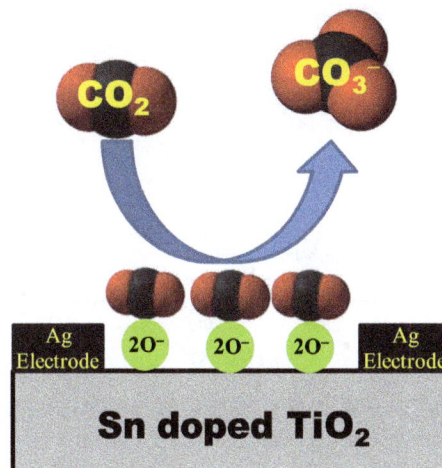

Fig. 6 Plausible sensing mechanism for CO_2 detection

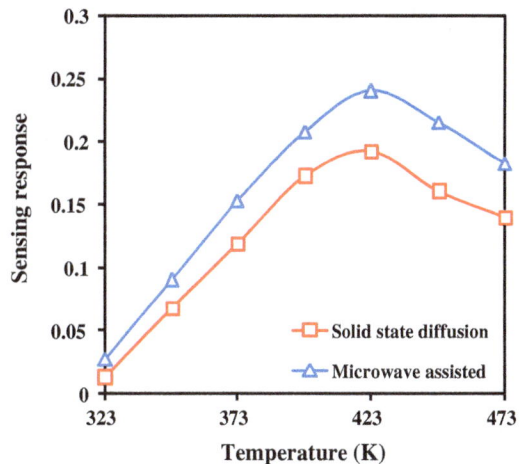

Fig. 5 Gas sensing response of Sn-doped TiO_2 nanoparticles synthesized by microwave-assisted and solid-state diffusion route

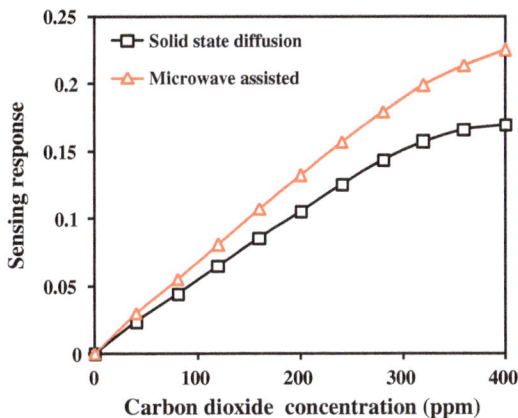

Fig. 7 Operating temperature response of Sn-doped TiO_2 nanoparticles to 200 ppm CO_2 gas

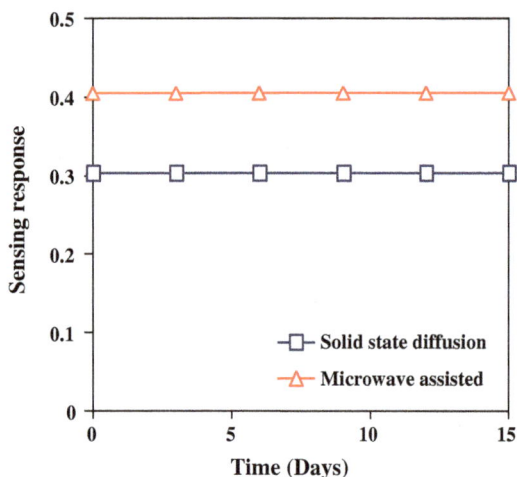

Fig. 8 Stability response of Sn-doped TiO_2 nanoparticles to 400 ppm concentration of CO_2 at 423 K

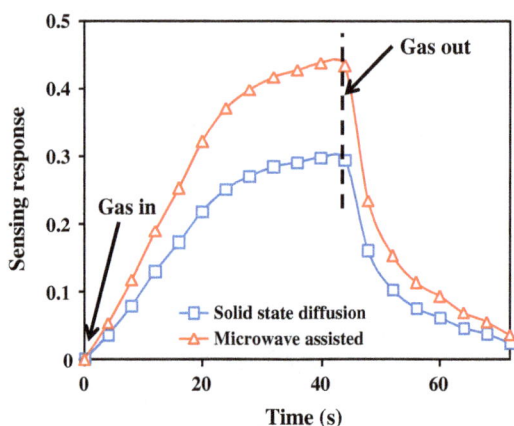

Fig. 9 Transient response of Sn-doped TiO_2 nanoparticles to 400 ppm concentration of CO_2 at 423 K

response. This shows that stability of both sensors is good against CO_2.

The transient response characteristic of Sn-doped TiO_2 nanoparticles based on both sensors for 400 ppm CO_2 at 423 K was studied and is shown in Fig. 9. For this measurement, gas was introduced in the chamber and resistance of sensors was measured in background gas, that is air and in the presence of CO_2 gas. Both sensors show fast response time towards the CO_2 of the order 42 s. For measuring the recovery time, sensors were exposed to air. Both the sensors achieve fast recovery in 30 s.

Conclusions

We have comparatively demonstrated CO_2 gas sensing by Sn-doped TiO_2 nanoparticles, which were synthesized by microwave-assisted and solid-state diffusion route. The structural analysis by XRD, pointed out the formation samples. The Sn-doped TiO_2 nanoparticles synthesized by microwave-assisted route found to be efficient for the good sensing response of CO_2 gas, due to sufficient sensing response at room temperature and relatively low operating temperature. The sensors realized the detection of CO_2 with fast response and recovery. Both the sensors were found to exhibit good stability against CO_2.

Acknowledgments One of the authors, K.R. Nemade is very much thankful to Shri Sanjayji Agrawal, Chairman, Jaidev Education Society and Shri Ajayji Agrawal, Secretary, Jaidev Education Society for providing necessary facilities.

mechanism for CO_2 detection is represented as Fig. 6. This formation of surface carbonates was responsible for the CO_2 detection through resistance change.

Figure 7 shows the operating temperature characteristics of as-synthesized materials for 30 ppm CO_2 gas. In the measured temperature domain, response to 200 ppm CO_2 increases up to 423 K and finally drops. This decrease in response value may be due to desorption of atmospheric oxygen ions from sensing surface due to thermal vibrations (Nemade and Waghuley 2013). Therefore, highest value of the sensing response at 423 K may be considered as operating temperature of Sn-doped TiO_2 gas sensing materials.

Figure 8 shows stability characteristics of as-synthesized Sn-doped TiO_2 nanoparticles for 400 ppm concentration of CO_2 at 423 K. To check the stability of sensor, its response was measured for 15 days, at an interval of 3 days. Both the sensors have almost constant sensing

References

Al-Homoudi I, Thakur JS, Naik R, Auner GW, Newaz G (2007) Anatase TiO_2 films based CO gas sensor: film thickness, substrate and temperature effects. Appl Surf Sci 253:8607–8614

Benkara S, Zerkouta S, Ghamrid H (2013) Synthesis of Sn doped ZnO/TiO_2 nanocomposite film and their application to H_2 gas sensing properties. Mater Sci Semi Process 16:1271–1279

Choi HG, Yong S, Kim DK (2012) Synthesis and photocatalytic properties of SnO_2-mixed and Sn-doped TiO_2 nanoparticles. Kor J Mater Res 22:352–357

Duan Y, Fu N, Liu Q, Fang Y, Zhou X, Zhang J, Lin Y (2012) Sn-doped TiO_2 photoanode for dye-sensitized solar cells. J Phys Chem C 116:8888–8893

Fan K, Qin H, Wang L, Ju L, Hu J (2013) CO_2 gas sensors based on $La_{1-x}Sr_xFeO_3$ nanocrystalline powders. Sens Actuators B 177:265–269

Lopez IA, Vazquez A, Gomez I (2013) Microwave assisted synthesis of CdS nanoparticles and their size evolution. Rev Mex Fis 59:160–164

Nemade KR, Waghuley SA (2013a) Chemiresistive gas sensing by few-layered graphene. J Electro Mater 42:2857–2866

Nemade KR, Waghuley SA (2013b) Strontium oxide quantum dot decorated graphene composites for liquid petroleum gas sensing. J Chinese Adv Mater Soc 1:219–228

Nemade KR, Waghuley SA (2014a) Role of defects concentration on optical and carbon dioxide gas sensing properties of Sb_2O_3/graphene composites. Optical Mater 36:712–716

Nemade KR, Waghuley SA (2014b) Low temperature synthesis of semiconducting α-Al_2O_3 quantum dots. Ceram Int 40:6109–6113

Radecka M, Jasinski M, Klich-Kafel J, Rekas M, Lyson B, Czapla A, Lubecka M, Sokolowski M, Zakrzewska K, Heel A, Graule TJ (2010) TiO_2-based nanopowders for gas sensor. Ceram Mater 62:545–549

Raji P, Binitha HS, Kumar KB (2011) Synthesis and humidity sensing properties of Sn-doped nano- TiO_2. J Nanotechnol 2011:569036–569042

Sui R, Young JL, Berlinguette CP (2010) Sol–gel synthesis of linear Sn-doped TiO_2 nanostructures. J Mater Chem 20:498–503

Sun B, Shi T, Peng Z, Sheng W, Jiang T, Liao G (2013) Controlled fabrication of Sn/TiO_2 nanorods for photoelectrochemical water splitting. Nanoscale Res Lett 8:462–470

Surati MA, Jauhari S, Desai KR (2012) A brief review: microwave assisted organic reaction. Arch Appl Sci Res 4:645–661

Wu JM (2012) Tin-doped rutile titanium dioxide nanowires: luminescence, gas sensor, and field emission properties. J Nanosci Nanotechnol 12:1434–1439

Xiufeng Z, Juan L, Lianghai L, Zuoshan W (2011) Preparation of crystalline Sn-doped TiO_2 and its application in visible-light photocatalysis. J Nanomater 2011:432947–432952

14

Capsaicin-capped silver nanoparticles: its kinetics, characterization and biocompatibility assay

Nagoth Joseph Amruthraj ·
John Poonga Preetam Raj ·
Antoine Lebel

Abstract Capsaicin was used as a bio-reductant for the reduction of silver nitrate to form silver nanoparticles. The formation of the silver nanoparticles was initially confirmed by color change and Tyndall effect of light scattering. It was characterized with UV–visible spectroscopy, FTIR and TEM. Hemagglutination (H) test and H-inhibition assay were performed in the presence of AgNPs–capsaicin conjugates. The silver colloid solution after complete reduction turned into pale gray color. The characteristic surface plasmon resonance of silver nanoparticles (SNPs) was observed at 450 nm. Time taken for complete bio-reduction of silver nitrate and capping was found to be 16 hours. The amount of capsaicin required to reduce 20 ml of 1 mM silver nitrate solution was found to be 40 μg approximately. The FTIR results confirmed the capping of capsaicin on the silver metal. The particle size was within the range of 20–30 nm. The hemagglutination and H-inhibition test was negative for all the blood groups. The capsaicin-capped silver nanoparticles were compatible with blood cells in hemagglutination test implying biocompatibility as future therapeutic drug.

Keywords Capsaicin · Silver nanoparticles (SNPs) · AgNO$_3$ · Bio-reductant · Biocompatibility · Hemagglutination

Introduction

Nanotechnology has a massive impact on therapeutic drugs. It significantly improves the performance of drugs in terms of efficacy and safety. Wide range of nano materials can be designed with surface modification to tag variety of chemical, molecular and biological entities (Kumari et al. 2012; Baimark 2012). Capping agents that bind to the nanoparticles surface will improve stability, water solubility, biocompatibility, specificity and prevent aggregation (Sing et al. 2009).

Silver has been used for medicinal purpose in the field of health care from ancient period (Priyaragini et al. 2013). The reduction of silver nitrate to silver nanoparticles was described by (Jin et al. 2001). "Green synthesis" of nanoparticles is a beneficial method that makes use of non-toxic reagents. Silver nanoparticles have tremendous scope, because of their important applications in antimicrobial, catalysis and surface-enhanced Raman scattering (Li et al. 2006; Chen et al. 2005; Setua et al. 2007). Silver-based drugs have proved as antimicrobial agents. The rapid breakdown of silver nanoparticles releases ionic silver that inactivates vital bacterial enzymes by interacting with essential thiol groups. Silver ions can inhibit bacterial DNA replication, damage bacterial cytoplasmic membranes, deplete the levels of intracellular adenosine triphosphate (ATP) and finally cause cell death (Feng et al. 2000).

Plant-derived drugs have served as the base for an enormous fraction of the current pharmacopoeia (Kingston and Newman 2002). Plant-based therapeutic molecules as drugs have limitations due to reduced solubility, permeability and bioavailability (Kumari et al. 2012). In spite of the challenges, the following natural products like artemisinin, curcumin, triptolide and capsaicin have been

N. J. Amruthraj · J. P. Preetam Raj (✉) · A. Lebel
Department of Plant Biology and Biotechnology, Loyola College, Chennai 600 034, Tamil Nadu, India
e-mail: preetamraj.jp@gmail.com

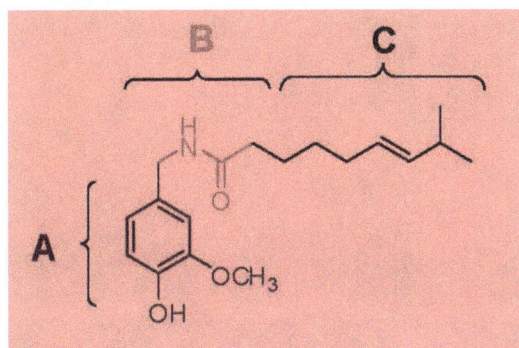

Fig. 1 Chemical structure of capsaicin

broadly studied and have entered into clinical trials (Corson and Crews 2007). In chemotherapy, plant origin drugs cause damage equally to both normal and malignant cells. By carefully designing the nanocarriers, natural product formulations can be targeted to specific cells (Sahoo and Labhasetwar 2003).

Capsaicin (8-Methyl-*N*-vanillyl-*trans*-6-nonenamide) is an active component of capsicum species. It is responsible for the characteristic pungent smell. It has several potential applications as anticarcinogenic [14] antioxidant (Rosa et al. 2002), suppressor of fat accumulation (Yoshioka et al. 1999), anti-inflammatory (Fraenkel et al. 2004; Fernando Spiller et al. 2008) and as antimicrobial agent (Kurita et al. 2002; Zeyrek and Oguz 2005).

Pure capsaicin is a hydrophobic, colorless, odorless, and crystalline to waxy compound. It has the chemical formula $C_{18}H_{27}NO_3$, with a molar mass of 305.40 g/mol ("ChemSpider"). The structure of the capsaicin molecule can be divided into three distinct regions, [**A**] as a vanillyl group, [**B**] an amide bond and [**C**]a fatty acid chain (Fig. 1) (Hopur2 2009). Due to the presence of the hydrophobic side chain, capsaicin is lipophilic and, therefore, highly soluble in fat, oil, and alcohol (Reyes-Escogido et al. 2011). It has poor bioavailability due to its hydrophobic nature like many other natural phytomedicines which is a major limitation in drug delivery. Therefore, in the present study, the capping of capsaicin on AgNPs by the direct reduction of silver nitrate in the aqueous phase, without the use of any other reducing agents was undertaken with the aim of developing a potential therapeutic drug.

Materials and methods

Synthesis of silver nanoparticles and UV–vis spectroscopy studies

Capsaicin (8-methyl-*N*-vanillyl-*trans*-6-nonenamide) was purchased from Sigma Chemical Co, St. Louis, MO, USA. The silver nitrate was purchased from Qualigens, Mumbai.

Glasswares were rinsed with deionised water before starting the synthesis. 20 ml of 1 mM $AgNO_3$ was rapidly stirred at 60 °C to which 100 μl of 1 mg/ml concentrated solution of capsaicin was added. Silver nanoparticles gradually formed as the capsaicin reduces and binds to the surface of silver nanoparticles (SNPs). Further heating was stopped when the reaction mixture turned purple. The presence of colloidal nanoparticles in the solution was detected by passing laser beam. The bio-reduction was monitored at different time intervals of 30 min, 1, 2, 16 and 24 h by color change and UV spectroscopy measurement.

Kinetics

UV–vis spectroscopy measurements of the silver nitrate solution, capsaicin and capsaicin-capped silver nanoparticles were carried out on a UV–vis spectrophotometer UV-2450 (Shimadzu). The kinetics of bio-reduction of silver nano particles at different time intervals was recorded. The spectra from 200 to 800 nm were recorded for the synthesized silver nanoparticles—pellet and remaining capsaicin in the supernatant after centrifugation.

Fourier transforms infrared (FTIR) spectroscopy measurements

After complete reduction, the colloidal silver nanoparticles were centrifuged at 12,000 rpm for 10 min to isolate the silver nanoparticles. The silver nanoparticles pellet obtained after centrifugation were resuspended in deionised water and again centrifuged to remove the traces of free capsaicin present in the solution prior to FTIR analysis. FTIR spectrum was taken to assess functional groups and the involvement of possible capping of capsaicin on silver nanoparticles was carried out on a Perkin-Elmer FTIR spectrum one spectrophotometer.

TEM measurements

TEM samples of the silver nanoparticles were prepared by placing a drop over carbon-coated copper grids and allowing the solvent to evaporate. TEM measurements were performed on a JEOL model 1200EX instrument operated at an accelerating voltage between 80 and 200 kV.

Hemagglutination test

This test was performed in the microtiter plate. To each well, 50 μl of PBS was added. The following solutions were added in the given order. In column 1 (A–F) 50 μl of PBS that serves as control without any treatment was added. To columns 2, 3 and 4 50 μl of Anti-A, Anti-B and

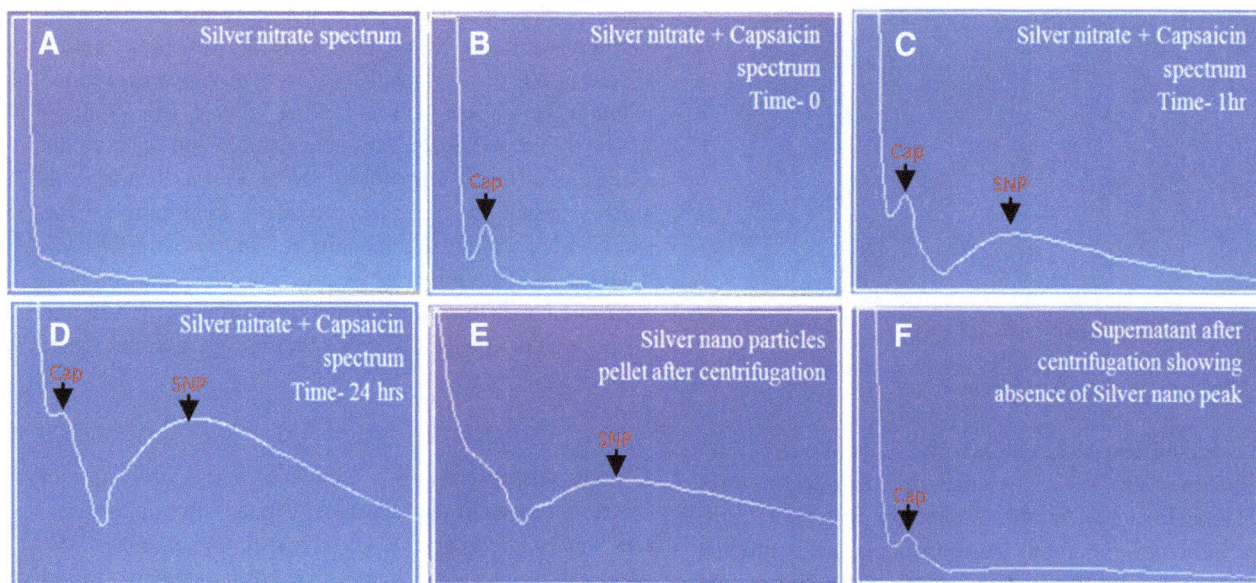

Fig. 2 Time dependent UV–vis spectra during the formation of capsaicin-capped silver nanoparticles

Anti-D were added, respectively. To the column 5, 50 μl of PHA (Himedia, Mumbai) was added. In column 6 of A, C and F 50 μl of capsaicin-capped silver nanoparticles was added, mixed well and 50 μl transferred to the next well on its right. Repeated mixing and transfer of 50 μl down the length of the plate up to column 9 was done; final 50 μl from the last well was discarded into a bleach solution. Agglutination inhibition test was performed by mixing the anti-A, B and D with nanoparticles in columns 10–12. To each well, 50 μl of 0.5 % red blood cells was added. It was incubated at room temperature for 30–60 min. Negative results appeared as dots in the center of round-bottomed plates. Positive results formed a uniform reddish color across the well.

Results

Synthesis and UV–vis spectroscopy studies

Silver nitrate solution mixed with capsaicin solution showed no significant change in the UV–vis spectra at 0 h, only the characteristic peak of capsaicin at 220 nm was observed in the curve Fig. 2b and found to be 100 μg calibrated using linear regression equation ($Y = 0.00919X - 0.0084$) of capsaicin. As the time increased decrease in the capsaicin peak at 220 nm, with an emerging peak around 458 nm as well as a change in color of the solution to pale gray color indicated the reduction of silver nitrate to silver nanoparticles as shown in (Fig. 3a). Absorption spectra of silver nanoparticles formed in the reaction mixture had absorbance peak at 450 nm, arising

due to the excitation of surface plasmon vibrations in the silver nanoparticles as seen in (Fig. 2d).

Kinetics

Time taken for complete bio-reduction of silver nitrate and capping was found to be 16 h even after which the reduction was stationary up to 24 h (Fig. 2d). The UV spectrum of the nano pellet after centrifugation was devoid of capsaicin peak (Fig. 2e). But free capsaicin peak was found in the supernatant and measured about 60 μg (Fig. 2f). The amount of capsaicin required to reduce 20 ml of 1 mM silver nitrate solution was found to be 40 μg approximately.

FTIR

FTIR measurements were carried out to identify the possible biomolecules in the capsaicin-reduced nanoparticles and also the capping agents responsible for the stability of the biogenic nanoparticles solution. FTIR spectrum of capsaicin showed prominent absorption bands at 3,435, 2,932, 2,351, 1,637, 1,538, 14,437, 1,026 and 812 cm^{-1}. Capsaicin structure has an aromatic ring, a polar group and a hydrophobic moiety (Fig. 4a). FTIR was done to interpret the structural modifications that had occurred in capsaicin upon formation of capsaicin-capped SNPs. We got conspicuous peaks around 3,447, 2,922, 2,352, 1,635, and a peak around 1,025 cm^{-1} followed by not-so conspicuous peaklets (Fig. 4b). The bands around 3,400 cm^{-1} are likely due to the presence of residual water molecules or the presence of OH group in capsaicin. The bands around

Fig. 3 SNPs colloidal solution
(a), Tyndall effect (b) and TEM
image of capsaicin-capped
SNPs (c)

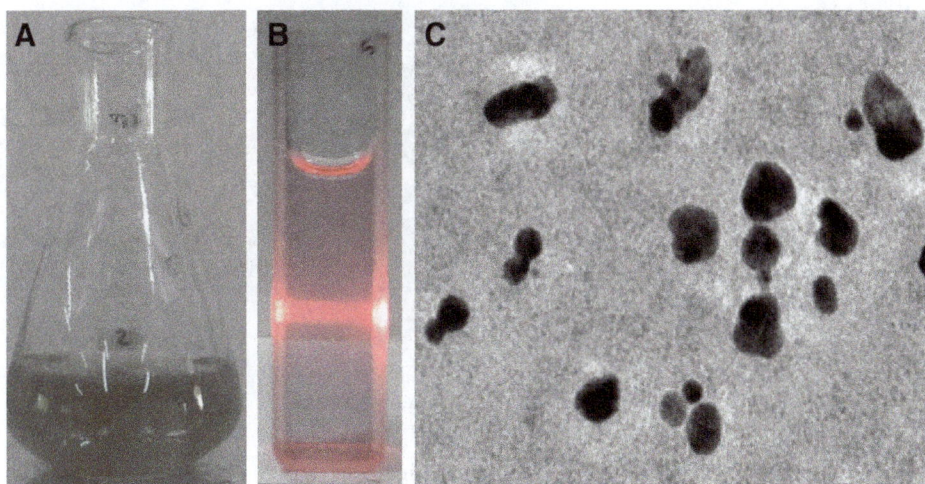

Fig. 3 SNPs colloidal solution (a), Tyndall effect (b) and TEM image of capsaicin-capped SNPs (c)

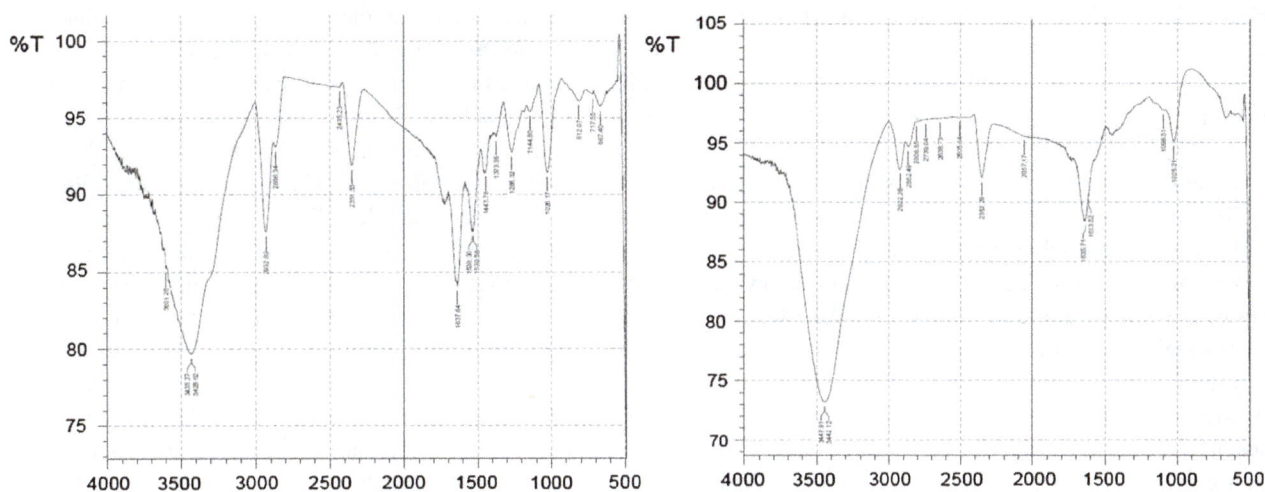

Fig. 4 FTIR transmittance spectra of capsaicin (a) and SNPs capped by capsaicin (b)

$2,750–3,000$ cm^{-1} are likely to originate from carbonyl C=O group present in the capsaicin. The bands around $1,600–1,750$ cm^{-1} are likely due to the amide present in capsaicin molecule and the presence of peaklets at the lower wave numbers following it could also likely suggest the presence of aromatic C=C, which is present in capsaicin molecule, and the disappearance of the peaklets at the lower wavenumber region after the formation of SNPs suggests the possible structural change of the aromatic group. It is known that aromatic groups reduce Ag ions into metallic NPs, which peak around $1,646$ cm^{-1} which is likely to be aliphatic (alkene) C=C and whose absorption is converged with aromatic and amide absorption bands.

HR-TEM measurements

TEM measurements divulged that the resultant product after 24 h was composed of nearly spherical AgNPs with 20–30 nm diameter (Fig. 3c).

Hemagglutination test

One of the main aims of the current study is to increase the bioavailability of capsaicin for biological applications. The hydrophobic capsaicin is now conjugated with SNPs which is in water, hence becoming more available. The SNPs act as carrier of drug capsaicin. Before utilizing the SNPs–capsaicin conjugates for clinical application, it is compulsory to confirm their biocompatibility particularly with human blood. Hence, we have performed hemagglutination (H) and H-inhibition assay in the presence of SNPs–capsaicin conjugates. The test was conducted with A, B and O blood groups. The tested blood groups showed positive results to respective test antigens. The capsaicin-capped silver nanoparticles showed negative results for the screened blood groups. But "O" positive blood group showed light agglutination when compared to "A" and "B" blood groups. The hemagglutination inhibition test was negative for all the blood groups. The assay result

Fig. 5 Hemagglutination and hemagglutination inhibition test for the capsaicin-capped silver nanoparticles for ABO blood groups. *C* Control, *A* Anti-A, *B* Anti-B, *D* Anti-D, *PHA* phytohemagglutinin, *NN3* nanoparticles dilution, *A+NP* Anti-A and nanoparticles, *B+NP* Anti-B and nanoparticles, and *D+NP* Anti-D and nanoparticles

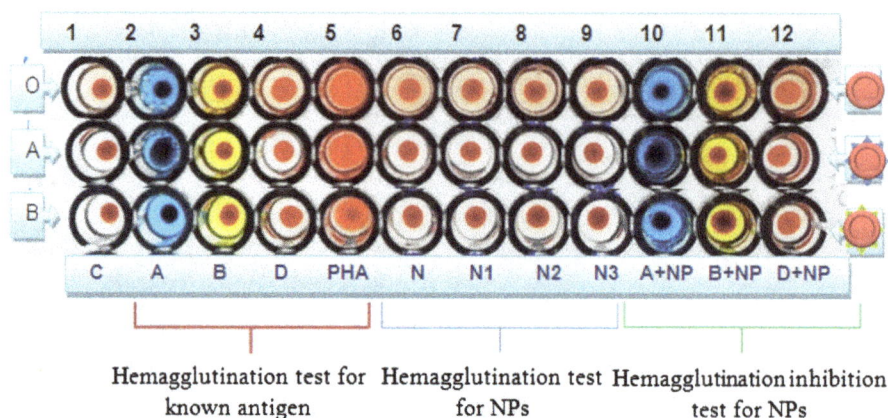

Hemagglutination test for   Hemagglutination test   Hemagglutination inhibition
        known antigen   for NPs   test for NPs

confirms the compatibility of AgNPs–capsaicin conjugate with normal blood cells, as RBC disruption was not observed which would lead to H-agglutination (Fig. 5).

Discussion

In ayurvedic medicine, silver is used in petite amounts as a tonic, elixir or rejuvenating agent. Currently, these are also available in oral, injectable, and topical forms (Kishore Madhukar Paknikar 2008). The ash of silver, also known as Raupya Bhasma, is used to treat diseases and disorders like pain, neuralgias, inflammation, anxiety, convulsions, memory loss etc. (Nadeem et al. 1999; Hamilton et al. 1972). Moreover, nanosize of silver particle is possibly responsible for improving the absorption of silver in brain for the treatment of various pain and neurological conditions, (Khanna et al. 1997; Wang et al. 2005; Chou et al. 2001). Capsaicin has been included in topical treatments aimed at relief of different neuropathic pain syndromes such as post-herpetic neuralgia, musculoskeletal pain, diabetic neuropathy, osteoarthritis and rheumatoid arthritis (Backonja et al. 2010; Tesfaye 2009; Watanabe et al. 1987; Sawynok 2005; Derry et al. 2009). Hence our study on capsaicin-capped silver nanoparticles has a huge impact in the field of pharmacology as active drug for neurological disorders.

In vitro studies in human skin showed that capsaicin adsorption and biotransformation were found to be slow and that most capsaicin remained unaffected, while a small fraction was metabolized to vanillylamine and vanillyl acid (Suresh and Srinivasan 2010; Chanda et al. 2008; Kawada et al. 2008; Kawada and IwaiK 1985). The reason for this is varying structures of the compounds, their fair aqueous solubility, poor permeability and instability. The natural products undergo fast oxidation under basic conditions and first pass metabolism before reaching systemic circulation (Haslam 1996; Anders 2002). Many pharmaceutical companies synthesize several million compounds, but only a

few of those new chemical entities become leads for pre-clinical development as a result of issues associated with their solubility. Nanotechnology paves way for insoluble compounds to be attached or encapsulated with highly soluble nanoparticles, offering number of drugs entry into clinical trials (McNeil 2005).

In this present study, capsaicin was found to be a potential lipophilic alkaloid which reduced the silver nitrate to silver nanoparticles. The absorption maxima of silver nanoparticles were reported in the range of 400–500 nm (Prathna et al. 2011). The plasmon resonance results of the present study were in accordance with previous reports of Vijayakumar et al. (2012) on *Artemisia nilagirica*, indicating occurrence of a silver band at 340–400 nm. According to the finding of Geethalakshmi and Sarada (2012), FTIR analysis of the silver nanoparticles showed absorption peaks of reduced silver at 1,653.96 and 1,027.44 cm^{-1} for capsaicin-capped silver nanoparticles. In FTIR, the shifting of peaks was observed which may be attributed to the difference in capping and nature of co-ordination with the metal surface of silver (Sing et al. 2009).

Protective role of capsaicin in the gastrointestinal tract has produced controversial results and clearly indicate the need for further clinical trials to better define effective dosages (Holzer and Lippe 1988). Studies have been carried to elucidate the pharmacokinetics of curcumin as it is poorly absorbed from the gastrointestinal (GI) tract after oral administration due to its low water solubility and low stability in GI fluids. But nanoparticle formulations of curcumin exhibited increased solubility in water compared to free curcumin (Bisht et al. 2007). The orally administered capsaicin was absorbed by the tissues and concentrated in the blood. About 24.4 % of administered capsaicin was circulated to liver, kidney and intestine through blood stream (Suresh and Srinivasan 2010). In our study, the capsaicin-capped silver nanoparticles tested for hemagglutination proved that they are highly biocompatible with all three types of blood groups. Hence, the capsaicin-capped silver nanoparticles would be a promising drug after high throughput in vivo screening.

Conclusion

Pure capsaicin lacks in bioavailability and permeability, mainly due to its poor solubility in water. To improve the bioavailability of capsaicin, it is conjugated to the surface of metal nanoparticles. In the present study, we report the binding of capsaicin to the surface of silver nanoparticles (SNPs). The nanoparticles were characterized by FTIR and TEM. The synthesized gold nanoparticles were found to be 20–30 nm in size. We found that capsaicin acts both as a reducing and capping agent, stabilizing the silver solution for 6 months. The synthesized SNPs were highly compatible with ABO blood groups in hemagglutination test. In vivo studies on the capsaicin-capped silver nanoparticles can be carried out to test its efficacy in the topical applications.

Acknowledgments Our grateful thank to Rev. Sr. Mercy Margret B.ICM, Mrs. Sargunam Amma and Ameer for their encouragement. Special thanks to Mr. Lourd Xavier IIT Madras Nanotechnology Unit, for his intellectual guidance and valuable suggestion. Authors thank Loyola College Management for their sustained support.

References

Anders B (2002) Interaction of plant polyphenols with salivary proteins. Crit Rev Oral Biol Med 13:184–196

Backonja MM, Malan TP, Vanhove GF et al (2010) Tobias (NGX-4010, a high-concentration capsaicin patch, for the treatment of postherpetic neuralgia: a randomized, double-blind, controlled study with an open-label extension. Pain Med 11:600–608

Baimark Y (2012) Preparation of surfactant-free linear and star-shaped poly (l-lactide)-b-methoxy polyethylene glycol nanoparticles for drug delivery. J Appl Sci 12:263–270

Bisht S, Feldmann G, Soni S, Ravi R, Karikar C,Maitra A, Maitra A (2007) Polymeric nanoparticle-encapsulated curcumin ("nanocurcumin"): a novel strategy for human cancer therapy. J Nanobiotechnol 5. 10.1186/1477-3155-5-3

Chanda S, Bashir M, Babbar S, Koganti A, BleyK (2008) In vitro hepatic and skin metabolism of capsaicin. Drug Metab Dispos 36:670–675

Chen YY, Wang CA, Liu HY, Qiu JS, Bao XH (2005) Ag/SiO2: a novel catalyst with high activity and selectivity for hydrogenation of chloronitrobenzenes. Chem Commun 42:5298–5300

Chou CW, Chu SJ, Chiang HJ, Haung CY, Lee CJ, Sheen SR et al (2001) Temperature programmed reduction study on calcinations of nano palladium. J Phys Chem B 38:9113–9117

Corson TW, Crews CM (2007) Molecular understanding and modern application of traditional medicines: triumphs and trials. Cell 130:769–774

Derry S, Lloyd R, Moore RA, McQuay HJ (2009) Topical capsaicin for chronic neuropathic pain in adults. Cochrane Database Syst Rev CD007393

Feng QL, Wu J, Chen GQ, Cui FZ, Kim TN, Kim JO (2000) A mechanistic study of the antibacterial effect of silver ions on *Escherichia coli* and *Staphylococcus aureus*. J Biomed Mater Res 52:662–668

Fernando Spiller, Márcia K, Alves et al (2008) Anti-inflammatory effects of red pepper (*Capsicum baccatum*) on carrageenan- and antigen-induced inflammation. J Pharm Pharm 60(4):473–478

Fraenkel L, Bogardus ST Jr, Concato J, Wittink DR (2004) Treatment options in knee osteoarthritis: the patient's perspective. Arch Intern Med 164:1299–1304

Geethalakshmi R, Sarada DVL (2012) Gold and silver nanoparticles from trianthema decandra: synthesis, characterization, and antimicrobial properties. Intern J Nanomedicine 7

Hamilton EJ, Minski MJ, Cleary JJ (1972) The concentration and distribution of some stable elements in healthy human tissues from United Kingdom. Sci Total Environ 1:341–374

Haslam E (1996) Natural polyphenols (vegetable tannins) as drugs: possible modes of action. J Nat Prod 59:205–215

Holzer P, Lippe IT (1988) Stimulation of afferent nerve endings by intragastric capsaicin protects against ethanol-induced damage of gastric mucosa. Neuroscience 27:981–987

Hopur2 (2009) Pharmacophore structure of capsaicin. Accessed 3 November 2009

Jin R, Cao YW, Kelly KL et al (2001) Photoinduced conversion of silver nanospheres to nanoprisms. Science 294:1901–1903

Kawada T, IwaiK (1985) In vivo and in vitro metabolism of dihydrocapsaicin, a pungent principle of hot pepper, in rats. Agric Biol Chem 49:441–448

Khanna AT, Silvaraman R, Vohora SB (1997) Analgesic activity of silver preparations used in Indian system of medicine. Indian J Pharmacol 29:393–398

Kingston DG, Newman DJ (2002) Mother nature's combinatorial libraries: their influence on the synthesis of drugs. Curr Opin Drug Discov Dev 5:304–316

Kishore Madhukar Paknikar (2008) Anti-microbial activity of biologically stabilized silver nano particles. EP1753293 A4

Kumari V, Kumar, Yadav SK (2012) Nanotechnology: a tool to enhance therapeutic values of natural plant products. Trends Med Res 7:34–42

Kurita S, Kitagawa E, Kim CH, Momose Y, Iwahashi H (2002) Studies on the antimicrobial mechanisms of capsaicin using yeast DNA microarray. Biosci Biotechnol Biochem 66:532–553

Li Z, Lee D, Sheng XX et al (2006) Two-level antibacterial coating with both releasekilling and contact-killing capabilities. Langmuir 22:9820–9823

McNeil SE (2005) Nanotechnology for the biologist. J Leukocyte Biol 78:585–594

Nadeem A, Khanna T, Vohora SB (1999) Silver preparations used in Indian system of medicine: neuropsychobehavioural effects. Indian J Pharmacol 31:214–221

Prathna TC, Raichur AM, Chandrasekaran N, Mukherjee A (2011) Biomimetic synthesis of silver nanoparticles by Citrus limon (lemon) aqueous extract and theoretical prediction of particle size. Colloids Surf B Biointerfaces 82:152–159

Priyaragini S, Sathishkumar SR, Bhaskararao KV (2013) Biosynthesis of silver nanoparticles using actinobacteria and evaluating its antimicrobial and cytotoxicity activity. Int J Pharm Pharm Sci 5(2):709-712

Reyes-Escogido ML, Gonzalez-Mondragon EG, Vazquez-Tzompantzi E (2011) Chemical and pharmacological aspects of capsaicin. Molecules 16:1253–1270

Rosa A, Deiana M, Casu V et al (2002) Antioxidant activity of capsinoids. J Agric Food Chem 50:7396–7401

Sahoo SK, Labhasetwar V (2003) Nanotech approaches to drug delivery and imaging. Drug Discov Today 8:1112–1120

Sawynok J (2005) Topical analgesics in neuropathic pain. Curr Pharm Des 11:2995–3004

Setua P, Chakraborty A et al (2007) Synthesis, optical properties, and surface enhanced Raman scattering of silver nanoparticles in nonaqueous methanol reverse micelles. J Phys Chem C 111:3901–3907

Sing S, Patel P, Jaiswal S, Prabhune AA, Ramana CV, Prasad BLV (2009) A direct method for the preparation of glycolipid-metal nanoparticle conjugates: sophorolipids as reducing and capping agents for the synthesis of water redispersible silver nanoparticles and their antibacterial activity. New J Chem 33:646–652

Suresh D, Srinivasan K (2010) Tissue distribution and elimination of capsaicin, piperine and curcumin following oral intake in rats. Indian J Med Res 131:682–691

Tesfaye S (2009) Advances in the management of diabetic peripheral neuropathy. Curr Opin Support Palliat Care 3:136–143

Vijayakumar M, Priya K, Nancy FT, Noorlidah A, Ahmed ABA (2012) Biosynthesis, characterization and anti-bacterial effect of plant-mediated silver nanoparticles using Artemisia nilagirica. Indus Crops Prod 41:235–240

Wang X, Zhuang J, Peng Q, Li Y (2005) A general strategy for nanocrystal synthesis. Nature 431:3968

Watanabe T, Kawada T, Yamamoto M, Iwai K (1987) Capsaicin, a pungent principle of hot red pepper, evokes catecholamine secretion from the adrenal medulla of anesthetized rats. Biochem Biophys Res Commun 142:259–264

Yoshioka M, St-Pierre S et al (1999) Effects of red pepper on appetite and energy intake. Br J Nutr 82:115–123

Zeyrek FY, Oguz E (2005) In vitro activity of capsaicin against Helicobacter pylori. Ann Microbiol 55:125–127

15

2-Aminoethanol-mediated wet chemical synthesis of ZnO nanostructures

Tehmina Naz · Adeel Afzal · Humaira M. Siddiqi ·
Javeed Akhtar · Amir Habib · Mateusz Banski ·
Artur Podhorodecki

Abstract The synthesis of ZnO nanostructures via co-precipitation of $Zn(NO_3)_2 \cdot 2H_2O$ in 2-aminoethanol under different reaction conditions is presented. The effect of temperature and time on crystal structure, size, morphology, and optical properties of ZnO nanopowders is studied. XRD analyses demonstrate that single crystalline wurtzite ZnO nanostructures are instantaneously formed at higher temperature, or at low temperature with growth times equal to 2 h. However, the mean crystallite size increases as a function of reaction temperature and growth time. XRD and SEM results reveal that ZnO nuclei grow along favored crystallographic planes [wurtzite (101)] in 2-aminoethanol to form single crystalline nanorods. The optical band-gap energies of ZnO crystallites measured from their UV absorption spectra increase from 3.31 to 3.52 eV with decreasing particle size. ZnO nanopowders also exhibit good photoluminescent characteristics with strong UV and weak visible (violet, blue) light emissions corresponding to surface defects and oxygen vacancies in ZnO products.

Keywords 2-Aminoethanol · Band gap · Crystallite size · Photoluminescence · ZnO nanostructures

Introduction

In the past few years, ZnO-based nanostructured materials have emerged as the first-choice materials not only for semiconductor devices such as sensors (Chougule et al. 2012; Afzal et al. 2012) and solar cells (Kim et al. 2013; Patra et al. 2014), but also in other fields as UV emitters (Zhang et al. 2013), optical waveguides (Jiang et al. 2012), and biomedical (Yang et al. 2012) and optoelectronic devices (Zhang et al. 2012). The nontoxic nature, low cost, stability, easy processability, and nanostructuring are the chief distinctions of ZnO (Schmidt-Mende and MacManus-Driscoll 2007).

A wide range of synthetic methods are currently available for developing different morphologies of ZnO nanostructures, which result in a variety of material applications such as those mentioned above (Djurišić et al. 2012). These include hydrothermal, sol–gel, sonochemical, organometallic, pyrolysis, laser ablation, and vapor phase epitaxial growth (Schmidt-Mende and MacManus-Driscoll 2007; Djurišić et al. 2012; Wang 2004; Ismail et al. 2011; Samanta and Bandyopadhyay 2012; Zhan et al. 2012) synthetic methods. It is well known that physical properties of

T. Naz · A. Afzal (✉) · H. M. Siddiqi (✉)
Department of Chemistry, Quaid-i-Azam University,
Islamabad 45320, Pakistan
e-mail: aa@aafzal.com

H. M. Siddiqi
e-mail: humairas@qau.edu.pk

A. Afzal
Affiliated Colleges at Hafr Al-Batin, King Fahd University of Petroleum and Minerals, P.O. Box 1803, Hafr Al-Batin 31991, Saudi Arabia

J. Akhtar
Department of Physics, COMSATS Institute of Information Technology, Islamabad Campus, Chak Shahzad, Islamabad 44000, Pakistan

A. Habib
School of Chemical and Materials Engineering, National University of Science and Technology, H-12, Islamabad 44000, Pakistan

M. Banski · A. Podhorodecki
Institute of Physics, Wroclaw University of Technology,
Wybrzeze Wyspianskiego 27, 50-370 Wroclaw, Poland

thus obtained ZnO nanostructures radically depend on the choice of the synthetic method.

In the recent past, researchers have also used different structure-directing agents to prepare ZnO nanomaterials and to optimize their structure, morphology, and properties (Shouli et al. 2010; Aimable et al. 2010; Veriansyah et al. 2010; Rai and Yu 2012; Foe et al. 2013). These structure-directing agents include simple organic molecules such as methanol, ethanol, amines, and urea, specialized surfactants such as cetyltrimethylammonium bromide (CTAB), trisodiumcitrate, and sodium dodecyl sulfate (SDS), and polymers such as polyethylene glycol (PEG), polyvinylpyrrolidone (PVP), and poly(acrylic acid). These reagents principally control the rate of reaction, growth of ZnO crystals, morphology (size, shape, and distribution), and physical properties of the final ZnO products.

In this bargain, the researchers successfully prepared different forms of ZnO nanostructures by precipitating aqueous zinc salt solutions with NaOH in the presence of small amounts of aminoethanol (Costa and Baptista 1993; Wang et al. 2011). Wang et al. (2011) specially studied the effect of aminoethanol and NaOH concentration on the morphology of ZnO nanostructures synthesized via one-pot hydrothermal synthesis. They concluded that varying ethanolamine concentration had a pronounced effect on the morphology and structure of ZnO nanopowders due to competitive adsorption of aminoethanol and $[Zn(OH)_4]^{2-}$ on ZnO nuclei.

Herein, we further exploit 2-aminoethanol as the structure-directing agent in the wet chemical synthesis of ZnO nanopowders. While the concentration of aminoethanol is fixed during these experiments, reaction conditions such as temperature and time are varied to study their effects on the crystalline nature, crystallite size, morphology, and optical properties including UV emission, band gap, and photoluminescent characteristics of the so formed ZnO nanopowders. Furthermore, in our experiments, 2-aminoethanol is used as the solvent for $Zn(NO_3)_2$, while it also acts as a surfactant and selectively adsorbs on growing ZnO nuclei to facilitate preferred growth on certain crystallographic planes, thus resulting in the formation of hexagonal ZnO nanorods. The temperature and time are found to increase the crystallite size. It is found that ZnO nanopowders show weak violet and blue emissions along with stronger UV emission.

Experimental

Chemicals and reagents

All chemicals were used as received without further purification. Zinc nitrate hexahydrate $(Zn(NO_3)_2 \cdot 6H_2O$;

crystallized, ≥ 99.0 %) was purchased from Sigma-Aldrich. 2-Aminoethanol (ACS reagent, ≥ 99.0 %) was obtained from Fluka. Liquid ammonia (33 %) was received from Lab-Scan. Ethanol (anhydrous) was obtained from Sigma-Aldrich.

Synthesis of ZnO nanostructures

ZnO nanopowders of variable sizes and morphology were synthesized by co-precipitation of $Zn(NO_3)_2 \cdot 6H_2O$ in 2-aminoethanol. For this purpose, 2-aminoethanol (20 mL) was placed in a 250-mL round-bottom flask. $Zn(NO_3)_2 \cdot 6H_2O$ (14 mmol; 4.15 g) was then added into the flask and dissolved through ultrasound sonication for several minutes. After the clear solution of $Zn(NO_3)_2 \cdot 6H_2O$ in 2-aminoethanol was obtained, liquid NH_3 (10 mL) was added slowly to the mixture. The mixture was subsequently diluted with a small amount of deionized water to attain a pH of 11.2. From this point onward, the temperature and time of the reaction were controlled to study the effect of increasing temperature and time on crystallinity, geometry, surface morphology, and optical properties of ZnO nanostructures. The temperature of the reaction was varied between 60 and 94 °C, whereas growth time was varied in the range of 15–120 min.

Table 1 provides details of the reaction times and temperatures adapted in different procedures to prepare ZnO nanopowders. White precipitates of ZnO nanopowders were obtained after the completion of these reactions. The solvent (2-aminoethanol) and excess water were decanted off after centrifuge, and the products (white powders) were dried in an oven at 100 °C. The colloidal suspensions of

Table 1 The optimization of reaction conditions (time and temperature) to obtain different ZnO nanostructures

Experiment	Growth time (min)	Reaction temperature (°C)	Crystallite size (nm)	Sample ID
Series A	60	60	12.50 ± 0.92	ZnO-A60
		70	33.82 ± 1.20	ZnO-A70
		75	40.28 ± 1.20	ZnO-A75
		80	43.15 ± 1.29	ZnO-A80
		94	65.40 ± 1.97	ZnO-A94
Series B	15	60	9.80 ± 0.50	ZnO-B15
	30		10.66 ± 0.65	ZnO-B30
	60		12.50 ± 0.92	ZnO-B60
	120		15.26 ± 1.05	ZnO-B120
Series C	15	94	35.62 ± 1.23	ZnO-C15
	30		40.28 ± 1.54	ZnO-C30
	60		65.40 ± 1.97	ZnO-C60
	120		69.50 ± 1.54	ZnO-C120

different ZnO nanopowders were prepared in anhydrous ethanol by ultrasound sonication for 15 min for the purpose of analytical characterization.

Analytical methods

X-ray diffraction (XRD) patterns of as-synthesized ZnO nanostructures were obtained by Siemens D500 XRD instrument equipped with Cu Kα irradiation source (wavelength: 1.5406 Å), at 40 kV and 30 mA. XRD patterns were obtained at a scanning rate of $10°$ min^{-1} from $10°$ to $80°$ of 2θ.

The surface morphology of synthesized ZnO nanostructures was studied on Jeol JFM 5910 SEM instrument equipped with tungsten filament electron emitter at an accelerating voltage of 5–10 kV.

The selected ZnO nanoparticle suspensions in ethanol were also characterized by UV–vis spectroscopy at 25 °C. A Schimadzu UV–vis spectrophotometer, model Pharma Spec UV-1700, and quartz cuvette were used for this purpose. The photoluminescence (PL) spectra were also measured at room temperature at an excited wavelength of 296 nm.

Results and discussion

The structural characterization of ZnO nanopowders

The structural characterization of ZnO nanopowders to determine their crystallinity, crystal phase characteristics, and purity was performed with the help of XRD. Figure 1 shows the XRD patterns of different ZnO samples belonging to series A. These nanopowders were prepared in 2-aminoethanol at different temperatures (60–94 °C) with a constant growth time of 60 min. The reference XRD patterns and relative peak intensity of phase-pure hexagonal wurtzite and cubic ZnO are also provided in Fig. 1 for comparison.

It is evident that ZnO nanopowders obtained at lower temperatures, e.g., ZnO-A60, ZnO-A70, and ZnO-A75 prepared at 60, 70, and 75 °C, respectively, do not exhibit phase-pure crystalline structure. Instead, these ZnO nanopowders show the characteristics of both hexagonal wurtzite and cubic phases along with some impurities such as $Zn(OH)_2$ (Yang et al. 2006; Rai and Yu 2012). It is due to the fact that as-synthesized ZnO nanopowders at low temperature are not treated thermally at higher

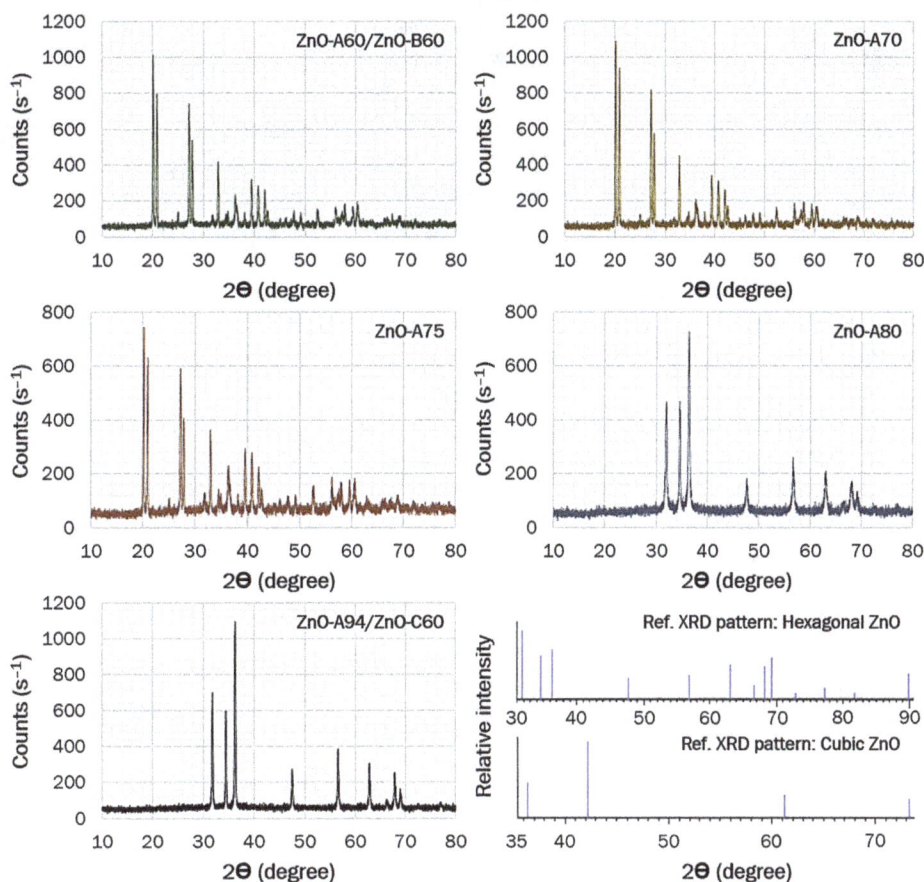

Fig. 1 XRD patterns of ZnO nanopowders (series A) prepared in 2-aminoethanol at different temperatures. The reference XRD patterns of crystalline hexagonal wurtzite and cubic ZnO are also given for comparison

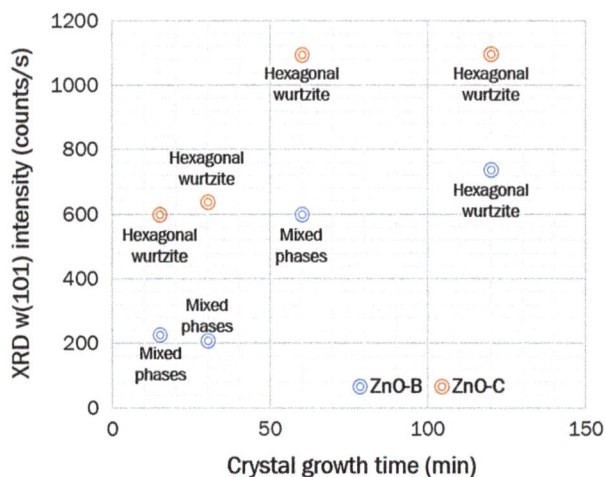

Fig. 2 The primary wurtzite diffraction peak w(101) intensity plotted as a function of reaction time. The crystalline nature of ZnO nanostructures is also indicated

temperature, thus they retain some functional hydroxyl groups on the surface.

However, ZnO nanopowders synthesized at higher temperatures, e.g., at 80 °C (ZnO-A80) and at 94 °C (ZnO-A94) show crystalline structure and their XRD reflections can be indexed as hexagonal wurtzite ZnO phase (Guo et al. 2002; Geng et al. 2004), i.e., the results are in agreement with the standard hexagonal wurtzite ZnO crystal structures with space group P63mc. The respective XRD patterns indicate that these ZnO nanopowders obtained at higher temperature were phase-pure single crystalline powders with no observable peaks relevant to impurities such as $Zn(OH)_2$. The sharp XRD peaks in ZnO-A94 also suggest that ZnO-A94 nanopowders are well crystallized as compared to ZnO-A80.

The XRD patterns of ZnO nanopowders corresponding to series B and C were also observed. The diffraction peaks around 31.8°, 34.5°, and 36.3° 2θ positions, which are indexed as w(100), w(002), and w(101) planes of wurtzite crystal, respectively, were observed in each sample indicating the existence of a wurtzite crystal structure. However, the samples of series B (prepared at 60 °C) also exhibit XRD peaks corresponding to the cubic phase and $Zn(OH)_2$ impurities, except ZnO-B120 that has wurtzite crystalline structure. On the other hand, all samples of ZnO nanopowders (belonging to series C) have single crystalline wurtzite structure.

Figure 2 shows the intensity of w(101) XRD peak, i.e., major wurtzite diffraction in series B and C samples of ZnO nanopowders plotted as a function of reaction/growth time. The crystalline nature of all samples is also indicated in Fig. 2. It is found that higher reaction temperature or prolonged reaction time is required to prepare hexagonal wurtzite ZnO nanostructures in 2-aminoethanol as the

reaction medium. Therefore, we conclude that slightly higher temperature is needed for the growth of wurtzite crystal structures, although the formation of ZnO nanophases is thermodynamically favored under ambient conditions. Likewise, for reactions carried out at lower or room temperatures, additional time is needed for crystal growth.

XRD analyses also offer further structural information on the crystallite size of ZnO nanostructures. The mean crystallite size of various ZnO nanopowders can be calculated by using the Scherrer's formula that is given in the following equation: $(D = 0.9\lambda/B\cos\theta)$, where D is the mean crystallite size of the ZnO nanopowder, λ is the wavelength of Cu Kα radiation, i.e., $\lambda = 1.5406$ Å, B is the full-width-at-half-maximum (FWHM) intensity of the diffraction peak in radian, and θ is the Bragg's diffraction angle (Mahmood et al. 2013). Consequently, the mean crystallite size of different samples of ZnO nanopowders was measured and reported in Table 1. It is found that the mean crystallite sizes of different ZnO nanopowders synthesized under different reaction conditions of time and temperature lie in the range 9.8–69.5 nm.

Effect of temperature on mean crystallite size of ZnO nanopowders

The mean (or average) crystallite size of ZnO nanopowders (series A) prepared in 2-aminoethanol at different temperatures (60–94 °C) and constant reaction time of 60 min are determined from the respective XRD patterns. These crystallite sizes are plotted as a function of reaction temperature in Fig. 3. It is obvious that a straight trendline is obtained that shows a regular increase in the mean crystallite size as a function of reaction temperature.

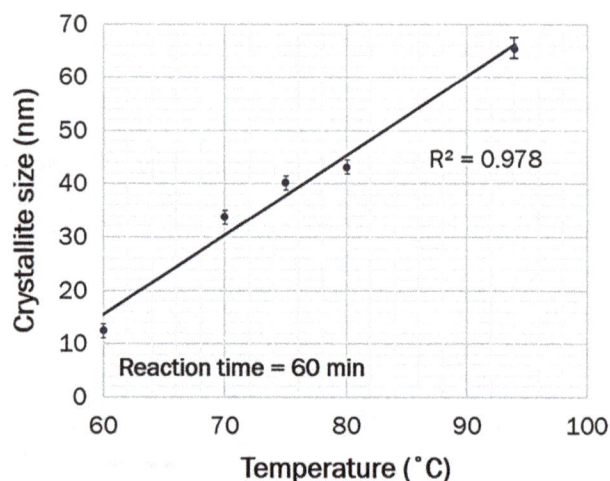

Fig. 3 The mean crystallite size of ZnO nanopowders synthesized in 2-aminoethanol at different reaction temperature (60 °C) and constant growth time of 60 min

Amin et al. (2011) reported that the surface area of hydrothermally grown ZnO nanostructures increases as a function of temperature within the temperature range of 50–90 °C, which is an indication of decreasing particle size with increasing temperature. However, we observed different results during the wet chemical synthesis of ZnO nanopowders involving 2-aminoethanol as a solvent as well as a surfactant, which indicates a regular growth in the mean crystallite size of the so formed ZnO nanopowders as a function of increasing temperature in the range of 60–94 °C. This may be attributed to an increased growth rate of wurtzite ZnO crystals at higher temperature that leads to relatively larger mean crystallite size.

Effect of growth time on mean crystallite size of ZnO nanopowders

The mean crystallite size of ZnO nanopowders belonging to the experimental series B and C were also measured from the respective diffraction patterns to study the effect of growth (reaction) time on crystallite size. These ZnO nanopowders were prepared in 2-aminoethanol at a constant temperature of 60 (series B) or 94 °C (series C), while growth time was varied from 15 to 120 min. Figure 4 shows the crystallite size of various ZnO samples as a function of growth time. It is obvious from the linear trendline that the mean crystallite size of ZnO nanopowders consistently increases with the increasing growth time.

These results are well in agreement with previous reports on hydrothermal and microwave-assisted synthesis of ZnO nanostructures (Amin et al. 2011; Barreto et al. 2013). For ZnO nanopowders prepared in 2-aminoethanol, it is found that with an increase in reaction time while maintaining constant temperature, concentration and pH,

ZnO nanocrystallites continue to grow and form bigger ZnO nanocrystallites. Furthermore, the nucleation and growth of ZnO nanocrystallites is faster at higher temperature (94 °C) leading to much bulkier single crystalline ZnO nanostructures, whereas the growth of ZnO nanocrystallites is slow at low temperature (60 °C) giving rise to smaller and mixed phase ZnO powders (Rai et al. 2011).

The surface morphology of ZnO nanopowders

The suspensions of ZnO nanopowders prepared in ethanol by sonication were coated on quartz slides to monitor the surface morphology, i.e., size, shape, and distribution, of ZnO nanopowders. Figure 5 shows the SEM micrograph of ZnO-A60 or ZnO-B60 sample corresponding to ZnO nanopowders prepared at 60 °C and with 60 min of growth time. The image demonstrates that ZnO nanopowders are composed of small ZnO nanoparticles of sizes equal to $\sim$20 nm. These ZnO nanoparticles are non-agglomerated and uniformly distributed on the surface with a narrow particle size range. The size of ZnO nanoparticles obtained from the SEM image is slightly greater than the mean crystallite size (12.5 nm) determined from the respective XRD patterns.

Figure 6 shows the SEM image of ZnO-A94 or ZnO-C60 sample corresponding to ZnO nanopowders prepared at 94 °C and with 60 min of growth time. The micrograph shows completely different surface morphology as compared to ZnO nanopowders prepared at lower temperature. In this case, ZnO-A94 nanopowders are composed of hexagonal nanorods, with edge-to-edge distance of $\sim$200 nm. In addition, these nanorods are aligned on quartz surface in the shape of a flower with six nanorods surrounding a single vertically aligned central nanorod,

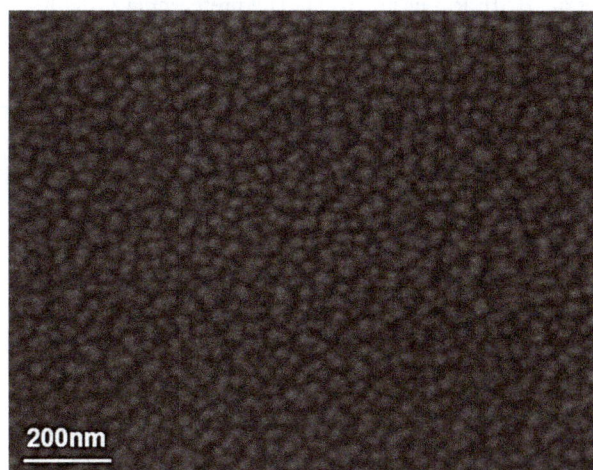

Fig. 4 The mean crystallite size of ZnO nanopowders synthesized in 2-aminoethanol at constant reaction temperatures of 60 and 94 °C and different growth times varying between 15 and 120 min

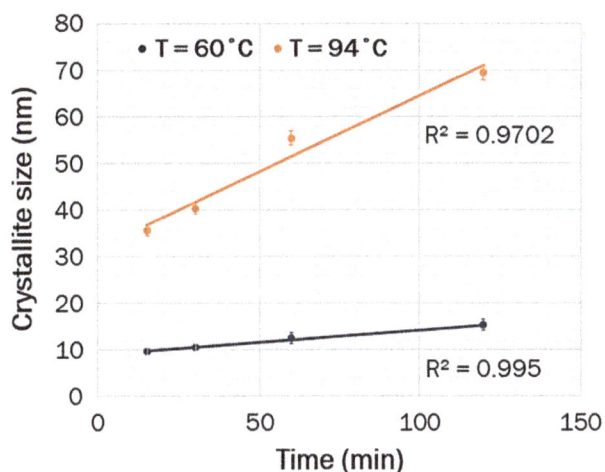

Fig. 5 The surface morphology of ZnO nanopowders synthesized in 2-aminoethanol at 60 °C with 60 min of growth time (sample ID: ZnO-A60/ZnO-B60)

Fig. 6 The surface morphology of ZnO nanopowders synthesized in 2-aminoethanol at 94 °C with 60 min of growth time (sample ID: ZnO-A94/ZnO-C60)

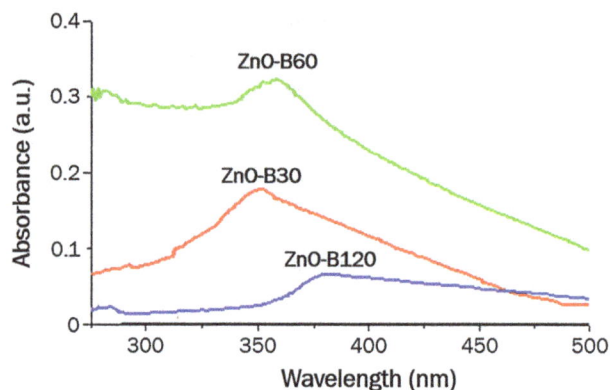

Fig. 7 UV–vis spectra of ZnO nanopowders (series B) prepared in 2-aminoethanol at constant temperature (60 °C)

thus forming a narcis-like architecture (Kajbafvala et al. 2010).

The difference in surface morphology of ZnO nanopowders synthesized under different conditions of temperature is obvious, and it may be attributed to faster and selective crystal growth on ZnO nuclei at higher temperature. In the respective diffraction patterns of ZnO nanopowders such as that of ZnO-A94 (in Fig. 1), it is clear that the diffraction peak corresponding to the w(101) plane of wurtzite crystal structure is the most intense peak, whereas the reference XRD pattern of hexagonal wurtzite structure shows the diffraction peak corresponding to w(100) plane as the most intense diffraction signal.

This fact supports the preferential growth mechanism of ZnO nanocrystallites along a particular lattice plane, i.e., w(101) that leads to the formation of ZnO nanorods (Nirmala et al. 2010; Rai et al. 2011). The preferential growth of ZnO nanocrystallites in a particular crystallographic plane may be attributed to the presence of 2-aminoethanol, which hereby acts as the structure-directing agent. 2-aminoethanol, the solvent used in the synthesis of ZnO nanopowders, also acts as a surfactant or capping agent, which may alter the relative growth of ZnO crystallites by adsorbing on specific crystallographic planes of the growing ZnO nuclei, thus inhibiting growth along these planes (Harunar Rashid et al. 2009). This may ultimately result in the formation of single crystalline wurtzite nanorods.

UV–vis spectroscopy and optical band-gap calculations

In recent years, wide band-gap semiconductors such as ZnO have attracted the scientific community due to increasing commercial demands of short wavelength light-emitting devices. ZnO is a wide band-gap (3.3 eV) semiconductor with a high excitation binding energy (60 meV) (Caglar et al. 2010). In general, bulk ZnO can absorb UV radiations with the wavelength ≤ 385 nm (Nirmala et al. 2010). The UV–vis absorption spectra of ZnO nanopowders prepared in 2-aminoethanol under controlled reaction conditions are shown in Fig. 7. The UV–vis absorption spectra were obtained from ZnO nanopowder suspensions in anhydrous ethanol. For this purpose, we have chosen ZnO nanopowders having smaller mean crystallite size, i.e., those belonging to the series B, as shown in Table 1. These ZnO nanopowders were prepared at a constant temperature of 60 °C, but growth time was varied between 15 and 120 min.

The mean crystallite size of these nanopowders varies in the range of 9.8–15.3 nm. The UV–vis absorption spectra of ZnO nanopowders exhibit well-defined excitation bands in the range of 350–375 nm showing a significant shift as compared to the bulk ZnO (Nirmala et al. 2010). The stronger UV emission of ZnO nanopowders in this range corresponds to the excitation of electrons from the valence band to the conduction band. At room temperature, these excellent UV emission properties of ZnO nanopowders may be attributed to the crystalline nature and small size of the synthesized ZnO nanopowders (Kajbafvala et al. 2010).

Furthermore, these UV absorptions can be used to calculate the band gap ZnO nanopowders using the following equation: $(E = hc/\lambda)$. While bulk ZnO has a band-gap energy of 3.3 eV at room temperature, the band gap of nanoscale ZnO generally increases with decreasing particle size due to quantum confinement (Ceylan et al. 2013). Therefore, we have chosen (series B) ZnO nanopowders for their smaller mean crystallite size. The band-gap energies calculated from their UV absorptions are found to be 3.52, 3.46, and 3.31 eV for ZnO-B30, ZnO-B60, and ZnO-B120 samples, respectively. Thus, we found that ZnO nanopowders with the mean crystallite size larger than 16 nm would have same band-gap energies as bulk ZnO.

It was also observed that a decrease in ZnO crystallite size from 15.3 to 10.7 nm enhanced the band-gap energy from 3.31 to 3.52 eV. This is expected since with the decrease in crystallite size, the energy gap between the valence band and the conduction band increases and more energy is required for electronic transitions. Thus, smaller size particles absorb a higher amount of energy (corresponding to lower wavelengths) and show maximum UV absorption at comparatively shorter wavelength as compared to larger size particles and bulk ZnO.

Photoluminescence characteristics

The photoluminescent behavior of ZnO nanopowders draws particular concern from the academic and applied research viewpoints. The luminescent ZnO nanostructures are promising for their applications in UV lasing devices, sensors, and bio labeling due to their cost-effectiveness and nontoxic characteristics (Xiong et al. 2011). We studied the photoluminescent properties of the selected samples of ZnO nanopowders (series A) due to a variety of crystallite sizes, difference in crystallinity, and phase purity. Figure 8 shows the typical photoluminescence (PL) spectrum of ZnO nanopowders prepared at 60 (ZnO-A60/ZnO-B60) and 94 °C (ZnO-A94/ZnO-C60). The mean crystallite size of the ZnO nanopowders obtained at 60 and 94 °C is 12.5 and 65.4 nm, respectively.

Evidently, ZnO-A60 and ZnO-A94 nanopowders demonstrate a sharp near-band-edge UV emission at 365 and 376 nm. The sharp near-band-edge emissions may be attributed to the well-known recombination of excitonic transitions (Bekeny et al. 2006; Buyanova et al. 2007). In addition, a few weak emission bands can be observed in the range of 420–430 nm corresponding to violet emission in both of these spectra. The source of violet emission band has been identified by Zeng et al. (2006) as the electronic transition from the high-concentration Zn interstitial defect levels to the valence band.

Furthermore, ZnO-A60 shows another weak emission band at 476 nm corresponding to blue emission, which is not observed in ZnO-A94 nanopowders. Blue emission band is also attributed to the crystal defects in ZnO nanopowders and electronic transitions from the lower levels of oxygen vacancies to the valence band (Xue et al. 2002; Zeng et al. 2010). Thus, it is concluded that ZnO nanopowders produced in 2-aminoethanol inherently possess zinc interstitials and oxygen vacancies, which give rise to such photoluminescent characteristics. The smaller ZnO crystallites produce relatively stronger and multiple emissions (violet and blue) in the visible region as compared to bigger ZnO crystallites.

Conclusions

The structural characterization, optical band gap, and photoluminescent characteristics of ZnO nanopowders prepared in 2-aminoethanol under different reaction conditions are presented in this article. The effects of reaction temperature, growth time, and 2-aminoethanol (solvent and surfactant) on crystallinity, mean crystallite size, morphology, and properties of synthesized ZnO nanopowders are investigated. We conclude that crystallite size is positively affected by increasing temperature due to increased growth rate as well as by increasing growth time, which end up in the formation of bulkier, single crystalline wurtzite ZnO nanostructures. When provided higher temperature or sufficient growth time, 2-aminoethanol acts as a structure-directing agent leading to preferred growth on the w(101) plane of wurtzite crystal and to the formation of nanorods. These ZnO nanorods align on quartz surfaces in the form of a narcis giving rise to a flower-like structure. The optical band gap of ZnO nanopowders is enhanced to 3.52 from 3.3 eV of bulk ZnO. Furthermore, photoluminescent studies demonstrate that ZnO nanopowders emit both UV and visible (violet and blue) radiation owing to some inherent oxygen vacancies and crystal defects in these ZnO nanostructures.

Acknowledgments The authors gratefully acknowledge the Higher Education Commission (HEC, Pakistan) for financial support under the National Research Program for Universities (NRPU) grant # 1308.

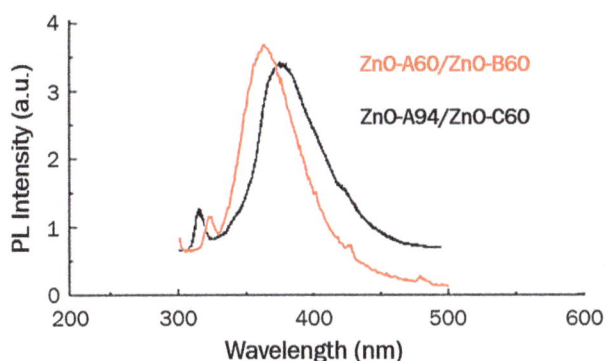

Fig. 8 Typical photoluminescence spectra of ZnO nanoparticles and ZnO nanorods prepared in 2-aminoethanol at 60 and 94 °C, respectively

References

Afzal A, Cioffi N, Sabbatini L, Torsi L (2012) NOx sensors based on semiconducting metal oxide nanostructures: progress and perspectives. Sens Actuators B Chem 171–172:25–42.

Aimable A, Buscaglia MT, Buscaglia V, Bowen P (2010) Polymer-assisted precipitation of ZnO nanoparticles with narrow particle size distribution. J Eur Ceram Soc 30:591–598.

Amin G, Asif MH, Zainelabdin A et al (2011) Influence of pH, precursor concentration, growth time, and temperature on the morphology of ZnO nanostructures grown by the hydrothermal method. J Nanomater.

Barreto GP, Morales G et al (2013) Microwave assisted synthesis of ZnO nanoparticles: effect of precursor reagents, temperature, irradiation time, and additives on nano-ZnO morphology development. J Mater

Bekeny C, Voss T, Gafsi H et al (2006) Origin of the near-band-edge photoluminescence emission in aqueous chemically grown ZnO nanorods. J Appl Phys 100:104317.

Buyanova IA, Bergman JP, Pozina G et al (2007) Mechanism for radiative recombination in ZnCdO alloys. Appl Phys Lett 90:261907.

Caglar M, Caglar Y, Aksoy S, Ilican S (2010) Temperature dependence of the optical band gap and electrical conductivity of sol–gel derived undoped and Li-doped ZnO films. Appl Surf Sci 256:4966–4971.

Ceylan H, Ozgit-Akgun C, Erkal TS et al (2013) Size-controlled conformal nanofabrication of biotemplated three-dimensional TiO2 and ZnO nanonetworks. Sci Rep.

Chougule MA, Sen S, Patil VB (2012) Fabrication of nanostructured ZnO thin film sensor for NO2 monitoring. Ceram Int 38:2685–2692.

Costa MEV, Baptista JL (1993) Characteristics of zinc oxide powders precipitated in the presence of alcohols and amines. J Eur Ceram Soc 11:275–281.

Djurišić AB, Chen X, Leung YH, Ng AMC (2012) ZnO nanostructures: growth, properties and applications. J Mater Chem 22:6526–6535.

Foe K, Namkoong G, Abdel-Fattah TM et al (2013) Controlled synthesis of ZnO spheres using structure directing agents. Thin Solid Films 534:76–82.

Geng C, Jiang Y, Yao Y et al (2004) Well-aligned ZnO nanowire arrays fabricated on silicon substrates. Adv Funct Mater 14:589–594.

Guo L, Ji YL, Xu H et al (2002) Regularly shaped, single-crystalline ZnO nanorods with wurtzite structure. J Am Chem Soc 124:14864–14865.

Harunar Rashid M, Raula M, Bhattacharjee RR, Mandal TK (2009) Low-temperature polymer-assisted synthesis of shape-tunable zinc oxide nanostructures dispersible in both aqueous and non-aqueous media. J Colloid Interface Sci 339:249–258.

Ismail RA, Ali AK, Ismail MM, Hassoon KI (2011) Preparation and characterization of colloidal ZnO nanoparticles using nanosecond laser ablation in water. Appl Nanosci 1:45–49.

Jiang DY, Zhao JX, Zhao M et al (2012) Optical waveguide based on ZnO nanowires prepared by a thermal evaporation process. J Alloys Compd 532:31–33.

Kajbafvala A, Zanganeh S, Kajbafvala E et al (2010) Microwave-assisted synthesis of narcis-like zinc oxide nanostructures. J Alloys Compd 497:325–329.

Kim H, Jeong H, An TK et al (2013) Hybrid-type quantum-dot cosensitized ZnO nanowire solar cell with enhanced visible-light harvesting. ACS Appl Mater Interfaces 5:268–275.

Mahmood Q, Afzal A, Siddiqi HM, Habib A (2013) Sol–gel synthesis of tetragonal ZrO2 nanoparticles stabilized by crystallite size and oxygen vacancies. J Sol Gel Sci Technol 67:670–674.

Nirmala M, Nair MG, Rekha K et al (2010) Photocatalytic activity of ZnO nanopowders synthesized by DC thermal plasma. Afr J Basic Appl Sci 2:161–166

Patra AK, Dutta A, Bhaumik A (2014) Self-assembled ultra small ZnO nanocrystals for dye-sensitized solar cell application. J Solid State Chem

Rai P, Yu Y-T (2012) Citrate-assisted hydrothermal synthesis of single crystalline ZnO nanoparticles for gas sensor application. Sens Actuators B Chem 173:58–65.

Rai P, Jo J-N, Wu X-F et al (2011) Synthesis of well dispersed, regular shape ZnO nanorods: effect of pH, time and temperature. J Nanosci Nanotechnol 11:647–651.

Samanta PK, Bandyopadhyay AK (2012) Chemical growth of hexagonal zinc oxide nanorods and their optical properties. Appl Nanosci 2:111–117.

Schmidt-Mende L, MacManus-Driscoll JL (2007) ZnO: nanostructures, defects, and devices. Mater Today 10:40–48.

Shouli B, Liangyuan C, Dianqing L et al (2010) Different morphologies of ZnO nanorods and their sensing property. Sens Actuators B Chem 146:129–137.

Veriansyah B, Kim J-D, Min BK et al (2010) Continuous synthesis of surface-modified zinc oxide nanoparticles in supercritical methanol. J Supercrit Fluids 52:76–83.

Wang ZL (2004) Nanostructures of zinc oxide. Mater Today 7:26–33.

Wang X, Zhang Q, Wan Q et al (2011) Controllable ZnO architectures by ethanolamine-assisted hydrothermal reaction for enhanced photocatalytic activity. J Phys Chem C 115:2769–2775.

Xiong H-M, Ma R-Z, Wang S-F, Xia Y-Y (2011) Photoluminescent ZnO nanoparticles synthesized at the interface between air and triethylene glycol. J Mater Chem 21:3178.

Xue ZY, Zhang DH, Wang QP, Wang JH (2002) The blue photoluminescence emitted from ZnO films deposited on glass substrate by r.f. magnetron sputtering. Appl Surf Sci 195:126–129

Yang M, Pang G, Jiang L, Feng S (2006) Hydrothermal synthesis of one-dimensional zinc oxides with different precursors. Nanotechnology 17:206–212.

Yang C, Xu C, Wang X et al (2012) A displacement assay for the sensing of carbohydrate using zinc oxide biotracers. Electrochim Acta 60:50–54.

Zeng H, Cai W, Hu J et al (2006) Violet photoluminescence from shell layer of Zn/ZnO core–shell nanoparticles induced by laser ablation. Appl Phys Lett 88:171910.

Zeng H, Duan G, Li Y et al (2010) Blue luminescence of ZnO nanoparticles based on non-equilibrium processes: defect origins and emission controls. Adv Funct Mater 20:561–572.

Zhan X, Chen F, Salcic Z et al (2012) Synthesis of ZnO submicron spheres by a two-stage solution method. Appl Nanosci 2:63–70.

Zhang Z, Bian J, Sun J et al (2012) High optical quality ZnO films grown on graphite substrate for transferable optoelectronics devices by ultrasonic spray pyrolysis. Mater Res Bull 47:2685–2688.

Zhang C, Zhou S, Li K et al (2013) Color-tunable ZnO quantum dots emitter: size effect study and a kinetic control of crystallization. Mater Focus 2:11–19.

16

Synthesis, characterization and evaluation of silver nanoparticles through leaves of *Abrus precatorius* L.: an important medicinal plant

Bhumi Gaddala · Savithramma Nataru

Abstract Biologically synthesized nanoparticles have been widely used in the field of medicine. The present study reports the green synthesis of silver nanoparticles using *Abrus precatorius* leaf extract with silver nitrate solution as reducing agent. The synthesized silver nanoparticles were analyzed through UV–Visible spectroscopy, X-ray diffraction, scanning electron microscopy, energy-dispersive X-ray analysis, atomic force microscopy and Fourier transform infrared. The synthesized silver nanoparticles were disk shaped with an average size of 19 nm. These silver nanoparticles were evaluated for antibacterial activity. The diameter of inhibition zones around the disk of *Pseudomonas aeruginosa* and *Staphylococcus aureus* are resistant to silver nanoparticles, whereas *Escherichia coli* and *Bacillus thuringiensis* are susceptible when compared with the other two species. The results were compared with the ciprofloxacin-positive control and silver nitrate. It is concluded that the green synthesis of silver nanoparticles is very fast, easy, cost-effective and eco-friendly and without any side effects.

Keywords *Abrus precatorius* · Antibacterial activity · Biological synthesis

Introduction

Nanoparticles are being viewed as fundamental building blocks of nanotechnology. The most important and distinct property of nanoparticles is that they exhibit larger surface to volume ratio. Silver has long been recognized as having an inhibitory effect toward many bacterial strains and microorganisms commonly present in medical and industrial processes (Mostafa et al. 2011). The most widely used and known applications of silver and silver nanoparticles include topical ointments and creams containing silver to prevent infection of burns and wound (Murphy 2008). Many attempts have been made to use silver nanoparticles as an anti-cancer agent and they have all turned up positive (Vaidyanathan et al. 2009). The role of silver nanoparticles as an anti-cancer agent should open new doors in the field of medicine. Production of nanoparticles can be achieved through different methods, for example reduction in solutions, chemical and photochemical reactions in reverse micelles, thermal decomposition of silver compounds (Plante and Zeid 2010), radiation-assisted (Cheng et al. 2011), electrochemical (Hirsch and Zharnikov 2005) and recently via green chemistry methods (Sivakumar 2012). Biological synthesis is cost-effective, environmental friendly and easily scaled up for large-scale synthesis. In this method there is no need to use high pressure, energy, temperature and toxic chemical that may have adverse effect in medical applications. Biosynthesis of silver nanoparticles has been performed using a number of plants: *Svensonia hyderobadensis* (Linga Rao and Savithramma 2011), *Shorea tumbuggaia* (Venkateswarlu et al. 2010) and *Thespesia populnea* (Bhumi et al. 2013). The potential of the plants as biological materials for the synthesis of nanoparticles is currently under exploitation.

Abrus precatorius Linn. belongs to the family Fabaceae, commonly known as rosary pea and ratti is a medicinal plant used for various diseases. The plant parts are purgative, emetic, toxic, antiphlogistic, aphrodiasiac

B. Gaddala (✉) · S. Nataru
Department of Botany, Sri Venkateswara University, Tirupati, Andhra Pradesh, India
e-mail: bhumi.gaddala10@gmail.com

S. Nataru
e-mail: prof.savithri@yahoo.in

(a) **(b)**
Plant extract **Treated with Ag(No₃)₂**

Fig. 1 Synthesis of SNPs (color change) using leaf extract of *Abrus precatorius*

and anti-ophthalmic (Manoharan et al. 2010). In India hot water extract of dried leaves and roots are applied to treat eye diseases. In Brazil, water extract of dried leaves and roots are taken orally as nerve tonic (Ivan 2003).

In the present study, we have explored the synthesis of silver nanoparticles and characterized them using UV–Visible spectroscopy, XRD, SEM, EDAX, AFM and FI-IR. Furthermore, the antibacterial activity of synthesized silver nanoparticles was evaluated against *E. coli, Bacillus, Pseudomonas* and *Staphylococcus*.

Materials and methods

All the chemicals and reagents used in the present study were of analytical grade. Silver nitrate was purchased from Sigma-Aldrich Chemicals. The glassware was washed with dilute nitric acid, thoroughly washed with double-distilled water and dried in hot air oven.

Preparation of plant extract

Abrus precatorius leaves were collected from S.V.U. Botanical Garden, Tirupati, Andhra Pradesh, India. The leaves were washed thoroughly thrice with distilled water and shade dried for 10 days. The fine powder was obtained from dried leaves by using kitchen blender. The leaf powder was sterilized at 121 °C for 5 min. 5 g of powder was taken into a 250-ml conical flask and 100 ml of sterile distilled water and boiled for 15 min at 100 °C. Then the leaf extract was collected in a separate conical flask by a standard filtration method.

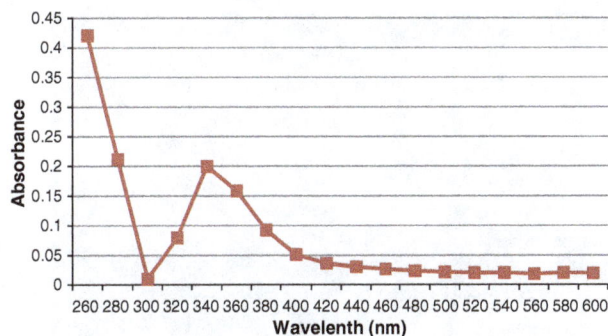

Fig. 2 UV–Vis spectra of silver nanoparticles synthesized from leaf extract of *Abrus precatorius*

Synthesis of silver nanoparticles

1 mM AgNO₃ solution was prepared and stored in amber color bottle. The leaf extract was added to 1 mM AgNO₃ solution. The color change of the solution from yellow to brown indicated that the silver nanoparticles were synthesized from the leaf for characterization and antibacterial activity.

Antibacterial assay

The following bacterial strains were used in this study, viz., *Bacillus thuringiensis* (ATCC10792) *Escherichia coli* (ATCC25922), *Staphylococcus aureus* (ATCC6538) and *Pseudomonas aeruginosa* (ATCC15442). Disc diffusion assay method was carried out by using standard protocol (Anonymous 1996). Overnight, bacterial cultures (100 μl) were spread over Muller Hinton Agar (Hi Media Laboratories Private Limited, Mumbai, India) plates with a sterile glass L-rod. 100 μl of each extract were applied to each filter paper disc, Whatman No. 1 (5 mm dia), and allowed to dry before being placed on the agar. Each extract was tested in triplicate and the plates were inoculated at 37 °C for 24 h after incubation, the diameter of inhibition zones was measured with the help of MIC scale and the results were tabulated.

UV–Vis spectra analysis

The reduction of pure silver ions was monitored by measuring the UV–Vis spectrum of the reaction medium at 5 h after diluting. A small aliquot analysis was done using UV–Vis spectrophotometer UV-2450 (Shimadzu).

SEM analysis of silver nanoparticles

Scanning electron microscope (SEM) analysis was carried out by using Hitachi S-4500 SEM Machine. Thin films of the sample were prepared on a carbon-coated copper grid

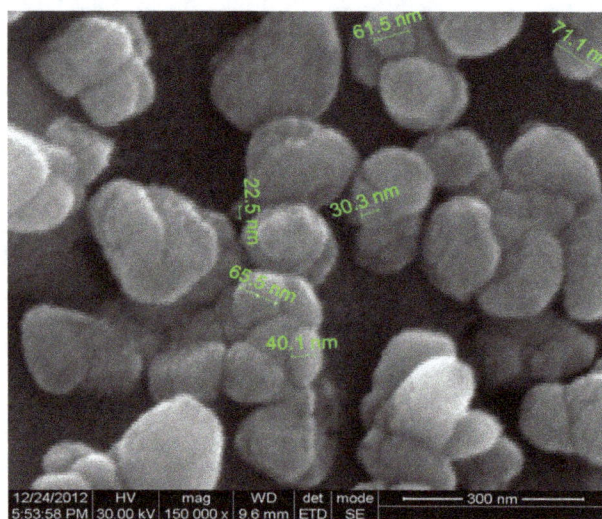

Fig. 3 SEM images of SNPs synthesized using leaf extract of *A. precatorius*

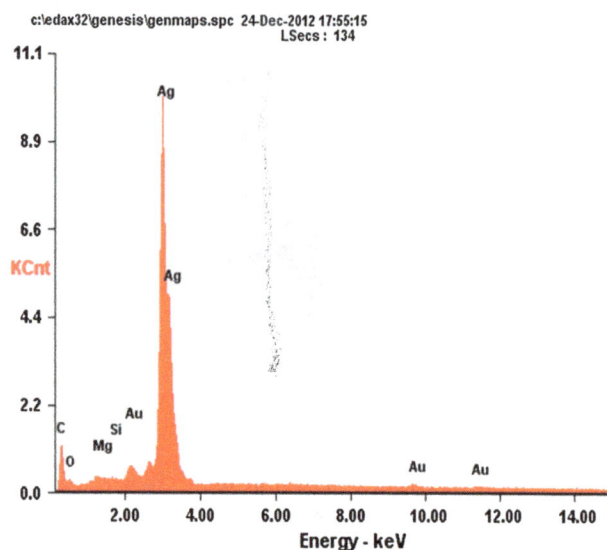

Fig. 4 EDAX images of SNPs of leaves of *A. precatorius*

by just dropping a very small amount of the sample on the grid, extra solution was removed using a blotting paper and then the film on the SEM grid was allowed to dry.

EDAX measurements

The drop of leaf extract with reduced silver nanoparticles was dried on a film coated with carbon and EDAX analysis was performed on Hitachi S-3400N SEM instrument equipped with thermo EDAX attachments.

AFM measurements

The silver nanoparticles extracted through the above protocol were visualized with an atomic force microscope. A thin film of the sample prepared on a glass on the slide was allowed to dry for 5 min and the slides were then scanned with the AFM (Nano Surf® AG, Switzerland, Product: BTO 2089, BRO).

X-ray diffraction (XRD) analysis

The practice size and nature of the silver nanoparticles were determined using XRD. This was carried out using Shimadzu XRD-6000/6100 model with 30 kV, 30 mA with Cukα radians at 2θ angle. X-ray powder diffraction is a rapid analytical technique primarily used for phase identification of a crystalline material and can provide information on unit cell dimensions. The analyzed material is finely ground and average bulk composition is determined. The particle or grain size of the particles on the silver nanoparticles was determined using Debye Sherrer's equation:

Table 1 EDAX of synthesized element during the formation of silver nanoparticles through the leaves of *Abrus precatorius*

Plant name	*Abrus precatorius*	
Element	Weight	Atomic (%)
CK	22.60	62.68
OK	06.75	14.07
MgK	1.54	2.12
SiK	0.19	0.23
AgL	66.22	20.45
AuL	2.70	0.46

$$D = k\lambda/\beta(\cos\theta)$$

FT-IR

The functional group of silver nanoparticles was identified by FT-IR using (Thermo Nicolet-nexus 670 spectrometer of resolution 4 cm^{-1}) one drop of sample placed between the plates of sodium chloride. The drop forms a thin film between the plate and sodium chloride is transparent to infrared light.

Results and discussion

Green synthesis of silver nanoparticles using 1 mM AgNO$_3$ is shown in Fig. 1. The fresh suspension of leaves of *A. precatorius* was yellow in color. However, after addition of Ag(NO$_3$)$_{2and}$ accelerated the reaction at 60 °C for 5 min, the suspension turned dark brown, roughly indicating that the formation of silver nanoparticles was confirmed. The time duration of change in color and

Nanosurf Easyscan 2 - Measurement Report

File: E:\samples\2013\APR-13\05-04-13-SV University-silverAD\10-04-2013\AB\AB1-7um.nid

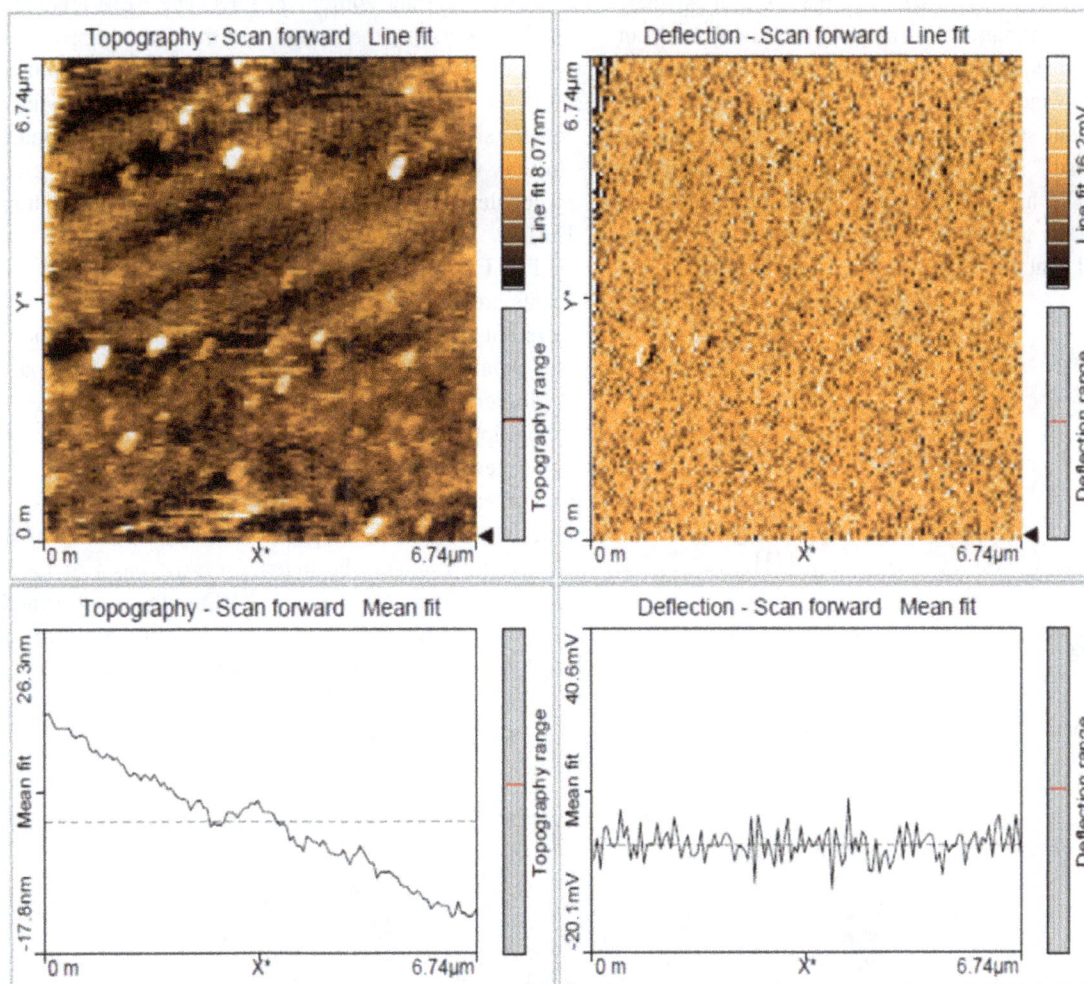

Fig. 5 AFM image of SNPs of *A. precatorius*

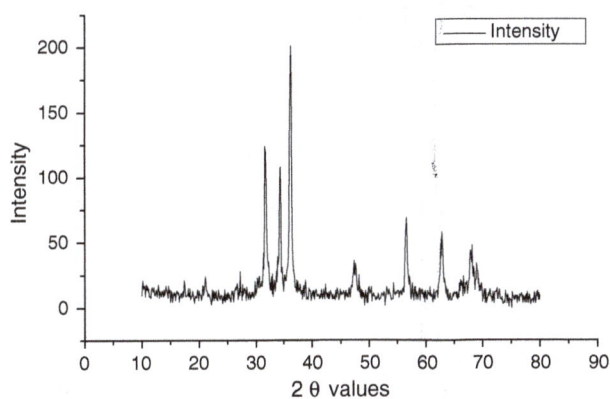

Fig. 6 XRD pattern of SNPs of *A. precatorius*

thickness of the color varies from plant to plant. The reason could be that NAD^+ is a coenzyme found in all living cells. NAD is a strong reducing agent. NAD^+ is involved in

redox reactions, carrying electron from one reaction to another. The coenzyme is therefore found in two forms in cells. NAD^+ is an oxidizing agent. It accepts electrons from other molecules and becomes reduced. This reaction results in NADH, which can donate electrons. These electrons transfer reactions are the main function of NAD:

$$AgNo_3 \rightarrow Ag^+ + No_3$$

$$NAD + e^- \rightarrow NAD$$

$$NAD + H^+ \rightarrow NADH + e^-$$

$$e^- + Ag^+ \rightarrow Ag^0$$

NAD continues to get reoxidised and constantly regenerated due to redox reactions. This might have led to the transformation of Ag ions to Ag^0 (zero-valence state). Ascorbic acid is further responsible for the reduction of Ag-NPs. Ascorbic acid is present at high levels in all parts of plants. Ascorbic acid is a reducing agent and can reduce, and thereby neutralize reactive oxygen species, leading to the formation of ascorbate radical and an electron. This free electron reduces the Ag^+ ion to Ag^0. The formation and stability of silver nanoparticle in aqueous colloidal solution were confirmed using UV–Vis spectral analysis. The UV–Vis spectrum of colloidal solution of SNPs has the characteristic absorbance peaks ranging from 320 to 400 (Fig. 2). The broadening of the peak indicated that the particles were polydispersed. The

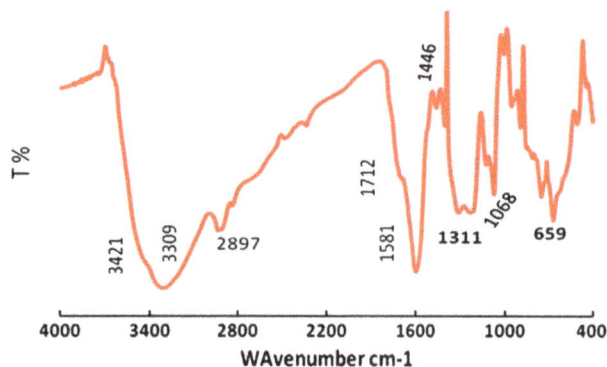

Fig. 7 FT-IR spectra of SNPs of leaf extract of *A. precatorius*

weak absorbance peaks at shorter wavelengths are due to the presence of several organic compounds which are known to interact with silver ions. The SEM analysis was used to determine the structure of the reaction products that were formed. The SEM image shows individual silver particles as well as a number of aggregates. The SEM image showed relatively spherical-shaped nanoparticles formed with diameter ranging 19–35.4 nm. Aggregated molecules were formed in the range of 20 nm, as shown in Fig. 3.

The EDAX spectrum (Fig. 4) reveals that various elements are identified, such as C, O, Mg, Si, Ag and Au with different percentages (Table 1). The AFM analysis of silver nanoparticles showed that the silver nanoparticles agglomerated and formed distinct nanostructures. The topographical image of irregular silver nanoparticles and agglomeration are clearly observed in Fig. 4.

The particle size of the silver nanoparticles ranged from 35 to 40 nm. The XRD pattern showed a number of Bragg reflections that may be indexed on the basis of face-centered cubic structure of silver. XRD analysis confirmed that the silver particles formed in the experiments were in the form of nanostructures, as evidenced by the peaks at 2θ values of 38.28°, 44.04°, 64.34° and 77.28 °, corresponding to (111), (200), (220) and (311) Bragg reflections,

Fig. 8 Antibacterial activity of leaf extract of *A. precatorius*

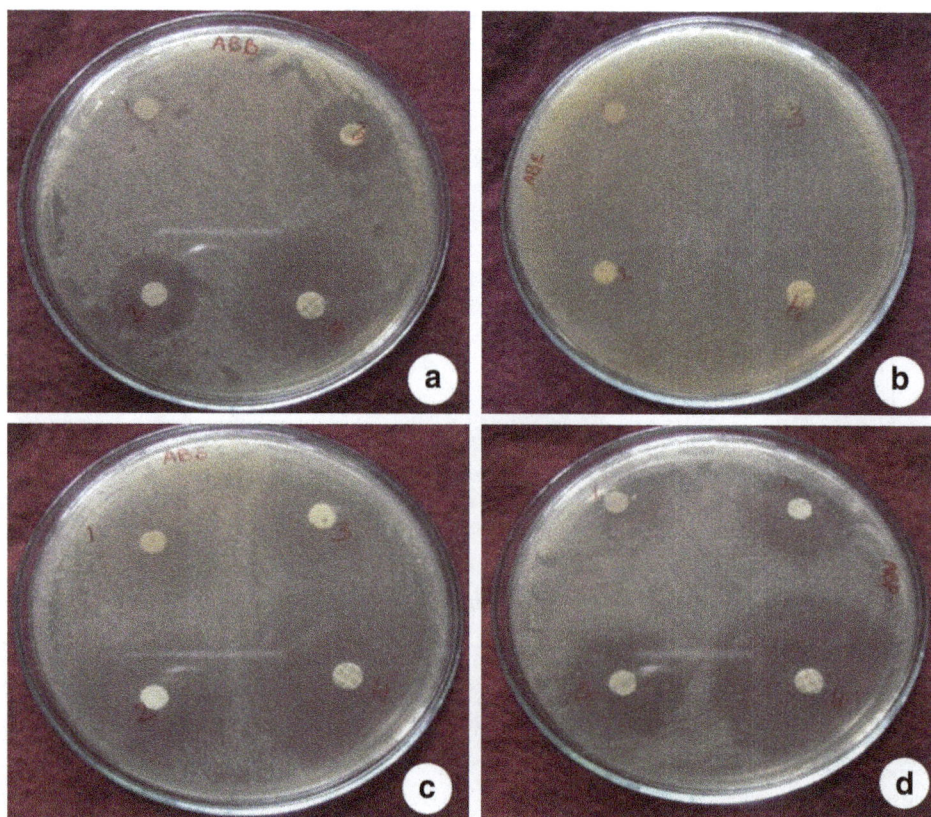

Table 2 In vitro antibacterial activity of SNPs of leaf extract of *A. precatorius*

S. no.	Name of the tested bacteria	Zone of inhibition (mm)		
		Positive control	Negative control	Treated
1.	*Pseudomonas*	43	21	24
2.	*Staphylococcus*	40	18	24
3.	*E. coli*	36	20	22
4.	*Bacillus*	28	18	19

respectively, in Fig. 5, related to crystalline and amorphous organic phases. It was found that the average size from XRD data and using the Debye–Scherrer equation was approximately 38 nm.

The FT-IR spectra in Fig. 6 shows that sharp absorption peaks located at 1,311, 1,446 and 1,581 cm^{-1}. The absorption peak at 1,331 may be assigned to the amide I bond of proteins arising from carbonyl stretching in proteins, and the peak at 1,446 is assigned to OH stretching in alcohols and phenolic compounds. The absorption peak at 1,581 cm^{-1} is close to that reported for native proteins, which suggests that proteins interact with biosynthesized silver nanoparticles and also their secondary structure is not affected during reaction with Ag^{+} ions or after binding with Ag nanoparticles. This FT-IR spectroscopic study confirmed that the carbonyl group of amino acid residues had a strong binding ability with silver, suggesting the formation of a layer covering silver nanoparticles and acting as a capping agent to prevent agglomeration and provide stability to the medium (Table 2). These results confirm the presence of possible proteins as reducing stabilizing agents (Fig. 7).

Biosynthesized silver nanoparticles of *A. precatorius* were analyzed for their antibacterial activity against two Gram-negative bacteria (*E. coli*—ATCC 25922, *Pseudomonas aeruginosa* ATCC 15442) and two Gram-positive bacteria (*Staphylococcus aureus* ATCC 6538, *Bacillus thuringiensis* ATCC 10792) by disk diffusion method (Fig. 8). The results showed that the highest antibacterial activity was observed against *Pseudomonas aeruginosa* followed by *Staphylococcus aureus, E. coli* and *Bacillus thuringiensis*. Among the four bacterial species, *P. aeruginosa* and *S. aureus* showed higher zone of inhibition, 24 mm each, respectively. The present study revealed that leaves of *Abrus precatorius* are ideal material for the synthesis of silver nanoparticles and also to extract novel biochemical compound.

Acknowledgments The first author is highly thankful to DST for sanctioning the INSPIRE Fellowship.

References

Anonymous, Pharmacopiea of India (The Indian Pharmacopeia) (1996) 3rd edn. Ministry of Health and Family Welfare, Delhi

Bhumi G, Lingarao M, Savithramma N (2013) Biological synthesis of silver nanoparticles from stembark of *Thespesia populnea* (L.) Soland. Ind Stre Resj J 3(3):1–7

Cheng Y, Yin L, Lin S, Wiesner M, Bernhardt E, Liu J (2011) Toxicity reduction of polymer-stabilized silver nanoparticles by sunlight. J Phys Chem C 115:4425–4432

Hirsch T, Zharnikov M, Shaporenko A, Stahl J, Weiss D, Wolfbeis OS et al (2005) Size-controlled electrochemical synthesis of metal nanoparticles on monomolecular templates, Angew. Chem Int Ed 44:6775–6778

Ivan AR (2003) Medicinal plants of the world-chemical constituents tradition and modern medicinal uses, vol 1, 2nd edn. Humana Press, Totawa, p 16

Linga Rao M, Savithramma N (2011) Biological synthesis of silver nanoparticles synthesized by using stem extract of *Svensonia hyderobadensis* (Walp.) Mold—a rare medicinal plant. Res Biol 3:41–47

Manoharan S, Balaji R, Aruna A, Niraimathi V, Manikandan G, Babu MBV, Vijayan P (2010) Preliminary phytochemical and cytotoxic property of leaves of *Abrus precatorius* Linn. J Herb Med Toxicol 4:21–24

Mostafa A, Oudadesse H, Legal Y, Foad E, Cathelinean G (2011) Characteristics of silver hydroxyapatite/pvp nano composite. Bio Ceran Dev Appl 1:1–3

Murphy CJ (2008) Sustainability as a design criterion in nanoparticle synthesis and applications. J Mater Chem 18:2173–2176

Plante IJL, Zeid TW, Yangab P, Mokari T (2010) Synthesis of metal sulfide nanomaterials via thermal decomposition of single-source precursors. J Mater Chem 20:6612–6617

Sivakumar P, Nethara Devi C, Renganathan S (2012) Asian J Pharma Clin Res 5:19

Vaidyanathan R, Kalishwaralal K, Gopalram S (2009) Gurunathan nano silver—the burgeoning therapeutic molecule and its green synthesis. Biotechnol Adv 27(6):924–937

Venkateswarlu P, Ankanna S, Prsad TNVKV, Elumalai EK, Nagajyothi PC, Savithramma N (2010) Green synthesis of silver nanoparticles using *Shorea tumbuggaia* stem bark. Int J Drug Dev Res 2:720–723

17

Nanoparticle film deposition using a simple and fast centrifuge sedimentation method

Andrew R. Markelonis · Joanna S. Wang ·
Bruno Ullrich · Chien M. Wai · Gail J. Brown

Abstract Colloidal nanoparticles (NPs) can be deposited uniformly on flat or rough and uneven substrate surfaces employing a standard centrifuge and common solvents. This method is suitable for depositing different types of nanoparticles on a variety of substrates including glass, silicon wafer, aluminum foil, copper sheet, polymer film, plastic, and paper, etc. The thickness of the films can be controlled by the amount of the colloidal nanoparticle solution used in the preparation. The method offers a fast and simple procedure compared to other currently known nanoparticle deposition techniques for studying the optical properties of nanoparticle films.

Keywords Centrifuge · Films · Simple and fast · Nanoparticles · Deposition

Introduction

Semiconductor quantum dots (QDs) are attracting much attention recently due to their size-tunable properties and their wide range of potential applications such as light-emitting diodes (Colvin et al. 1994; Tessler et al. 2002), biological labeling (Bruchez et al. 1998; Michalet et al. 2005), sensors (Zhang et al. 2005), and solar cells (Huynh et al. 2002). For example, lead sulfide (PbS) QDs absorb

A. R. Markelonis · C. M. Wai
Department of Chemistry, University of Idaho, Renfrew Hall, Moscow, ID 83844, USA

J. S. Wang (✉) · B. Ullrich · G. J. Brown
Materials and Manufacturing Directorate, Air Force Research Laboratory, 3005 Hobson Way, Wright-Patterson Air Force Base, OH 45433, USA
e-mail: jswang@uidaho.edu

and emit light in the near-infrared (NIR) region, which makes them potentially important for telecommunication (Bakueva et al. 2004) and biotechnology applications (Medintz et al. 2005). Methods to deposit nanoparticles (NPs) on a variety of platforms are a key part of the technology development. Current procedures of depositing nanoparticles on solid surfaces include chemical vapor deposition (CVD) (Okada 2007), physical vapor deposition (PVD) (Kong et al. 2001), sputtering (Brodsky et al. 1977), lithography (Chou et al. 1997), spin coating (Chang et al. 2004), pulsed-laser deposition (PLD) (Aziz 2008), and supercritical fluid CO_2 (sc-CO_2) deposition (Smetana et al. 2008; Wang et al. 2010). Most nanoparticle deposition techniques require specific equipment and a high level of technical expertise. For example, vapor deposition techniques require high-temperature and vacuum chambers. Spin coating requires highly concentrated nanoparticle solutions and a viscous polymer matrix. A vacuum chamber is required in the sputtering coating process. Layer-by-layer (LBL) deposition uses alternating layers of charged particles and a charged substrate (Decher et al. 1998). An in-depth and comprehensive knowledge of chemical processes is required for completing the LBL process. A high-power pulsed-laser beam, in pulsed-laser deposition, is focused inside a vacuum chamber to strike a target of the material that is to be vaporized and deposited as a thin film on a substrate. This process requires an ultra-high vacuum or presence of a background gas (Aziz 2008). The supercritical CO_2 deposition method enables filling nanostructures of substrates with nanoparticles (Smetana et al. 2008; Wang et al. 2010; Ye et al. 2003) but involves operations at high pressure with specialized equipment and knowledge.

South et al. (South et al. 2009) fabricated microgels (water-soluble polymer cross-linked into a contiguous network)-based films using centrifuge deposition method. Ahmadi et al. (2013) used a stable suspension containing carbon nanotubes and polymethylmethacrylate (PMMA) (PMMA/carbon weight ratio = 0.05:1) in cyclohexane solution to centrifuge and to form a smooth high-density carbon layer on the surface of graphite sample. Centrifuge deposition method has been used and reported previously for making a thin film. However, little has been explored in using centrifuge deposition for nanoparticle film formation. In this study, a facile and swift nanoparticle deposition method is described which utilizes only a standard centrifuge instrument and common solvents for the fabrication of nanoparticle films. This centrifuge deposition method (CDM) is a very simple and efficient process that can be applied to a wide variety of substrates and different nanoparticles. It allows good coverage and thickness control of the nanoparticle films and layer-by-layer deposition of different nanoparticles in a very simple manner. This method also shows that the deposition can take place on flexible or rough surfaces, not limited to flat surfaces, such as flexible plastic, aluminum foil, copper sheet, etc. Gold (Au) and lead sulfide (PbS) NPs were arbitrarily selected for this study. Optical properties of Au and PbS NP films formed by this method are presented to demonstrate potential applications of the technique in the nanomaterials field.

Experimental section

Materials and substrates

Different substrates were chosen for the deposition of Au and PbS NPs, which included rigid, flexible, flat, rough, conductive, and insulating materials. The sizes of substrates usually were tailored to 5×5 mm^2. The materials used in this study include a plastic sheet [polyethylene terephthalate (PET)], aluminum foil, copper sheet, copper grid, flexible black or white plastic [6 μm, high-density polyethylene (HDPE)], glass slide, paper, and silicon. Figure 1 shows some representative substrates used in this study with and without PbS NP coating. Transmission electron microscope (TEM) images were obtained from TEM Cu grids which were prepared with these substrates under the same experimental conditions and deposited with Au, or PbS NPs. Scanning electron microscope (SEM) images of the Au and PbS NPs deposited on these substrates were also obtained. However, in this study, our emphasis is focused on the exploration of CDM method using Au and PbS NPs.

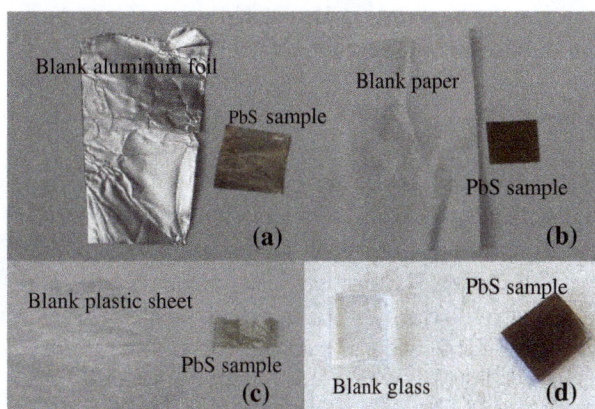

Fig. 1 PbS NPs deposited by the centrifuge method on different substrates. *Left*: blank substrates and *right*: PbS NPs deposited on **a** aluminum foil, **b** paper, **c** plastic sheet, and **d** glass

Preparation of Au and PbS nanoparticles

Au nanoparticles

Hydrogen tetrachloroaurate(III) trihydrate (99.99 %), NaCNBH$_3$ (95 %), hexanes, toluene, and sodium *bis*(2-ethylhexyl)sulfosuccinate (AOT, 98 %) were purchased from Aldrich and used as received.

Gold NPs were prepared by the reduction of gold ions suspended in AOT water-in-hexane microemulsions. Separate metal ion and reducing agent solutions were prepared by dissolving 0.0178 g of AOT in 2 mL of hexane and adding an aqueous solution of 7.2 μL (water/surfactant ratio, $W = 10$) of either the 0.4 M Au^{3+} gold ion or the 1.2 M NaCNBH$_3$ reducing agent. The W value was manipulated by the amount of aqueous solution added. These micellar solutions were stirred for 1 h before reduction to equilibrate the reagents to the reaction temperature. The reaction was carried out at the ambient temperature. The microemulsion containing the reducing agent was added dropwise over a time span of 60 s to the gold ion-containing microemulsion under vigorous stirring. Dodecanethiol (70 μL) was added to the reaction immediately after all of the reducing agents had been added to the solution. This solution was then allowed to stir at the reaction temperature for another hour. After this time, the gold particles were precipitated by adding a mixture of 6 mL of ethanol and 4 mL of methanol, followed by centrifugation. The supernatant was discarded, and the remaining particles were washed two more times with 5 mL of ethanol to remove AOT, spectator ions, and excess dodecanethiol. Multiple sets of the particle samples were combined, re-suspended in 1 mL of toluene or hexane, and the final concentration was approximately 0.005 M. All procedures were conducted on the bench-top without the need for inert environments. Detailed procedure can be found in the literature (Smetana et al. 2007).

PbS nanoparticles

Hexane, toluene, octadecene (ODE), *bis*(trimethyl-silyl)sulfide (TMS), oleic acid (OA), ethanol, and methanol were purchased from Aldrich and used as received.

Oleic acid-capped PbS nanoparticles were prepared, in principle, in a similar manner to Hines and Scholes (2003) but with some modifications. 0.25 mL (0.8×10^{-3} mol) oleic acid, 0.09 g PbO (0.4×10^{-3} mol), and 42 μL of TMS (0.2×10^{-3} mol) were used at a molar ratio of 4:2:1.

PbO (0.09 g), ODE (3.9 mL), and OA (0.25 mL) were added in a 3-neck reaction flask, in vigorous stirring, under a flow of continuous argon (Ar) gas at 150 °C for 1 h. After 1 h, TMS (42 μL) dissolved in 2 mL of ODE was quickly injected at 150 °C into the reaction flask, while the magnetic stirrer was vigorously working and Ar gas was continuously passed through. The solution color changed from colorless to dark brown immediately. The heating source was removed and the temperature reduced to 100–120 °C. The temperature of the system was kept at this temperature range and the stirring continued under Ar gas for another hour.

The sample was washed with absolute 200 proof ethanol for 3–4 times and methanol for 1 time using a 8 mL glass vial until the dark brown lower organic phase disappeared, and black PbS QDs were completely precipitated. The nanoparticles were dried with a stream of N_2 and then dissolved in toluene. The concentrations of PbS NP solutions varied from 2 mg mL^{-1} to 10 mg mL^{-1} for the particle sizes of 2.2, 2.7, 3.1, 4.8, and 14.4 nm.

Centrifuge deposition method

The centrifuge instrument was manufactured by Thermo Electron Corporation (model: IEC Centra CL2). A general procedure for depositing different colloidal nanoparticle samples using the centrifugation approach is described below. The amount of Au NPs used was calculated based on the amount desired for a specific deposition. In this study, 625 μL of Au NP solution (~0.005 M) in hexane was mixed with 5 mL of absolute ethanol. The desired substrate was placed in the bottom of a vial. The vial was then placed in a standard centrifuge with swinging basket and spun for 8 min at 3,400 rpm (Fig. 2). Before CDM, the sample was sonicated for a half min making sure the nanoparticles were dispersed well in hexane and methanol (or ethanol) mixed solutions. After centrifugation, the

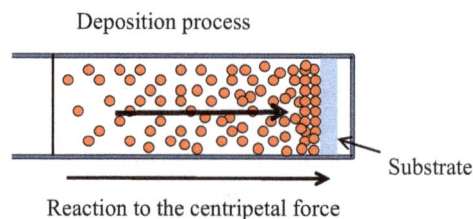

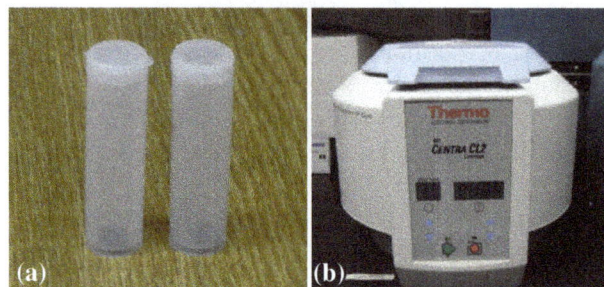

Fig. 2 *Top* schematic diagram to illustrate the centrifuge deposition process. *Bottom* **a** the vials used for the CDM procedure, **b** centrifuge instrument used in this study

colorless supernatant was pipetted out and the substrate with the deposited Au NPs was then removed and allowed to air dry for future measurements.

A procedure was also developed for depositing PbS NPs. The deposition method used is similar to the Au NP deposition described in the previous paragraph. 20 μL of PbS (depending on the concentration synthesized) plus 50 μL of toluene were added to prevent premature precipitation, which otherwise leads to uneven coverage. One mL of methanol was then added to make the particles less soluble in the solution. The vial was then placed in a centrifuge tube and spun, in which it deposited the particles on a substrate placed on the bottom of the vial. The time for the centrifuge was set at 8 min. Using a standard laboratory centrifuge sedimentation at 3,400 rpm which produces approximately 1,950 G, the deposition of 4.8 nm PbS NPs is complete but the 2.7 nm PbS NPs or 2.2 nm Au was only partially removed from the solution. An ultra-centrifuge at 10,000 rpm producing over 15,000 G was utilized for the deposition of the 2.7 nm PbS or 2.2 nm Au NPs.

For the deposition on the PET substrate, PbS NPs in toluene, with oleic acid as a capping reagent, were used. 12.5 mL of the PbS nanoparticle solution (4.7 nm, 40 mg mL^{-1}) was added to 5 mL ethanol in a 8 mL sample vial with the PET substrate. The deposition of the particles was accomplished by centrifuge sedimentation at 3,400 rpm (G = 1,950) for 8 min. After this, the PbS NP-coated film was obtained by removing the solution. Thus, we can directly employ PET as a carrying substrate for the PbS NP deposition.

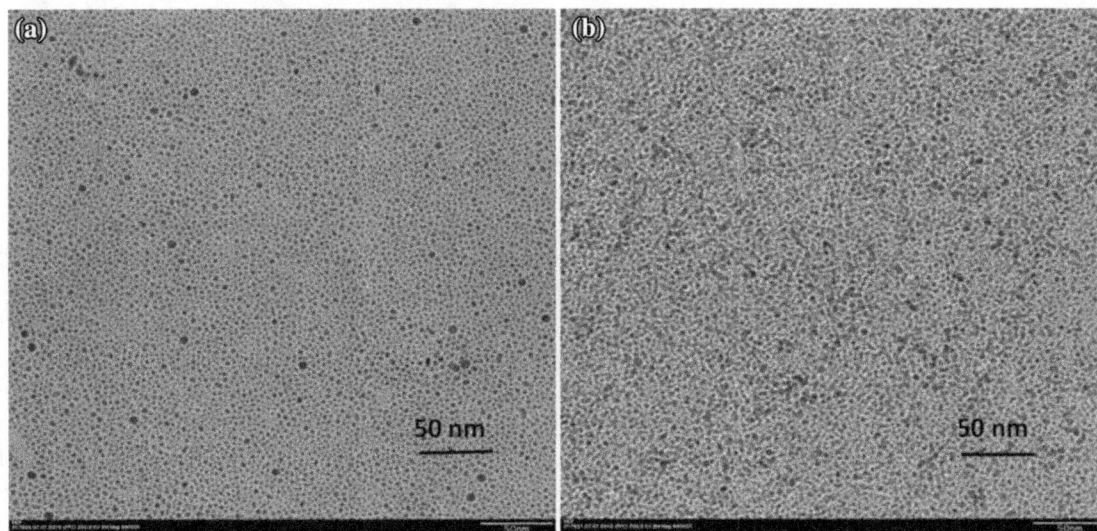

Fig. 3 TEM images of Au NPs (2.2 ± 0.37 nm) deposited on carbon-coated Cu grids by CDM. **a** monolayer and **b** multilayer

Characterization studies

TEM analysis

Carbon-coated copper grids purchased from Ted Pella were used to prepare samples for particle size determination from TEM images. The TEM samples were prepared with CDM. A Phillips CM 200 LaB_6 (lanthanum hexaboride cathode) transmission electron microscope operating at 200 kV was used for both low-and high-resolution imaging. The average size of the PbS NPs was obtained from the TEM images by measuring at least 300 particles using the ImageJ software.

SEM analysis

Scanning electron microscope images were collected with a Sirion instrument manufactured by FEI, Inc. The following parameters were used for the SEM field emission gun (FEG): an accelerating voltage of 10 kV, a spot size of 3.0, and a working distance of $\sim$4.5 mm.

To define the boundary of PbS layer when measuring the QD layer cross-section, SEM glue (Kleindiek NanoTechnik, SEM GLU) was injected onto a corner of the Si substrate, a small amount of which was then transferred onto the top layer of Au or PbS NPs . NOVA SEM (FEI NOVA NanoLab 600 FIB) instrument was employed for this injection. In this process, the SEM glue penetrated into the top layer of QDs to make the QD layer stiff. For some samples, a chemical vapor deposition platinum cap was also used to protect the QDs. The sample surfaces were milled using a FEI StrataDB235 focused ion beam (FIB) with a 30 keV Ga, to produce a few micrometers deep

cross-section, which includes the protection layer, nanoparticle layer, and Si surface. Furthermore, a FEI Sirion SEM instrument in conjunction with a back-scattering electron (BSE) detector was used, in which it offers a better definition and contrast between the Au and PbS NP layers. BSE images are very helpful for obtaining high-resolution compositional maps of a sample and for quickly distinguishing different phases. During the measurement of the cross-section of the films, dynamic focus and tilt correction were applied and specimen tilt was set at 45°. An ultra high-resolution (UHR) mode was turned off due to a far-away working distance (WD = 7.5 mm) between the sample and the pole piece of SEM instrument, in which the sample was set at 45°. The magnifications used in SEM images were 20,000×. Because UHR mode was off and magnification was not high enough, nanoparticles could not be observed in detail (Figs. 5, 8).

Surface profilometry, UV–visible, fluorescence, and photoluminescence spectroscopy analyses

The film thickness was also measured with a DekTak 6 M Stylus Profiler instrument. The scan length was 2,500 μm. UV–Vis–NIR spectra of PbS NPs were obtained using a Cary 5000 Varian UV–Vis–NIR spectrophotometer scanning from 400 to 1,600 nm. Fluorescence spectra were measured with a Horiba Jobin-Yvon Nanolog 916B spectrometer equipped with an IGA 512 InGaAs near-IR detector. The photoluminescence (PL) optical excitation of about 30 W/cm^2 was provided by the 532 nm continuous wave (CW) emission of a solid-state laser. The PL was measured by a double modulation Fourier transform infrared spectroscopy (FTIR) technique.

Results and discussion

Deposition of Au and PbS NPs using CDM method

Gold and PbS NPs can be homogeneously deposited on a TEM carbon-coated copper grid using the CDM method. The films show similar tightly packed particle arrangements as those prepared by the supercritical fluid CO_2 (sc-CO_2) deposition method (Smetana et al. 2008; Wang et al. 2010). As described by previous studies (Smetana et al. 2008; Wang et al. 2010), sc-CO_2 deposition requires a high-pressure chamber, specific skill, and a time-consuming process. It will take more than 3 h to complete the deposition if a volume of a 14 cm^3 high-pressure chamber was used. The results in Fig. 3 demonstrate that similar outcomes from the CDM method could be achieved as in sc-CO_2 deposition, but with a much faster and simpler procedure.

Figure 3 shows the TEM images of Au NPs (2.2 nm) deposited by CDM on carbon-coated copper grids with monolayer (Fig. 3a) and multilayer (Fig. 3b) arrangements. The thickness of the film can be varied by changing the concentrations or the volume of the nanoparticle solution. The surface morphology of these films is comparable to the PbS QD arrays achieved by a sc-CO_2 fluid deposition process (Smetana et al. 2008; Wang et al. 2010) including those reported on a GaAs substrate (Ullrich et al. 2010). In general, the images of the Au and PbS NP films deposited by the CDM procedure reveal fairly homogenous coverage on different substrates.

The deposition of mixed solutions of two different types, or different sizes, of nanoparticles using the CDM method was also studied. One experiment was the deposition of a mixture of small Au NPs (2.2 nm) and large PbS NPs (14.4 nm, in a low concentration) on copper grids by the CDM method. A similar experiment, i.e., a mixture of small PbS NPs (3.1 nm) and large PbS NPs (14.4 nm, in a low concentration) deposited on Cu grids, was also carried out by this method. Similar outcomes, revealed by TEM images, were observed for the above two experiments. The TEM image shows an even surface coverage by the Au or small PbS (3.1 nm) nanoparticle layer. The larger PbS NPs do not seem to cluster together but to be randomly dispersed among the Au or PbS (3.1 nm) nanoparticle layer. When Au or PbS NPs were deposited individually using CDM at a low concentration, the majority was in close proximity over large areas but there were some scattered void areas. When the concentration of nanoparticles was increased, the substrates showed fairly even coverage after deposition. When the films were deposited using a mixture of two different-sized nanoparticles, it was not known if the large particles would deposit first due to their larger mass being more affected by the reaction to the centripetal force. The TEM image showing the larger particles (PbS, 14.4 nm) dispersion within the smaller particles (Au, 2.2 nm, or PbS, 3.1 nm) suggests that this may not be the case (Fig. 4a). Figure 4b shows a TEM image of a mixture of 2.7 and 4.8 nm PbS NPs. For the mixed nanoparticles, as with Au and PbS NPs, or two different-sized PbS NPs, 50 μL of toluene was added to the PbS sample first before adding the second compound. Five mL of ethanol was then added to the particle mixture before centrifugation. This procedure seems to result in random distribution of both components in the film.

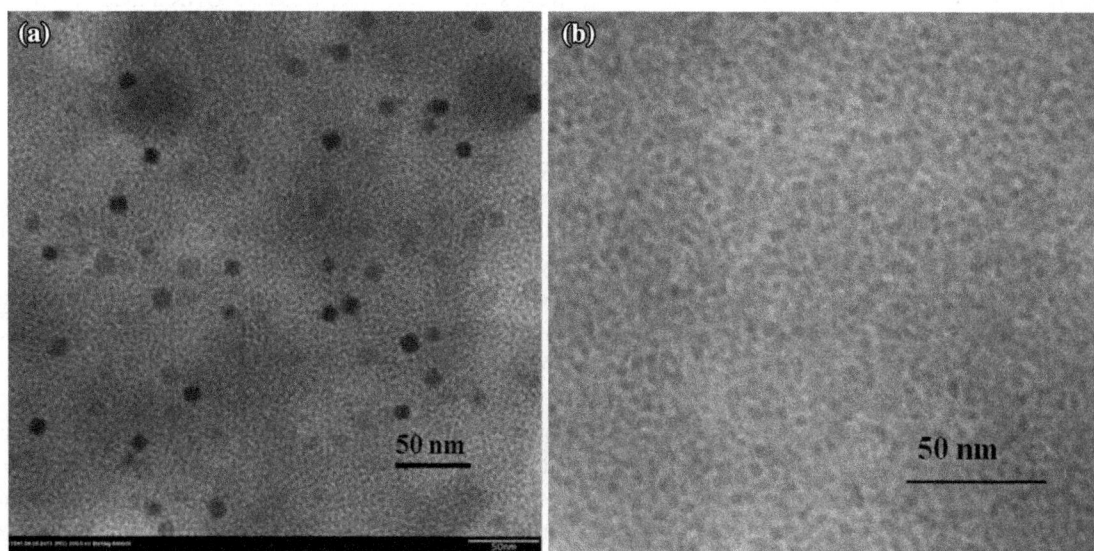

Fig. 4 **a** TEM image of PbS NPs (14.4 nm) and Au NPs (2.2 nm) mixed and then deposited, showing the dispersion of the large particles within the smaller particles. **b** TEM image of PbS deposited on a TEM grid using CDM method showing a mixture of 2.7 and 4.8 nm of PbS NPs

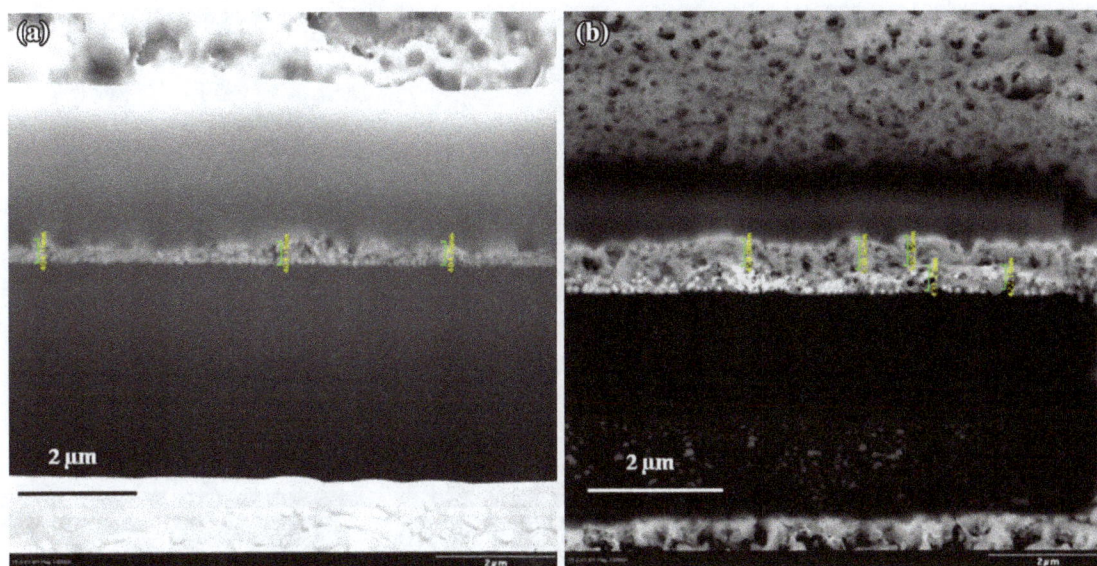

Fig. 5 SEM images of **a** a cross-section of a single Au layer film, the thickness is ∼400 nm, **b** a cross-section of a double-layer film, the thickness of Au layer (*bottom*) is about 430 nm, and PbS layer (*top*) is about 500 nm. The *green lines* shown on the cross-section in SEM images are measuring *bars*

To demonstrate the fabrication of double-layer films on a solid surface, a single layer of Au NPs was deposited on a piece of silicon, and the remaining liquid was then removed. The sample was left to dry overnight. Afterwards, the second solution containing PbS NPs was added and the CDM procedure was repeated to make the second layer. Figure 5 shows cross-sections of SEM images of a single Au NP layer (∼400 nm) deposited by CDM (Fig. 5a), and a double-layer film (Fig. 5b) made with a PbS NP layer (4.7 nm, thickness of ∼500 nm) deposited on the top of the Au NP layer (2.2 nm, thickness of ∼430 nm) applying the CDM procedure. The back-scattering electron detector shows the SEM image with distinct contrast between the two layers (Fig. 5).

Control of film coverage at low population of nanoparticles was demonstrated by TEM images. Figure 6 shows when the volume of Au NP added increases, the coverage or the thickness of the Au NP in the CDM process can be increased as well. As shown in Fig. 6a, when 10 μL of Au NP was used, the films (on TEM grids) showed randomly scattered islands. When 40 μL (Fig. 6b) and 80 μL (Fig. 6c) Au NPs were added separately in CDM, relatively dense (Fig. 6b) and monolayer (Fig. 6c) distributions could be observed. When more Au NPs, for example, 200 μL (Fig. 6d) was added in CDM, thick films can be formed. However, the multilayer creates a hindrance for clear observation of nanoparticles from TEM. For the image in Fig. 6d, when 200 μL NP solution was added, an area at the edge of uniform coverage was selected where the voids allow the film thickness to be imaged. Nevertheless, the process of the film formation by CDM with increasing amount of the Au NPs is clearly illustrated in Fig. 6.

Film thickness measurements

The film thickness was measured with a DekTak 6 M Stylus Profiler instrument as well. Figure 7 shows the thickness measurements of the Au NP films. The scan length was 2500 μm. The film thickness [average scan height (ASH)] of Au NP on Si substrate is 2.02 KÅ (202 nm, Fig. 7a), 5.10 KÅ (510 nm, Fig. 7b), and 10.32 KÅ (1,032 nm, Fig. 7c) for films a, b and c, respectively. The volumes of the Au NP solution used in films a, b and c were 20, 200 and 400 μL, respectively, corresponding to the ratio of 1:10:20. Based upon the thickness data, the thickness ratio for films b and c is quite proportional to the relative amounts of the Au NP solution added, i.e., when the volume increases by a factor of 2 (from 200 to 400 μL), the film thickness doubles from 510 nm to 1,032 nm. However, from 20 to 200 μL, the film thickness acquired is not proportional to the Au NP solution added. This is likely due to the fact that the population of the NPs is not high enough to cover the substrate surface and only island-like structures are formed. Only when the population of the Au nanoparticles is high enough to form at least a monolayer film, then the thickness of the film is proportional to the volume of NP added. The average scan height is between 200 and 1,000 nm, whereas the scan length is 2,500 μm (2.5 mm). It provides sufficient information for measuring the layer thickness. The reaction to the centripetal force allows a random and even scattering deposition on the substrates. If the scan height measured was 1,000 nm, the ratio of the scan length to array scan height will be a factor of 2,500.

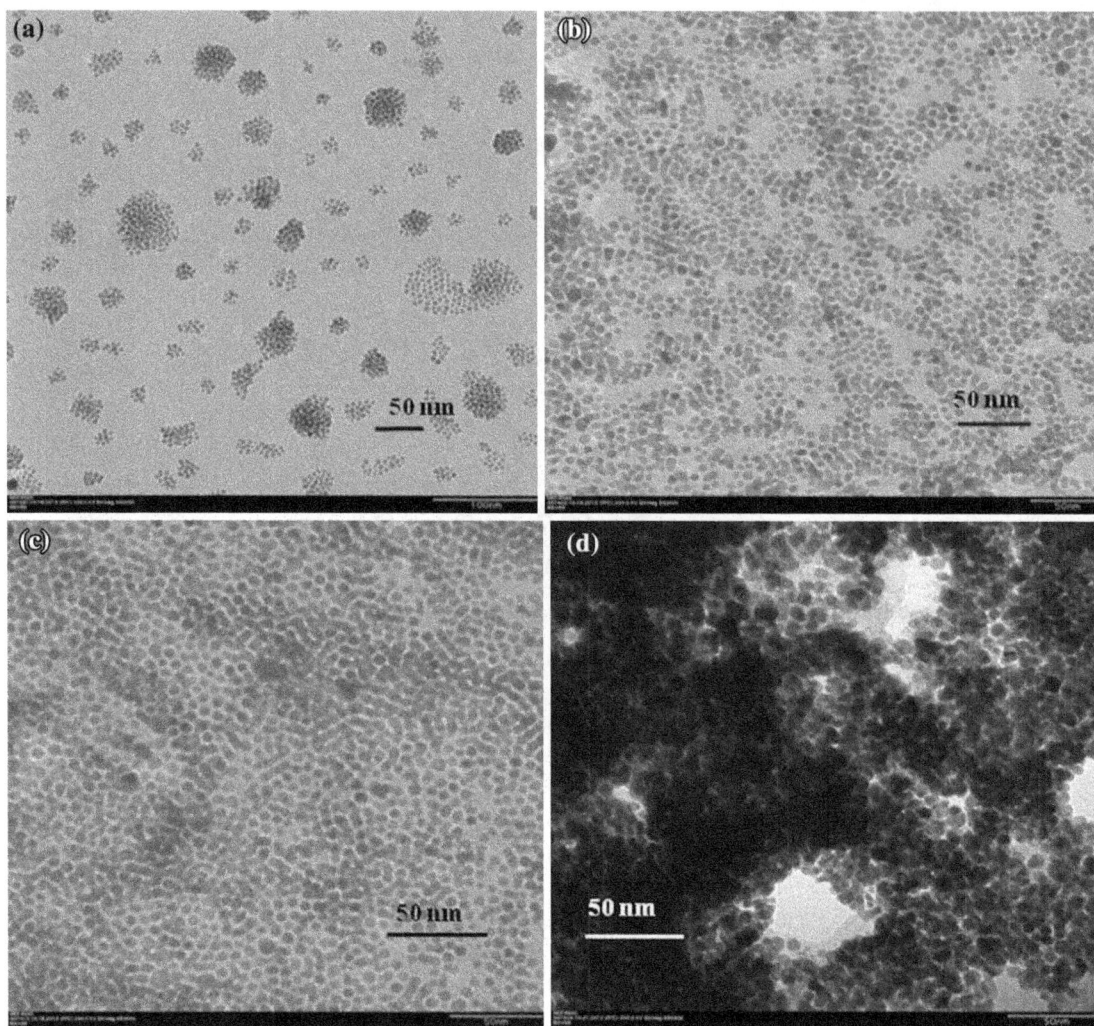

Fig. 6 TEM images of **a** 10 μL, **b** 40 μL, **c** 80 μL, and **d** 200 μL Au NPs added in CDM processes

To check the consistency of film thickness, the cross-sections of the Au film were also measured by SEM as shown in Fig. 8. Au NP volume ratios used for making these films are 1:10:20. The average cross-section thicknesses of the Au films measured with the Sirion SEM instrument displayed in Fig. 8a–c were 157, 404, and 804 nm, corresponding to the Au NP solution ratios of 20 μL:200 μL:400 μL, respectively. The thickness outcomes in relative values are consistent with the results obtained from DekTak 6 M Stylus Profiler instrument.

In previous works combining PET and PbS NPs, the PbS NPs were either embedded in polymers or the PET was used as a carrier for Fano filters containing PbS NPs (Binder et al. 2006, 2009), which is much more tedious than this simple centrifuge sedimentation method. The adhesion between NPs and substrate was tested by different methods. All the nanoparticles adhered firmly to the substrates, albeit with different strengths. The nanoparticle films formed in this way did not peel off during bending,

mechanical abrasion, and exposure to solvents. When deposited on a flexible substrate like plastic sheet, we simply bent the plastic by hand and the films did not visibly flake off or crack. The PbS NPs deposited on PET sheet could not be removed by Scotch tape. On the other hand, the PbS NPs deposited on glass could be removed somewhat using Scotch tape. However, when a layer of Au NPs was first applied on the glass, the PbS NP layer on top was not readily removed. When the metal nanoparticles were deposited on silicon wafers, there was also no noticeable removal of the Au NPs when the tape was peeled off from the wafer. The nanoparticles also cannot be removed from the wafer when exposed to solvents such as acetone and ethanol. We do not know the bonding specifics between the nanoparticles and PET, but nanoparticles can bind to polymeric surfaces through van der Waals forces (Min et al. 2008), hydrogen bonding (Binder et al. 2006), electrostatic linkage (Kinge et al. 2008), and covalent bonding (Li et al. 2010). Furthermore, ligand-stabilized

Fig. 7 Thickness measurements of Au NP films on silicon surface (5 × 5 mm²). The scan length was 2,500 μm. The plots on the *top* are the thickness [average scan height (ASH)] measurements and the measurements of **a** 202 nm, **b** 510 nm, and **c** 1,032 nm from the *top* correspond to films **a**, **b**, and **c** in the images, respectively

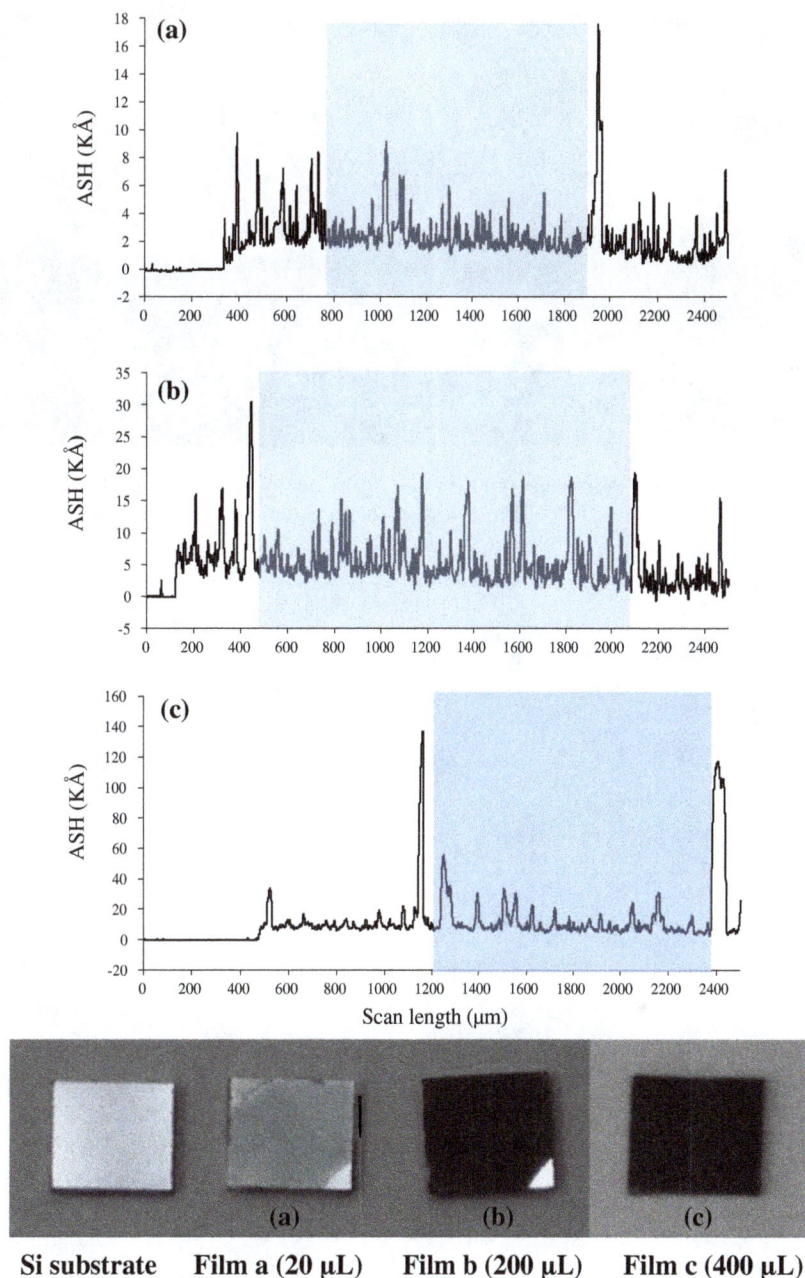

Si substrate   Film a (20 μL)   Film b (200 μL)   Film c (400 μL)

nanoparticles are able to bond to specific sites on monomers (Haryono and Binder 2006).

Spectroscopy studies

Photoluminescence studies of PbS/PET

For the PL experiments, the optical excitation of about 30 W/cm² was provided by the 532 nm continuous wave (cw) emission of a solid-state laser. The PL was measured by a double modulation Fourier transform infrared spectroscopy (FTIR) technique. The double modulation

technique was necessary due to the rather weak PL signal of the sample at room temperature. The PL spectra were recorded using a nitrogen cooled InGaAs detector and a quartz beamsplitter in the Bomem DA3 FTIR from 5 K up to 300 K using a closed-cycle refrigeration cryostat with a diamond window. Figure 9 shows the PL spectra of PbS/PET deposited using CDM measured at 5 K and 300 K. The broken line represents the measured PL intensity (I) as a function of the emitted photon energy (hv) and the solid line fits using the Gaussian intensity distribution,

$$I(hv) = A \times \exp\left(-\left(hv - hv_{\mathrm{p}}\right)^2 / \left(2w^2\right)\right) \tag{1}$$

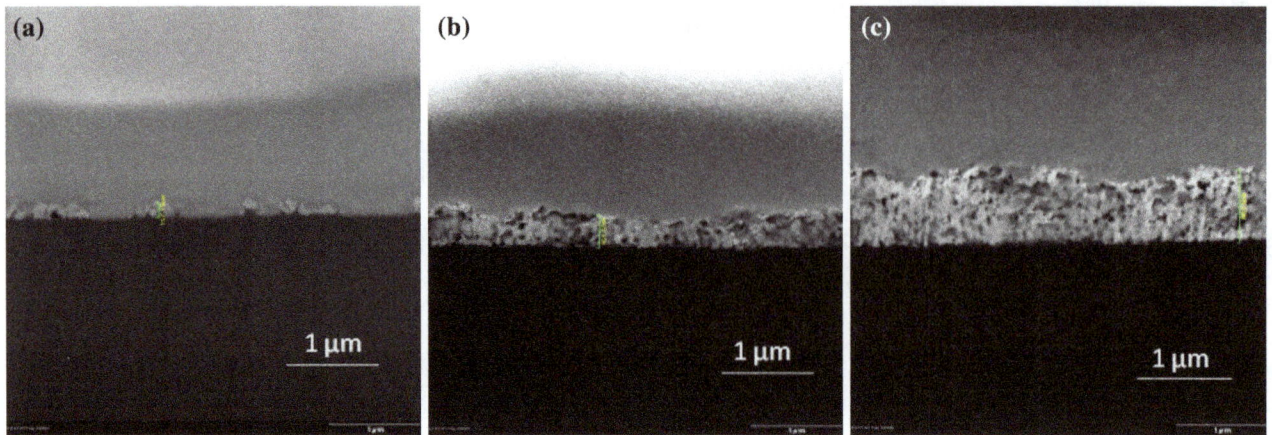

Fig. 8 SEM of Au layer cross-section, Au NP volume ratios added are 1:10:20. SEM magnifications are all ×20,000. Cross-sections in **a–c** are 157, 404, and 804 nm, respectively. The *green lines* shown on the cross-section in SEM images are measuring *bars*

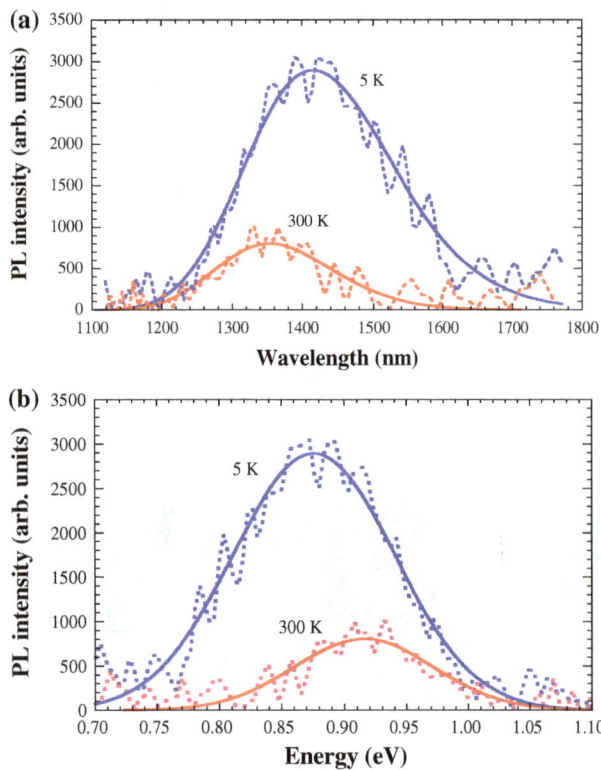

Fig. 9 The PL intensity **a** vs. wavelength and **b** vs. energy of PbS/PET deposited using CDM. The *broken lines* represent the measured PL intensity and *solid lines* represent the fits using Eq. (1)

Fig. 10 Fluorescence of a PbS nanoparticle solution before and after CDM

comparable to the 42 nm (30 meV) shift in the PbS/glass absorption peak (Ullrich et al. 2012). We also noticed that the PbS/PET data of PL peak position (eV) versus temperature (K) are more scattered than other fits reported previously (Ullrich et al. 2011). Therefore, it is possible that colloidal PbS QDs on PET substrates deposited by CDM are subject to photo-induced changes under laser irradiation more than PbS QDs on inert substrates such as glass, suggesting the established bonding linkage at the PET surface was built-up during CDM deposition.

Fluorescence measurements

Fluorescence spectroscopy was used to characterize the film and solution samples. A fluorescence spectrum of a PbS nanoparticle solution before deposition was measured. The spectrum shows a substantial emission peak around 1350 nm. Following deposition of PbS on glass by the CDM method, another spectrum of the centrifuged solution in the vial was taken to determine the concentration of the PbS NPs left in the solution (Fig. 10). As shown in Fig. 10, there is virtually no fluorescence feature for the solution

where A is the peak height, hv_p is the center energy of the PL peak, and w defines the peak width parameter, which is related to the full-width at half maximum (FWHM) by FWHM = 2.35 × w. The peak intensity at 5 K is significantly enhanced, while the PL emission is weaker at 300 K, as expected, the decrease is less than an order of magnitude. The 61 nm (40 meV) PL peak shift from 5 to 300 K in PbS QDs/PET sample observed (Fig. 9a, b) is

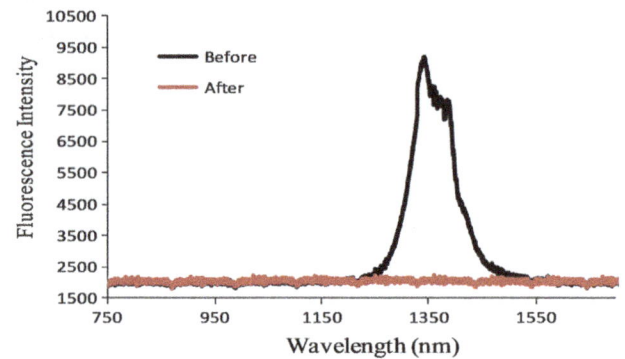

(a)

(b)

(c)

Fig. 12 Fluorescence spectra of PbS QD films made of individual and a mixed sample solutions with 2.7 and 4.8 nm QDs

surfaces are known to have a quenching effect on nano-particle fluorescence signals (Ganesh et al. 2008; Matsuda et al. 2008). The quenching of the fluorescence is probably due to energy transfer from the PbS NPs to the metal substrate resulting in a diminished emission (Dulkeith et al. 2001, 2002). Above 0.1 mg cm^{-2}, the fluorescence of the PbS NPs on the copper sheet can be detected and the intensity of the fluorescence continues to increase up to 0.54 mg cm^{-2}. For PbS NPs in contact with a conductive metal such as copper, the effective range of the quenching effect is about 0.1 mg cm^{-2}.

Energy transfer

Energy transfer between semiconductor nanoparticles of difference sizes is of current interest and more under-standing of this phenomenon is needed (Stephen et al. 2007). Most energy transfer studies are done in composites using solvent evaporation or crude deposition processes (Kagen et al. 1996; Stephen et al. 2007). In this study, we used the CMD method to deposit a mixture of two different sizes of PbS NPs (2.7 and 4.8 nm) on PET. The particles were deposited individually to determine the initial inten-sity of each size of PbS NPs separately. The two size particles were then mixed and deposited as a mixed film. The fluorescence spectra for these samples are given in Fig. 12. The individually deposited particles have an intensity ratio of 1 (2.7 nm):1.3 (4.8 nm). In the mixed nanoparticle film, the fluorescence peak of the 2.7 nm QDs decreased and the peak for the 4.8 nm QDs increased. Obviously, energy transfer occurred between the PbS QDs with different sizes, the small dots (2.7 nm) serving as energy donors and the large ones (4.8 nm) as acceptors. This energy transfer resulted in a new emission intensity ratio of 1 (2.7 nm):3 (4.8 nm) as shown in Fig. 12.

The energy transfer between PbS NPs appears to depend on the concentration per cm^2 of the film prepared by the CDM method. Figure 13 shows two cases, one sample with a mixed PbS nanoparticles (2.7 and 4.8 nm) at 0.05 mg cm^{-2} and the other at 0.25 mg cm^{-2}; both

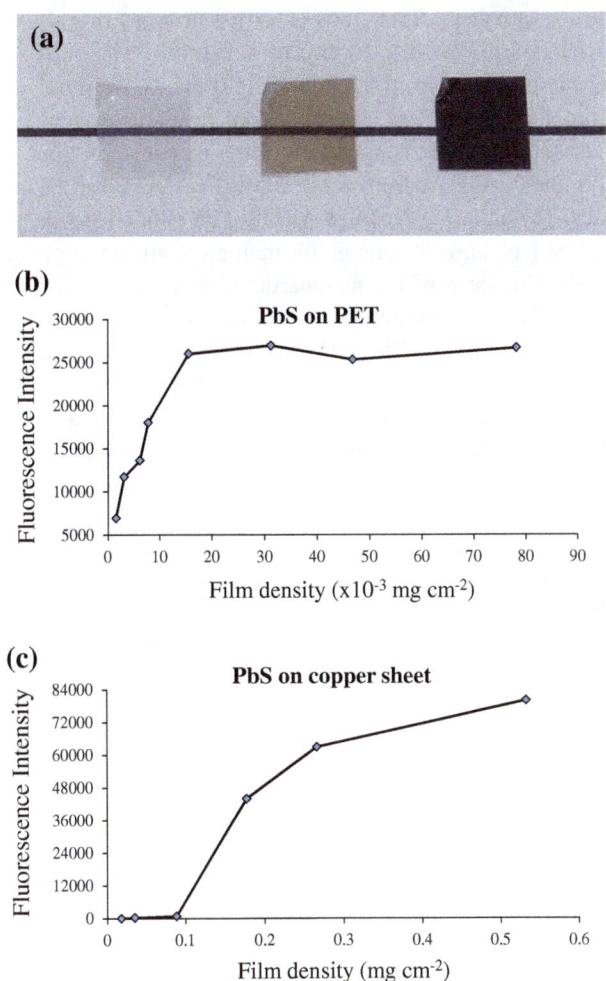

Fig. 11 a Optical images of PET with increasing coverage of PbS QDs (from *left* to *right*: 8.8 × 10^{-3}, 44 × 10^{-3}, and 440 × 10^{-3} mg cm^{-2}, respectively. Concentration ratio = 1:5:50). Fluorescence intensities as a function of PbS nanoparticle concentrations **b** on a PET film and **c** on a copper sheet

after the deposition indicating that essentially all PbS NPs are removed from the solution using the CDM method.

When the concentration of PbS NPs in solution is increased, the amount of particles deposited on the surface of the sample, after the CDM process, also increases. This increase in PbS NPs deposited on the PET substrate leads to a visually darker sample (Fig. 11a) and the expected increase in fluorescence intensities (Fig. 11b). At a film density of about 0.016 mg cm^{-2} or greater, the fluores-cence intensity ceases to increase indicating that this is the film thickness which contributes to the fluorescence of PbS NPs (Fig. 11b). With films of PbS NPs deposited on a copper sheet, the fluorescence of the PbS particles is completely quenched until the film density is about 0.1 mg cm^{-2} or greater (Fig. 11c). As noted by Guo (2008), substrates are known to play an active role in the fluorescence of semiconductor nanoparticles. Metal

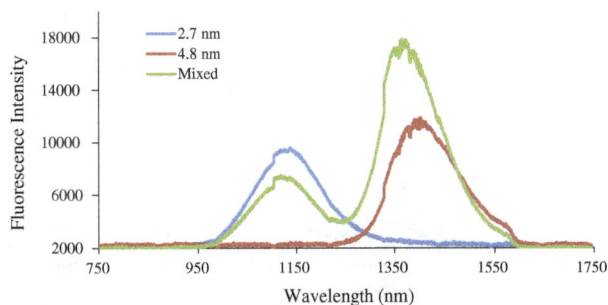

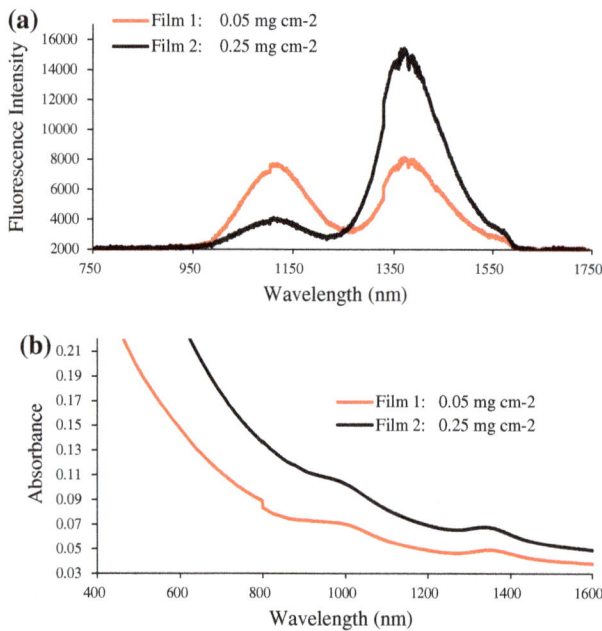

Fig. 13 **a** Fluorescence and **b** absorbance spectra of PbS nanoparticle films made of a mixture of 2.7 and 4.8 nm sized QDs. Film 2 (0.25 mg cm^{-2}) has 5 times more areal density than that of film 1 (0.05 mg cm^{-2})

is able to deposit uniform films of different types of colloidal nanoparticles on a variety of substrates. The substrates can be varied from flexible to rigid sheets, from flat to rough and uneven surfaces, and from insulator to conductive materials. The nanoparticle films prepared by this method show good stability and affinity to the substrate surfaces. The adhesion between Au and PbS is fairly good. This CDM technique is suitable for making relatively thick films. The thickness of the nanoparticle films can be controlled by the concentration or the volume of the colloidal nanoparticle solutions. The CDM method can also make films of mixed nanoparticles and layered nanoparticles. The centrifuge deposition technique appears to provide a simple method for making uniform nanoparticle films on different substrate surfaces for chemical and optical property studies.

Acknowledgments We are grateful to Scott Apt at Material Characterization Facility at Air Force Research Laboratories for his helpful assistance with acquiring SEM cross-section images.

samples were prepared from the same mixed solution. As shown in Fig. 13a, the fluorescence peak ratio for sample 1 is about 1 (2.7 nm):1 (4.8 nm) for the 1,130 nm/1,390 nm peaks, whereas in sample 2 this ratio is 1 (2.7 nm):8 (4.8 nm). Energy transfer from the smaller PbS particles (2.7 nm) to the larger PbS particles (4.8 nm) appears more efficient for the sample with more PbS NPs per square centimeter. A more closely packed PbS NP film would facilitate energy transfer from smaller (higher energy bandgap) to larger (lower energy bandgap) QDs. As expected, the absorption spectra of the two samples in the visible and NIR regions show about the same ratio for the two PbS NP peaks regardless of the amount deposited per square centimeter (Fig. 13b). The peak absorbance maxima centered around 1,000 and 1,350 nm, for 2.7 and 4.8 nm QDs, respectively (Fig. 13b). Details about the experiments performed and results obtained related to fluorescence resonance energy transfer (FRET) involving PbS QDs can be found in the literature published by us previously (Wang et al. 2012).

Conclusion

A simple, fast, and straightforward centrifuge deposition method has been developed for making uniform nanoparticle films from colloidal solutions. The deposition involves common laboratory equipment and solvents. This method

References

Ahmadi R, Ehsani N, Soltani AK (2013) Carbon coating graphite substrates using centrifugal deposition process: effect of centrifugal rotation speed and heat treatment on coating quality. Middle East J Sci Res 15(2):287–290

Aziz MJ (2008) Film growth mechanisms in pulsed laser deposition. Appl Phys A 93:579–587

Bakueva L, Konstantatos G, Levina L, Musikhin S, Sargent EH (2004) Luminescence from processible quantum dot-polymer light emitters 1100–1600 nm: tailoring spectral width and shape. Appl Phys Lett 84:3459–3461

Binder WH, Kluger C, Josipovic M, Straif CJ, Friedbacher G (2006) Directing supramolecular nanoparticle binding onto polymer films: film formation and influence of receptor density on binding densities. Macromolecules 39:8092–8101

Binder WH, Lomoschitz M, Sachsenhofer R, Friedbacher G (2009) Reversible and irreversible binding of nanoparticles to polymeric surfaces. J Nanomater, Article ID 613813

Brodsky MH, Cardona M, Cuomo JJ (1977) Infrared and Raman spectra of the silicon-hydrogen bonds in amorphous silicon prepared by glow discharge and sputtering. Phys Rev B 16:3556–3571

Bruchez M Jr, Moronne M, Gin P, Weiss S, Alivisatos AP (1998) Semiconductor nanocrystals as fluorescent biological labels. Science 281:2013–2016

Chang JF, Sun BQ, Breiby DW, Nielsen MM, Solling TI, Giles M, McCulloch I, Sirringhaus H (2004) Enhanced mobility of poly(3-hexylthiophene) transistors by spin-coating form high-boiling point solvents. Chem Mater 16(23):4772–4776

Chou SY, Krauss PR, Zhang W, Guo L, Zhuang L (1997) Sub-10 nm imprint lithography and applications. J Vac Sci Technol B 15(6):2897–2904

Colvin VL, Schlamp MC, Alivisatos AP (1994) Light-emitting diodes made from cadmium selenide nanocrystals and a semiconducting polymer. Nature 370:354–357

Decher D, Eckle M, Schmitt J, Struth B (1998) Layer-by-layer assembled multicomposite films. Curr Opin Colloid Interface Sci 3:32–39

Dulkeith E, Morteani A, Sonnichsen C, Feldmann J, Riethmuller S, Spatz JP, Moller M (2001) Fluorescence quenching in the vicinity of metal nanoparticles. American Physical Society annual March meeting, March 12–16, APS, Seattle, ID: MAR01, abstract # C14.006

Dulkeith E, Morteani AC, Niedereichholz T, Klar TA, Feldmann J, Levi SA, van Veggel FCJM, Reinhoudt DN, Moller M, Gittins DI (2002) Fluorescence quenching of dye molecules near gold nanoparticles: radiative and nonradiative effects. Phys Rev Lett 89:203002(1)–203002(4)

Ganesh N, Mathias PC, Zhang W, Cunningham BT (2008) Distance dependence of fluorescence enhancement from photonic crystal surfaces. J Appl Phys 103:083104(1)–083104(6)

Guo S, Tsai S, Kan H, Tsai D, Zachariah MR, Phaneuf RJ (2008) The effect of an active substrate on nanoparticle-enhanced fluorescence. Adv Mater 20:1424–1428

Haryono A, Binder WH (2006) Controlled arrangement of nanoparticle arrays in block-copolymer domains. Small 2:600–611

Hines MA, Scholes GD (2003) Colloidal PbS nanocrystals with size-tunable near-infrared emission: observation of post-synthesis self-narrowing of the particle size distribution. Adv Mater 15(21):1844–1849

Huynh WU, Dittmer JJ, Alivisatos AP (2002) Hybrid nanorod-polymer solar cells. Science 295:2425–2427

Kagen CR, Murray CB, Bawendi MG (1996) Long-range resonance transfer of electronic excitations in close-packed CdSe quantum-dot solids. Phys Rev B 54:8633–8644

Kinge S, Crego-Calama M, Reinhoudt DN (2008) Self-assembling nanoparticles at surfaces and interfaces. Chem Phys Chem 9:20–42

Kong YC, Yu DP, Zhang B, Fang W, Feng SQ (2001) Ultraviolet-emitting ZnO nanowires synthesized by a physical vapor deposition approach. Appl Phys Lett 78:407–409

Li S, Lin MM, Toprak MS, Kim DK, Muhammed M (2010) Nanocomposites of polymer and inorganic nanoparticles for optical and magnetic applications. Nano Rev 1:5214(1)–5214(19)

Matsuda K, Ito Y, Kanemitsu Y (2008) Photoluminescence enhancement and quenching of single CdSe/ZnS nanocrystals on metal surfaces dominated by plasmon resonant energy transfer. Appl Phys Lett 92:211911(1)–211911(3)

Medintz IL, Uyeda HT, Goldman ER, Mattoussi H (2005) Quantum dot bioconjugates for imaging, labelling and sensing. Nat Mater 4:435–436

Michalet X, Pinaud FF, Bentolila LA, Tsay JM, Doose S, Li JJ, Sundaresan G, Wu AM, Gambhir SS, Weiss S (2005) Quantum dots for live cells, in vivo imaging, and diagnostics. Science 307:538–544

Min Y, Akbulut M, Kristiansen K, Golan Y, Israelachvili J (2008) The role of interparticle and external forces in nanoparticle assembly. Nat Mater 7:527–538

Okada K (2007) Plasma-enhanced chemical vapor deposition of nanocrystalline diamond. Sci Technol Adv Mater 8:624–634

Smetana AB, Wang JS, Boeckl JJ, Brown GJ, Wai CM (2007) Fine-tuning size of gold nanoparticles by cooling during reverse micelle synthesis. Langmuir 23(21):10429–10432

Smetana AB, Wang JS, Boeckl JJ, Brown GJ, Wai CM (2008) Deposition of ordered arrays of gold and platinum nanoparticles with an adjustable particle size and interparticle spacing using supercritical CO_2. J Phys Chem C 112:2294–2297

South AB, Whitemire RE, Garcia AJ, Lyon LA (2009) Centrifugal deposition of microgels for the rapid assembly of nonfouling thin films. Appl Mater Interfaces 1(12):2747–2754

Stephen W, Clark J, Harbold M, Wise FW (2007) Resonant energy transfer in PbS quantum dots. J Phys Chem C 111:7302–7305

Tessler N, Medvedev V, Kazes M, Kan S, Banin U (2002) Efficient near-infrared polymer nanocrystal light-emitting diodes. Science 295:1506–1508

Ullrich B, Xiao XY, Brown GJ (2010) Photoluminescence of PbS quantum dots on semi-insulating GaAs. J Appl Phys 108:013525(1)–013525(5)

Ullrich B, Wang JS, Brown GJ (2011) Analysis of thermal band gap variations of PbS quantum dots by Fourier transform transmission and emission spectroscopy. Appl Phys Lett 99:081901(1)–081901(3)

Ullrich B, Wang JS, Xiao XY, Brown GJ (2012) Fourier spectroscopy on PbS quantum dots. Spectrosc Photonics Int Eng (SPIE) 8271:82710A(1)–82710A(6)

Wang JS, Smetana AB, Boeckl JJ, Brown GJ, Wai CM (2010) Depositing ordered arrays of metal sulfide nanoparticles in nanostructures using supercritical fluid carbon dioxide. Langmuir 26(2):1117–1123

Wang JS, Brown GJ, Hung WC, Wai CM (2012) Supercritical fluid deposition of uniform PbS nanoparticle films for energy-transfer studies. Chem Phys Chem 13(8):2068–2073

Ye XR, Wai CM, Zhang QD, Kranov Y, McIlroy DN, Lin YH, Engelhard M (2003) Immersion deposition of metal films on silicon and germanium substrates in supercritical carbon dioxide. Chem Mater 15:83–91

Zhang CY, Yeh HC, Kuroki MT, Wang TH (2005) Single-quantum-dot-based DNA nanosensor. Nat Mater 4(11):826–831

18

Evaluation of the effect of indigenous mycogenic silver nanoparticles on soil exo-enzymes in barite mine contaminated soils

Durga Prameela Gaddam · Nagalakshmi Devamma ·
Tollamadugu Naga Venkata Krishna Vara Prasad

Abstract The biosynthesis of nanoparticles has received increasing attention due to the growing need to develop safe, cost-effective and environmentally friendly technologies for nanoscale materials synthesis. In this report, silver nanoparticles (AgNPs) were synthesized by treating aqueous Ag^+ ions with the culture supernatants of indigenous fungal species of *Fusarium solani* isolated from barite mine contaminated soils. The formation of AgNPs might be an enzyme-mediated extracellular reaction process. The localized surface plasmon resonance of the formed AgNPs was recorded using UV–VIS spectrophotometer and was characterized using the techniques transmission electron microscopy, particle size analyzer, Fourier transform-infrared spectroscopy (FT-IR), particle size (dynamic light scattering) and zeta potential. The synthesized AgNPs were stable, polydispersed with the average size of 80 nm. FT-IR spectra reveals that proteins and carboxylic groups present in the fungal secrets might be responsible for the reduction and stabilization of the silver ions. Applied to the barite mine contaminated soils, concentration of AgNPs and incubation period significantly influences the soil exo-enzymatic activities, viz., urease, phosphatase, dehydrogenase and β-glucosidase. To the best of our knowledge, this is the first report on this kind of work in barite mine contaminated soils.

Keywords Silver nanoparticles · *Fusarium solani* · Biosynthesis · Soil exo-enzymes · Barite mine

Introduction

Nanotechnology includes the synthesis of nanoscale materials for versatile applications. The unique properties of nanomaterials differ substantially from bulk materials—in fact, at this scale (1–100 nm), matter behaves differently from their bulk counter parts and exhibit novel properties. Nanoparticles (NPs) are the best known nanomaterials. They have predominant surface effects for the high proportion of the atoms located on their surface that leads to a relevant increase in their reactivity. The biological behavior of NPs is determined by the chemical composition, including coatings on the surface, the decrease in size and associated increase in surface to volume ratio and the shape. In addition, aggregations of NPs may have an effect on their biological behavior as well. Many microorganisms, plant extracts and fungi have been shown to produce NPs through biological pathways (Abou ElNour et al. 2010; Mohanpuria et al. 2008; Ghorbani et al. 2011; Popescu et al. 2010; Dhoondia and Chakraborty 2012; Sastry et al. 2003; Prasad and Elumalai 2011). The use of fungi in the synthesis of metallic NPs is relatively exciting, and was identified as one of the potential microorganisms as they secrete significant amounts of enzymes that are easily handled at the laboratory level. Synthesis of silver NPs (AgNPs) has been investigated utilizing many ubiquitous fungal species including *Trichoderma*, *Fusarium* and *Penicillium* (Vahabi et al. 2011; Durán et al. 2005; Naveen et al. 2010). Recent research regarding the use of fungi has generally investigated potential redox systems

D. P. Gaddam · N. Devamma (✉)
Department of Botany, S.V. University, Tirupati,
A.P. 517 502, India
e-mail: devi.bot@gmail.com

T. N. V. K. V. Prasad (✉)
Nanotechnology Laboratory, Institute of Frontier Technology,
Regional Agricultural Research Station, Acharya N G Ranga
Agricultural University, Tirupati, A.P. 517 502, India
e-mail: tnvkvprasad@gmail.com

using silver nitrate as the source of silver ions (Vahabi et al. 2011; Naveen et al. 2010). Several enzymes, α-NADPH-dependent reductases and nitrate-dependent reductases, were implicated in AgNPs synthesis from *Fusarium oxysporum* (Mohanpuria et al. 2008; Durán et al. 2005). Nitrate reductase was also suggested to initiate NPs formation USING in a *Penicillium* species (Naveen et al. 2010). Soil fungi are known to tolerate heavy metals (Baldrian 2003; Tuomela et al. 2005), but the sensitivity of fungi to heavy metals can differ between species and strains (Baldrian 2003; Fomina et al. 2007). Fungi that produce sterile mycelium (no spores) would have little impact on air quality and also represent another avenue for a greener production alternative. Lenhard (1956) introduced the concept of determining the metabolic activity of microorganisms in soil and other habitats by measuring dehydrogenase activity. Enzymes are the direct mediators for biological catabolism of soil organic and mineral components. These catalysts provide a meaningful assessment of reaction rates for important soil processes. In addition, soil enzyme activities can be helpful in measuring the microbial activity, soil productivity and inhibiting activity of soil pollutants (Tate 1995). Silver nanoparticles have been well known for their inhibitory and antimicrobial effects. It is also noteworthy that the presence of silver ions enhances biochemical processes in entities, such as soil, plant, etc. The present investigation is taken up for the synthesis of silver nanoparticles using the secrets of indigenous fungal species isolated from barite mine contaminated soils (collected from Mangampeta barite mining area of YSR Kadapa District, Andhra Pradesh, India) and to assess the impact of these silver nanoparticles on soil exo-enzymatic activity in the same soils. To the best of our knowledge, this is the first report on this kind of work in barite mine contaminated soils.

Materials and methods

Production of mycogenic AgNPs (m-AgNPs)

The soil samples from barite mines (Mangampeta, YSR Kadapa District, AP, India) were collected and serially diluted and inoculated on to potato dextrose agar (PDA) media for identification of species. Among the various fungal species *Fusarium solani* (*F. solani*) mycelia were found as major strain. The *F. solani* strains were confirmed by morphological, biochemical, and microscopic examination of spores before experimentation. The *F. solani* isolates were grown freshly by inoculating on to sterile PDA medium separately and the flasks were

incubated at 28 °C and 2,000 rpm at pH 7.2 for 96 h. After the 96 h of incubation, the cultures containing dense quantity of mycelium were selected and used for synthesis. The cultures were centrifuged at 4,000 rpm for 30 min and the supernatant was used for the biosynthesis of AgNPs. De-ionized water was used as solvent for the biological synthesis of AgNPs. The supernatant of *F. solani* was collected and added separately as 1 % (v/v) to the reaction vessel containing silver nitrate (1 mM concentration) and incubated on an orbital shaker under dark conditions for 96 h at 30 °C (Logeswari et al. 2012). Silver nanoparticles were biologically synthesized by the culture supernatant of *F. solani*. A 10 ml of filtered (Whatman no. 1 filter paper) supernatant of *F. solani* was added to the 90 ml of 1 mM $AgNO_3$ solution and kept at room temperature for 24 h.

UV–visible spectral analysis

The absorption spectrum, eventually the localized surface plasmon resonance (LSPR) of the m-AgNPs was recorded using a UV–visible spectrophotometer (Shimadzu, UV-2450, double beam). A small aliquot analysis was done using UV–Vis spectrophotometer.

FT-IR analysis

The biosynthesized silver nanoparticles were mixed with potassium bromide powder to form a pellet. The pellet was further analyzed using the Fourier transform infrared spectrophotometer (FT-IR) using the diffuse reflectance mode (BRUKER-Tensor 27).

Dynamic light scattering (particle size) and zeta potential analysis

Dynamic light scattering (DLS) technique is one of the widely accepted techniques to measure the size of the particles in a hydrosol. The particle size measurements were carried out using nanopartica SZ-100 (HORIBA). Zeta potential also measured using electrical conducting cell.

High-resolution transmission electron microscopy (HRTEM)

The sample was characterized by Transmission Electron Microscopy (JEOL 3010; USA) and the sample for transmission electron microscopy (TEM) analysis was prepared by drop casting the nanoparticles suspension on the carbon coated Cu grids.

Inductively coupled plasma optical emission spectroscopy (ICP-OES)

Heavy metals present in soil sample was determined by digesting the filtered soil solution with 10 ml of digestion mixture ($[HNO_3]$ + [HCl] in 3:1 ratio). The mixture was heated at 90–95 °C until complete digestion. It was then filtered and transferred into a standard flask and the volume was made up to the mark with 1 % HNO_3. The sample was analyzed using ICP-OES for the estimation of the heavy metals concentration.

Preparation of soil samples with *F. solani* cultures

The test unit was prepared using plastic cups with 30 g of sterile soil. The AgNPs concentrations of 10, 50, 100, 150 ppm were prepared. The soil sample was incubated with *F. solani* for 0–30 days at 28 °C and different concentrations of m-AgNPs (10, 50, 100, 150 ppm) were added. The moisture content was adjusted to about 60 % of the water holding capacity (WHC), with the soils then incubated at room temperature for 0, 24 h, 7 and 30 days. The soil pH (6.4) was measured for each exposure concentration and duration, but no significant differences were observed.

Assay of the soil exo-enzymatic activities

Biochemical analyses of soil involved determinations of the activity of soil dehydrogenases (SDH) with TTC (2,3,5-triphenyltetrazolium chloride) as substrate (Öhlinger 1996), the activity of urease enzyme (URE, EC 3.5.1.5) determined according to Alef and Nannipieri (1998), and activities of acid phosphatase (ACP, EC 3.1.3.1) and alkaline phosphatase (ALP, EC 3.1.3.2) measured according to the method described by Alef and Nannipieri (1998). β-Glucosidase (EC 3.2.1.21) activity was determined according to Eivazi and Tabatabai (1988). The assay was performed by spectrophotometric measurement of *p*-nitrophenyl (*p*-NP) released after 1 h incubation of soil samples with *p*-nitrophenyl-β-D-glucopyranoside (*p*NPG) at 37 °C in modified universal buffer (pH 6.0) as the substrate. The enzyme activity was expressed as mM of *p*-NP released kg^{-1} dry soil h^{-1} (mM *p*-NP kg^{-1} h^{-1}).

Results and discussion

Microbial synthesis of AgNPs (m-AgNPs)

The appearance and color change of solution from yellow to brown in the silver nitrate-treated flask containing *F. solani* culture filtrate indicates the formation of silver

Fig. 1 Change of colour of broth of *F. solani* before (**a**) and after (**b**) addition of silver nitrate (change of colour is due to the localized surface plasmon resonance and confirms the formation of silver nanoparticles)

nanoparticles, whereas, no color change was observed in the culture supernatant without silver nitrate. Figure 1 shows Erlenmeyer flasks containing the cell-free filtrate of *F. solani* alone (a) and *F. solani* mixed with silver nitrate (b) after completion of reaction for 24 h of duration. Earlier reports evidenced that nitrates can induce nitrate reductase, while ammonium and glutamine inhibit the same enzyme can cause nitrate repression in fungi (Dunn-Coleman et al. 1984; Premakumar et al. 1979). If nitrate reductase is the sole responsible enzyme for AgNPs synthesis, then repression of the enzyme would either inhibit NPs formation or cause NPs formation by another pathway. The mechanism of the biosynthesized nanoparticles involves the reduction of silver ions by the electron shuttle through enzymatic metal reduction process. Earlier reports revealed that NADH and NADH-dependent enzymes are vital in the biosynthesis of metal nanoparticles (Kalimuthu et al. 2008). The microbes are well known to secrete the cofactors NADH and NADH-dependent enzymes, such as nitrate reductase might be responsible for the bioreduction of metal ions and the subsequent formation of silver nanoparticles (Logeswari et al. 2012).

UV–Vis spectral studies: recording localized surface plasmon resonance of m-AgNPs

UV–Vis spectroscopy is one of the most widely used techniques for structural characterization of AgNPs. The color change was observed from yellow to brown due to

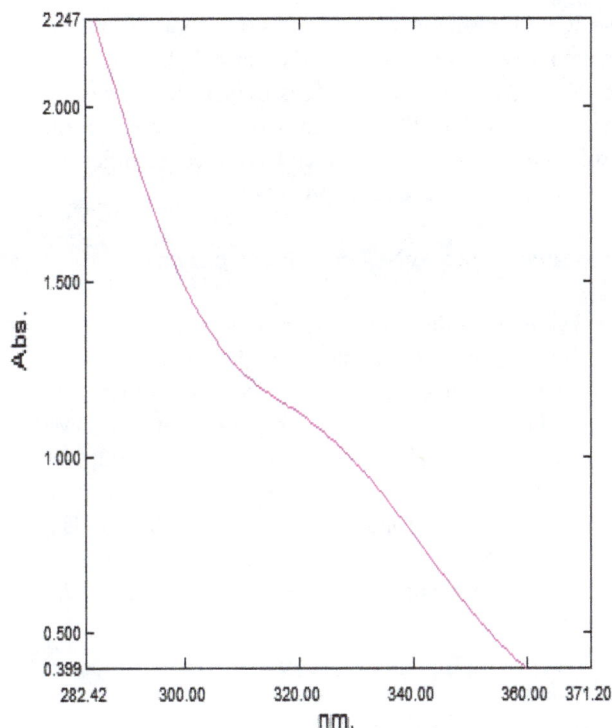

Fig. 2 UV–visible absorption spectra showing localized surface plasmon resonance (LSPR) of silver nanoparticles synthesized using *F. solani* at 330 nm

the bioreduction of the Ag^+ ions in the flask containing silver nitrate and fungal secrets. The extracellular components of the *F. solani* culture reduced the Ag^+ ions in the medium. The absorption spectra of nanoparticles showed highly symmetric single-band absorption with peak maximum at 330 nm (Fig. 2). The intensity of the biosynthesized AgNPs showed increased absorbance and broad peak observed at 330 nm, corresponds to the surface plasmon

resonance (SPR) of the silver nanoparticles. The broad spectrum clearly indicates the polydispersity of the formed AgNPs which is also confirmed with the TEM micrographs.

FT-IR analysis

The FT-IR spectrum was recorded from the potassium iodide pellet with the biosynthesized silver nanoparticles formed after 24 h of incubation with *F. solani*. Figure 3 shows the amide linkages between the amino acid residues in proteins giving rise to the well-known signatures in the infrared region of the electromagnetic spectrum. The bands observed at 3,343 cm^{-1}, which fall near 3,280 cm^{-1}, were assigned to the stretching vibrations of the primary amines, while the corresponding bending vibrations were seen at 1,635 cm^{-1}. The two small bands observed at 1,312 and 1,138 cm^{-1} represent the C–N stretching vibrations of aromatic and aliphatic amines, respectively. The bands observed in the FT-IR of the biosynthesized AgNPs using *F. solani* confirm the presence of proteins. The proteins present in the culture sample are able to bind nanoparticles either through free amine groups or cysteine residues in the proteins (Mandal et al. 2005) through electrostatic attraction of negatively charged carboxylate groups (Sastry et al. 2003) present in the enzyme produced by *F. solani*. The formation and stability of the biosynthesized AgNPs by proteins is a possibility (Sastry et al. 2003).

High-resolution transmission electron microscopy

The sample was characterized by Transmission Electron Microscopy (JEOL 3010; USA) and the sample for TEM analysis was prepared by drop casting the nanoparticles

Fig. 3 Fourier transform infrared spectra (FT-IR) micrograph showing the functional groups involved in the reduction and stabilization of silver nanoparticles synthesized using *F. solani*

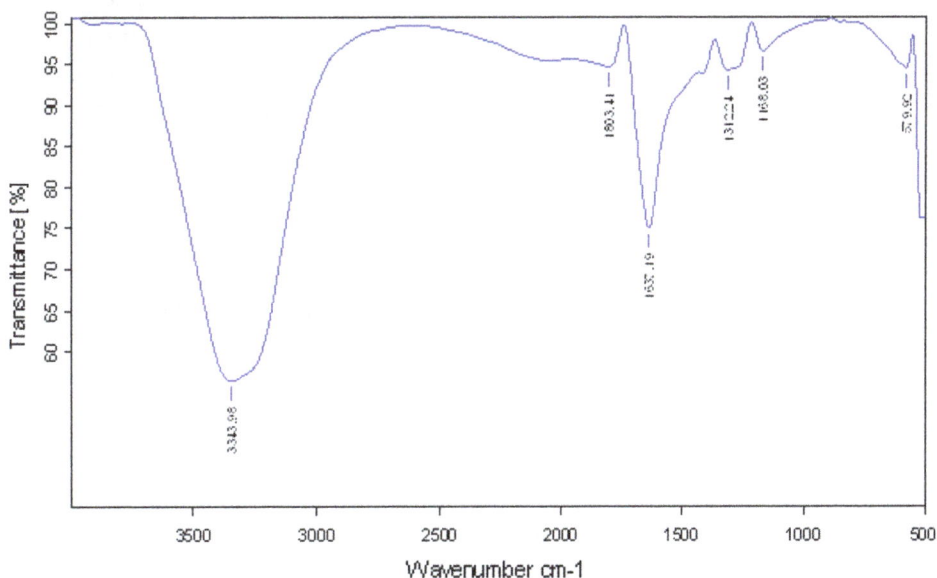

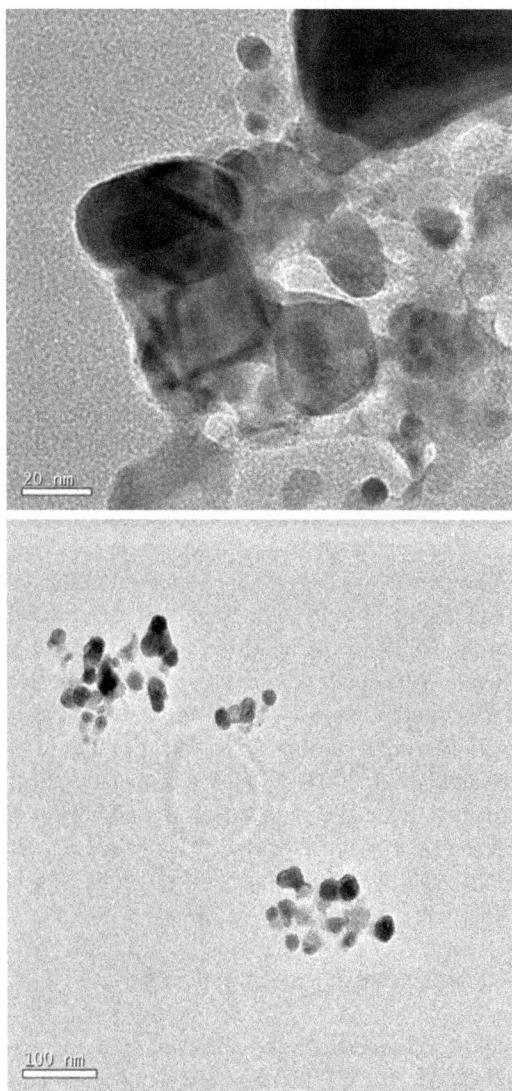

Fig. 4 Transmission electron microscopic (TEM) micrographs showing the silver nanoparticles synthesized using *F. solani*

suspension on the carbon coated Cu grids. The TEM micrographs show polydispersed (Fig. 4) relatively spherical AgNPs with an average size of 80 nm.

Dynamic light scattering and zeta potential measurements

The hydrodynamic diameter of the hydrosol measured using DLS technique also confirmed the size of the Ag-NPs as 87.2 nm (Figs. 5, 6) which is in good agreement with the TEM studies. The electrophoretic mobility ($-$0.000164 cm^2/Vs) coupled with negative zeta potential (21.1 mV) indicates the formed AgNPs are highly stable and receives greater repulsion in soil matrix since most of the soil colloids are negatively charged.

The ICP-OES analysis of the soils was done to reveal the quantitative presence of heavy metals (Table 1) in the barite mine contaminated soils. It is evident that a variety of heavy metals like, Pb, Zn, As, Cd, etc., were present with different concentrations which have great significance in containing soil biochemical processes.

Effect of mycogenic AgNPs on soil exo-enzymes

Soil enzyme activities have the potential to provide a unique integrative and reliable biological assessment of soils because of their relationship to soil biology and rapid response to changes in soil organic matter (SOM) and soil management (Dick 1994; Lulu and Insam 2000). Soil enzymatic activities have been used to ascertain different issues of environmental qualities (Table 2). The effect of m-AgNPs on the soil exo enzyme activities in barite mine contaminated soils were discussed (Figs. 7a, b, 8, 9, 10).

Phosphatases

In soil ecosystems, phosphatases are believed to play vital role in P-cycling (Speir and Ross 1978) as they are correlated to P stress and plant growth. Apart from being good indicators of soil fertility, phosphatase also plays a key role in the maintenance of soil system (Eivazi and Tabatabai 1977; Dick et al. 2000). The assay performed on ALP and ACP activities of soil with the application of m-AgNPs are depicted in Fig. 7a and b. It is evident from the results that enzymatic activity at zeroth day increased in all the treatments except for 50 ppm. However, inhibition of the activity was observed at first day, but on tenth day and 30 days of incubation at all the higher concentrations of AgNPs (50, 100, 150 ppm) the enzymatic activity increased. It has been reported in the literature that adsorption by SOM reduces the mobility of engineered nanoparticles (ENPs) in the soil matrix is curtailed and hence their influence on the microbial populations is drastically reduced. ENPs can be strongly sorbed to soil surfaces and SOM making them less mobile or are small enough to be trapped in the inter-spaces of soil particles and might therefore travel farther than larger particles before becoming trapped in the soil matrix. The strength of sorption would, however, depend on the size, chemistry, aggregation behavior, conditions under which it is applied, etc. In fact, whether an ENP can be hazardous in soil depends not only on its concentration, but also on the likelihood of it ever coming into contact with microbial cells. It may also be noted that natural colloids and ENPs in the environment can interact with one another and also with other larger particles (Simonet and Valcárcel 2009). Studies also demonstrate lack of toxicity from bulk

Fig. 5 Particle size distribution (measured using dynamic light scattering technique) of m-AgNPs synthesized using *F. solani*

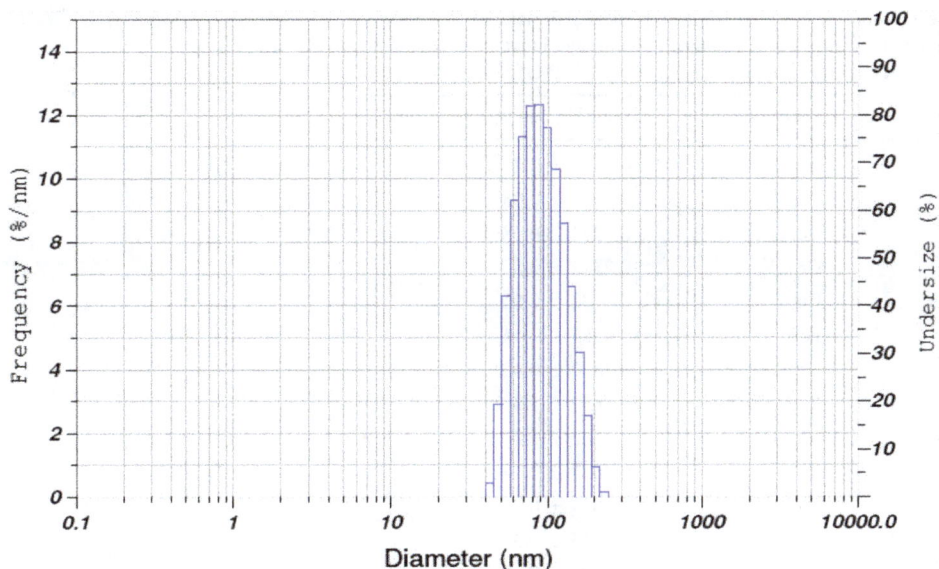

Calculation Results

Peak No.	S.P.Area Ratio	Mean	S. D.	Mode
1	1.00	95.9 nm	34.8 nm	87.2 nm
2	---	--- nm	--- nm	--- nm
3	---	--- nm	--- nm	--- nm
Total	1.00	95.9 nm	34.8 nm	87.2 nm

Fig. 6 Zeta potential of m-AgNPs synthesized using *F. solani*

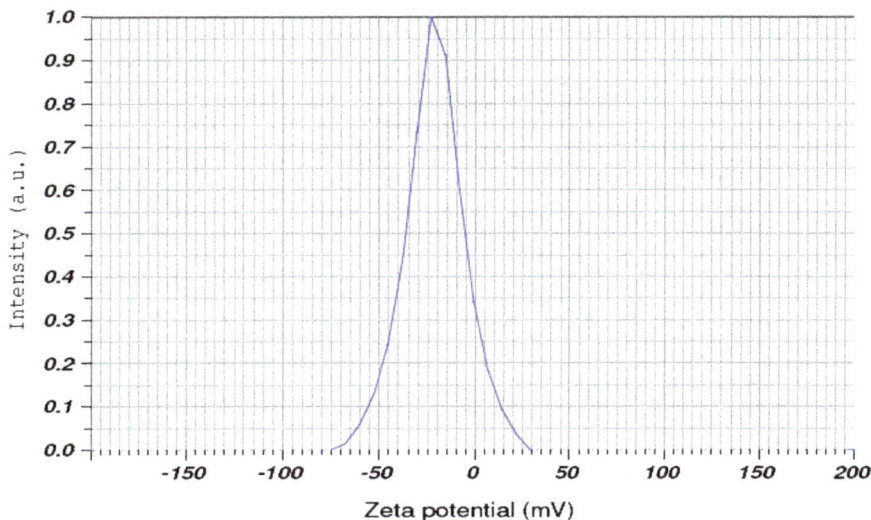

Calculation Results

Peak No.	Zeta Potential	Electrophoretic Mobility
1	-21.1 mV	-0.000164 cm2/Vs
2	--- mV	--- cm2/Vs
3	--- mV	--- cm2/Vs

Zeta Potential (Mean) : -21.1 mV
Electrophoretic Mobility mean : -0.000164 cm2/Vs

materials (300 nm) of Ag-, CuO- and ZnO-NP (Gajjar et al. 2009), suggesting that aggregation of the ENPs into larger particles, possibly by factors present in the soil (natural colloids, SOM, etc.) may reduce their antimicrobial activity.

β-Glucosidase

β-Glucosidase is a common and predominant enzyme in soils (Eivazi and Tabatabai 1988; Tabatabai 1994a, b). β-Glucosidase is characteristically useful as a soil quality

Table 1 The concentration of heavy metals present in the baryte mine contaminated soils estimated using the inductively coupled plasma-optical emission spectrophotometer (ICP-OES)

S. no.	Element	Metal ion concentration (ppm)
1.	As	0.6488 ± 0.09
2.	Ca	9.8372 ± 0.05
3.	Cd	0.1274 ± 0.08
4.	Cr	0.7317 ± 0.04
5.	Cu	0.8773 ± 0.06
6.	Fe	163.365 ± 2.54
7.	Hg	0.3630 ± 0.02
8.	K	41.3029 ± 1.8
9.	Mg	28.1182 ± 1.2
10.	Mn	0.4477 ± 0.04
11.	Mo	0.4039 ± 0.05
12.	Na	2.6413 ± 0.3
13.	P	1.0193 ± 0.07
14.	Pb	0.4850 ± 0.06
15.	Se	0.5300 ± 0.02
16.	Ni	0.8162 ± 0.1
17.	Zn	1.4555 ± 0.8
18.	B	0.4703 ± 0.09

The values presented are the ±SE of three replicates

Table 2 Important soil exo-enzymes as indicators of soil biochemical processes (Das and Varma 2011)

S. no.	Name of the enzyme	Enzyme reaction	Indicator of microbial activity/biochemical process
1.	Dehydrogenase	Electron transport system	C-cycling
2.	β-Glucosidase	Cellobiose hydrolysis	C-cycling
3.	Urease	Urea hydrolysis	N-cycling
4.	Phosphatase	Release of PO_4^-	P-cycling
5.	Soil enzymes	Hydrolysis	General organic matter Degradative enzyme activities

indicator, and largely reflects past biological activity, the capacity of soil to stabilize the SOM, and can be used to detect management effect on soils (Bandick and Dick 1999; Ndiaye et al. 2000). From Fig. 8, it is clear that enzymatic activity decreased in all the treatments at zeroth day as it is needed to disruption of the cell. However, from first day to 30 days the assay indicates the increased glucosidase activity in all treatments but relatively recorded low values at higher concentrations of AgNPs (100, 150 ppm) (Fig. 8). This enzyme has a strong relationship with the microbial biomass that suggests that the activity of

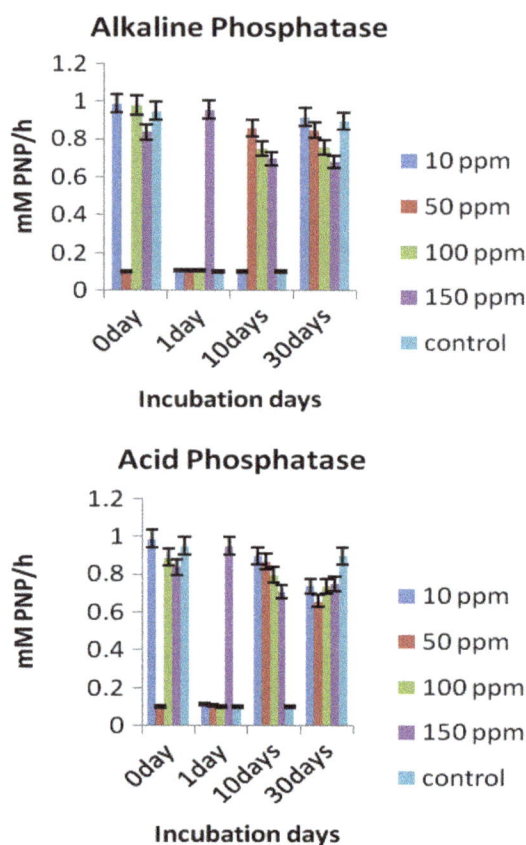

Fig. 7 Effect of m-AgNPs on alkaline phosphatase (ALP) and acid phosphatase (ACP) enzyme activities during 0–30 days of incubation in baryte mine contaminates soils

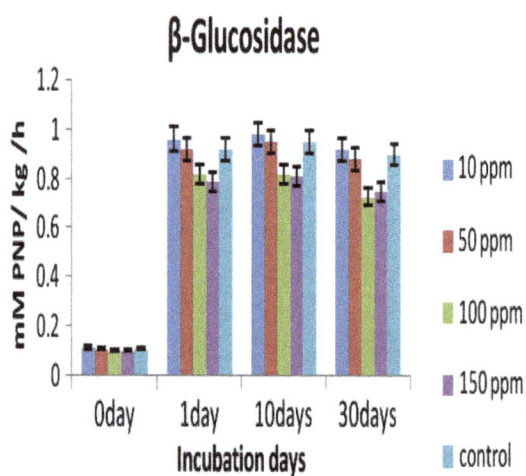

Fig. 8 Effect of m-AgNPs on β-glucosidase activity during 0–30 days of incubation in baryte mine contaminated soils

extracellular immobilized enzymes is negligible. However, several studies show a strong relationship between β-glucosidase activity and clay content which may reflect the potential for enzyme immobilization in the soil. Therefore, AgNPs at lower concentrations promotes the β-glucosidase activity in the soils.

Dehydrogenase

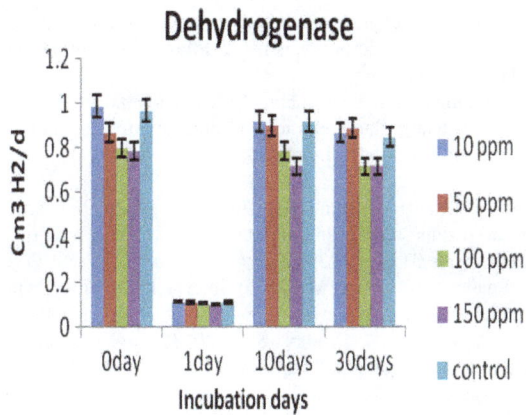

Fig. 9 Effect of m-AgNPs on dehydrogenase activity during 0–30 days of incubation in baryte mine contaminated soils

Urease

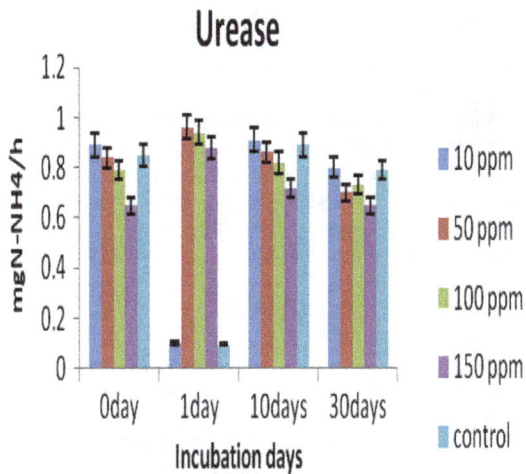

Fig. 10 Effect of m-AgNPs on soil urease activity during 0–30 days of incubation in baryte mine contaminated soils

Dehydrogenase

The dehydrogenase enzyme activity is also commonly used as an indicator of biological activity in soils (Burns 1978) to support biochemical processes that are essential to maintain soil fertility as well as soil health. This enzyme exists as an integral part of intact cells but does not accumulate extracellularly in the soil. Dehydrogenase enzyme is known to oxidize SOM by transferring protons and electrons from substrates to acceptors. Metal nanoparticles, such as Ag have been proven to be toxic to soil micro biota several researchers demonstrated the toxic effect of Ag on soil dehydrogenase activity as severe and bacterial colony growth was inhibited at levels between 0.1 and 0.5 mg Ag kg^{-1} soil. These finding also suggests that soil denitrifying bacteria are susceptible to inhibition by Ag. The effect of Ag was found that only

20 % due to denitrifying bacteria as compared to control remained in samples with the highest Ag concentration. Further during 14, 30 and 90 days of incubation, the inhibition patterns were similar, and no recovery was observed. In the present investigation, the mycogenic AgNPs showed similar inhibition activity up to 24 h, but after 1 up to 30 days inhibition activity decreased at all the concentrations but the recorded activity is less as compared to control and recovery of the activity was observed (Fig. 9). The surface of the m-AgNPs is coated with the secrets of *F. solani* which are organic in nature is the causative for the recovery of the activity over a period of time.

Urease

Urease enzyme is responsible for the hydrolysis of urea fertilizers applied to the soil into NH_3 and CO_2 with the concomitant rise in soil pH (Andrews et al. 1989; Byrnes and Amberger 1989). Urease activity in soils is influenced by many factors, which include cropping history, SOM, soil depth, soil amendments, heavy metals, and environmental factors, such as temperatures (Tabatabai 1977; Yang et al. 2006). For example, studies have shown that urease was very sensitive to toxic concentrations of heavy metals (Yang et al. 2006). It is evident from Fig. 10 that urease activity is low in control and at 10 ppm of m-AgNPs during day 1 and increased up to 10 days and almost similar up to 30 days. This indicates that m-AgNPs affects the N-cycling instantly, but has no significant affect over a period of time (30 days).

Conclusion

The fungi *F. solani* isolated from barytes mine contaminated soil was a potential candidate for the synthesis of AgNPs. On application, m-AgNPs were effective in improving soil quality and oxidising ability of SOM as assessed through the enzymatic processes, but there is no pattern observed though concentration dependent activity was recorded (10–150 ppm). As the properties of the NPs largely dependent on the genesis, surface coating and size the results obtained in the present study may not be extrapolated to other nanomaterial or any other soils contaminated with different pollutants. Thus, the effects of m-AgNPs on soil exo-enzymatic activity may be useful to reduce the concentration of the heavy metals leads to in situ bioremediation of contaminated soils.

References

Abou ElNour MM, Eftaiha A, AlWarthan A, Ammar RAA (2010) Synthesis and application of silver nanoparticles. Arab J Chem 3:135–140

Alef K, Nannipieri P (1998) Urease activity. In: Alef K, Nannipieri P (eds) Methods in applied soil microbiology and biochemistry. Academic Press, Harcourt Brace and Company, London, pp 316–320

Andrews RK, Blakeley RL, Zerner B (1989) Urease: a Ni(II) metalloenzyme. In: Lancaster JR (ed) The bioinorganic chemistry of nickel. VCH, New York, pp 141–166.

Baldrian P (2003) Interactions of heavy metals with white rot fungi. Enzyme Microbial Technol 32:78–91

Bandick AK, Dick RP (1999) Field management effects on soil enzyme activities. Soil Biol Biochem 31:1471–1479

Burns RG (1978) Enzyme activity in soil: some theoretical and practical considerations. In: Bums RG (ed) Soil enzymes. Academic, London, pp 295–340

Byrnes BH, Amberger A (1989) Fate of broadcast urea in a flooded soil when treated with N-(n-butyl) thiophosphoric triamide, a urease inhibitor. Fertil Res 18:221–231

Das SK, Varma A (2011) Role of enzymes in maintaining soil health. In: Shukla G, Varma A (eds) Soil enzymology, soil biology, chap 22. Springer, Berlin, pp 25–42.

Dhoondia ZH, Chakraborty H (2012) Lactobacillus mediated synthesis of silver oxide nanoparticles. In: Prete P (ed) Nanomaterials and nanotechnology, vol 2. InTech. ISBN: 1847-9804,

Dick RP (1994) Soil enzyme activities as indicators of soil quality. In: Doran JW, Coleman DC, Bezdicek DF, Stewart BA (eds) Soil enzymes. Soil Science Society of America Madison, Doran, pp 107–124

Dick WA, Cheng L, Wang P (2000) Soil acid and alkaline phosphatase activity as pH adjustment indicators. Soil Biol Biochem 32:1915–1919

Dunn-Coleman SN, Smarrell J, Garrett RH (1984) Nitrate assimilation in eukaryotic cells. Int Rev Cytol 92:1–50

Durán N, Marcato PD, Alves OL, de Souza IHG, Esposito E (2005) Mechanistic aspects of biosynthesis of silver nanoparticles by several *Fusarium oxysporum* strains. J Nanotechnol 3:1–7

Eivazi F, Tabatabai MA (1977) Phosphates in soils. Soil Biol Biochem 9:167–172

Eivazi F, Tabatabai MA (1988) Glucosidases and galactosidases in soils. Soil Biol Biochem 20:601–606

Fomina M, Charnock J, Bowen AD, Gadd GM (2007) X-ray absorption spectroscopy (XAS) of toxic metal mineral transformations by fungi. Environ Microbiol 9:308–321

Gajjar P, Pettee B, Britt DW, Huang W, Johnson WP, Anderson AJ (2009) Antimicrobial activities of commercial nanoparticles against an environmental soil microbe, *Pseudomonas putida* KT2440. J Biol Eng 3:1–13

Ghorbani HR, Safekordi AA, Attar H, Rezayat Sorkhabadi SM (2011) Biological and non-biological methods for silver nanoparticles synthesis. Chem Biochem Eng Q 25:317–326

Kalimuthu K, Babu RS, Venkataraman D, Bilal M, Gurunathan S (2008) Biosynthesis of silver nanocrystals by *Bacillus licheniformis*. Colloids Surf B Biointerfaces 65:150–153.

Lenhard G (1956) The dehydrogenase activity in soil as a measure of the activity of soil microorganisms. Z Pflanzenernaehr Dueng Bodenkd 73:1–11

Logeswari P, Silambarasan S, Abraham J (2012) Synthesis of silver nanoparticles using plants extract and analysis of their antimicrobial property. J Saudi Chem Soc.

Lulu B, Insam H (2000) Medium-term effects of a single application of mustard residues on soil microbiota and C content of vertisols. Biol Fertil Soils 31:108–113

Mandal S, Phadtare S, Sastry M (2005) Interfacing biology with nanoparticles. Curr Appl Phys 5:118–127

Mohanpuria P, Rana NK, Yadav SK (2008) Biosynthesis of nanoparticles: technological concepts and future applications. J Nanoparticle Res 10:507–517

Naveen H, Kumar G, Karthik L, Roa B (2010) Extracellular biosynthesis of silver nanoparticles using the filamentous fungus *Penicillium* sp. Arch Appl Sci Res 2:161–167

Ndiaye EL, Sandeno JM, McGrath D, Dick RP (2000) Integrative biological indicators for detecting change in soil quality. Am J Altern Agric 15:26–36

Öhlinger R (1996) Dehydrogenase activity with the substrate TTC. In: Schinner F, Öhlinger R, Kandeler E, Margesin R (eds) Methods in soil biology. Springer, Berlin, pp 241–243

Popescu M, Velea A, Lőrinczi A (2010) Biogenic production of nanoparticles. Dig J Nanomater Biostruct 5:1035–1040

Prasad TNVKV, Elumalai EK (2011) Biofabrication of Ag nanoparticles using *Moringa oleifera* leaf extract and their antimicrobial activity. Asian Pac J Trop Biomed 1:439–442

Premakumar R, Sorger GJ, Gooden D (1979) Nitrogen metabolite repression of nitrate reductase in *Neurospora crassa*. J Bacteriol 137:1119–1126

Sastry M, Ahmad A, Khan MI, Kumar R (2003) Biosynthesis of metal nanoparticles using fungi and actinomycete. Curr Sci 85:162–170

Simonet BM, Valcárcel M (2009) Monitoring nanoparticles in the environment. Anal Bioanal Chem 393:17–21

Speir TW, Ross DJ (1978) Soil phosphatase and sulphatase. In: Burns RG (ed) Soil enzymes. Academic, London, pp 197–250

Tabatabai MA (1977) Effects of trace elements on urease activity in soils. Soil Biol Biochem 9:9–13

Tabatabai MA (1994a) Soil enzymes. In: Weaver RW, Angle JS, Bottomley PS (eds) Methods of soil analysis, part 2. Microbiological and biochemical properties. Soil Science Society of America, Madison, pp 775–833

Tabatabai MA (1994b) Soil enzymes. In: Mickelson SH (ed) Methods of soil analysis, Part 2. Microbiological and biochemical properties. Soil Science Society of America, Madison, pp 775–833

Tate RL (1995) Soil microbiology. Wiley, New York

Tuomela M, Steffen KT, Kerko E, Hartikainen H, Hofrichter M, Hatakka A (2005) Influence of Pb contamination in boreal forest soil on the growth and ligninolytic activity of litter-decomposing fungi. FEMS Microbiol Ecol 53:179–186

Vahabi K, Mansoori GA, Karimi S (2011) Biosynthesis of silver nanoparticles by fungus *Trichoderma reesei*: a route for large scale production of AgNPs. Insci J 1:65–79

Yang Z, Liu S, Zheng D, Feng S (2006) Effects of cadmium, zinc and lead on soil enzyme activities. J Environ Sci 18:1135–1141.

19

Molecularly imprinted titania nanoparticles for selective recognition and assay of uric acid

Adnan Mujahid · Aimen Idrees Khan · Adeel Afzal ·
Tajamal Hussain · Muhammad Hamid Raza ·
Asma Tufail Shah · Waheed uz Zaman

Abstract Molecularly imprinted titania nanoparticles are successfully synthesized by sol–gel method for the selective recognition of uric acid. Atomic force microscopy is used to study the morphology of uric acid imprinted titania nanoparticles with diameter in the range of 100–150 nm. Scanning electron microscopy images of thick titania layer indicate the formation of fine network of titania nanoparticles with uniform distribution. Molecular imprinting of uric acid as well as its subsequent washing is confirmed by Fourier transformation infrared spectroscopy measurements. Uric acid rebinding studies reveal the recognition capability of imprinted particles in the range of 0.01–0.095 mmol, which is applicable in monitoring normal to elevated levels of uric acid in human blood. The optical shift (signal) of imprinted particles is six times higher in comparison with non-imprinted particles for the same concentration of uric acid. Imprinted titania particles have shown substantially reduced binding affinity toward interfering and structurally related substances, e.g. ascorbic acid and guanine. These results suggest the possible application of titania nanoparticles in uric acid recognition and quantification in blood serum.

Keywords Molecular imprinting · Nanoparticles · Spectroscopy · Titania · Uric acid

A. Mujahid (✉) · A. I. Khan · T. Hussain ·
M. H. Raza · W. uz Zaman
Institute of Chemistry, University of Punjab, Quaid-e-Azam
Campus, Lahore 54590, Pakistan
e-mail: adnanmujahid.chem@pu.edu.pk

A. Afzal (✉)
Affiliated Colleges at Hafr Al-Batin, King Fahd University of
Petroleum and Minerals, P.O. Box 1803, Hafr Al-Batin 31991,
Saudi Arabia
e-mail: aa@aafzal.com

A. Afzal · A. T. Shah
Interdisciplinary Research Centre for Biomedical Materials,
COMSATS Institute of Information Technology, Defence Road,
Off. Raiwind Road, Lahore 45600, Pakistan

Introduction

Molecular imprinting (Mosbach 1994; Chen et al. 2011) is a modern technique to create artificial receptor sites in a variety of materials such as polymers (Latif et al. 2011), nanoparticles (Lieberzeit et al. 2007), and others. In the last decade, molecular imprinting has proven itself a promising technology to build materials with several applications in analytical separations (Kempe and Mosbach 1995; Andersson 2000), enzyme-like catalysis (Ramström and Mosbach 1999; Wulff 2002), chemical sensors (Dickert and Hayden 1999; Mujahid et al. 2010b), and advance drug delivery systems (Hilt and Byrne 2004; Sellergren and Allender 2005; Kryscio and Peppas 2009). Molecularly imprinted polymers (MIPs) are robust, less expensive, can endure high temperature and oxidative environment, and are chemically stable as compared to their natural competitors (Vasapollo et al. 2011). Molecularly imprinted nanoparticles (MINPs), on the other hand, possess high surface area and a large number of interaction sites in addition to the aforementioned attributes of MIPs (Iqbal and Afzal 2013).

In general, imprinting can lead to fast and easy production of enzyme- or antibody-like binding sites (Poma et al. 2010; Mujahid et al. 2013), thus proving a potential breakthrough technique in the field of biomedical materials and clinical diagnostics (Piletsky et al. 2006). Imprinted materials can be synthesized by a variety of methods like suspension, emulsion polymerization (Pérez et al. 2001) and precipitation (Ye et al. 2000) and others to prepare MIPs directly (Ansell and Mosbach 1998; Zhang et al. 2003; Chen et al. 2005). Among various imprinting approaches, non-covalent imprinting is more advantageous due to simple and straightforward synthetic procedure, commonly available starting materials and chemicals, and relatively easy extraction of weekly bounded template molecules (Sellergren 2000; Zhang et al. 2006).

Uric acid is a final product of purine metabolism in human body (Kaur and Halliwell 1990). Elevated levels of uric acid can cause crystal formation in the joints, which is known as gout. It is reported that half of the antioxidant capacity of blood plasma in human comes from uric acid (Maxwell et al. 1997). The normal range of uric acid in human blood plasma is between 3.6 and 8.3 mg dL^{-1}, or 0.214 and 0.493 mmol L^{-1}. The elevated level of uric acid in blood may result in the formation of kidney stones, i.e. urate crystallizes in the kidney. These stones are radiolucent and do not appear in an abdominal plain X-ray scan, so their presence must be detected by ultrasound.

The main focus of this research is to generate molecularly imprinted titania nanoparticles for the selective recognition of uric acid. Since titania is resistive to oxidation and degradation processes (Mujahid et al. 2010a), therefore, it is suitable for working in complex matrices with imprinted affinity sites for uric acid. Since molecular imprinting accomplishes the desired affinity for target analyte molecules. Although, there are several reports focused on the detection and assay of uric acid using different types of sensing interfaces (Patel et al. 2009; Sun et al. 2011; Khasanah et al. 2012; Wang et al. 2013), imprinted nanomaterials such as metal oxide nanoparticles have not been considered for this purpose so far. Therefore, it would be interesting to develop highly sensitive and chemically stable recognition materials that can readily detect and monitor uric acid in blood plasma. This article presents our primary results showing the potential of imprinted titania nanoparticles toward real-time monitoring of uric acid in blood plasma.

Experimental section

Chemicals and reagents

All the chemicals and reagents used were obtained as follows: Titanium butoxide, uric acid, and iso-propanol from Sigma Aldrich, sodium hydroxide from Merck and methanol from United Laboratory Chemicals.

Characterization methods

Morphology of the imprinted titania nanoparticles is studied by Nanoscope III Atomic Force Microscope. Scanning electron microscopy (SEM) image of titania nanoparticles' film was recorded on Hitachi S-4700 Scanning Electron Microscope. The microscopic images were analyzed using WSxM v5.0 by NANOTEC Electronica (Horcas et al. 2007). Fourier transformation infrared spectroscopy (FTIR) spectra were recorded on Bruker Vector 22 FTIR Spectrometer. UV–vis spectroscopy was performed on Labomed Spectrophotometer model UVD-2950.

Synthesis of molecularly imprinted titania nanoparticles

Molecularly imprinted titania (TiO$_2$) nanoparticles were synthesized by sol–gel method using titanium butoxide as a precursor. A solvent system for dissolving uric acid was designed comprising of 60 % methanol and 40 % 0.1 N NaOH. Precisely, 0.284 g of titanium butoxide was dissolved in 10 mL of iso-propanol taken in a reaction flask placed on hot plate at 60 °C under 130 rpm. Following that 0.005 g of uric acid was dissolved in a 5 mL of the above-mentioned solvent and then added to above reaction mixture under constant stirring. The hydrolysis of titanium butoxide was carried out by the basic media of uric acid solvent. The nanoparticles started to form slowly. The stirring was kept on for almost 1 h. Then stirring and heating were turned off and the particles were subjected to centrifugation for 10 min at 5,000 rpm. Finally, the supernatant was removed and particles were dried in oven for maximum 1 h at 120 °C. As-prepared nanoparticles were washed with excess of solvent to remove any traces of template, i.e. uric acid present in them, and then were characterized for the successful removal of uric acid. Non-imprinted titania particles were also prepared following the same procedure except adding uric acid.

Re-inclusion/rebinding studies

Re-inclusion or rebinding studies of molecularly imprinted titania nanoparticles were carried out for different concentrations of uric acid as in human blood from normal to elevated levels. For binding studies, 0.02 g of thoroughly washed imprinted titania nanoparticles was added to uric acid solution of concentration 0.1 mmol of volume 20 mL under 130 rpm for 1 h at room temperature. After stirring, the mixture was centrifuged for 10 min at 5,000 rpm and the supernatant was separated and analyzed by spectrophotometer. The absorbance of uric acid solution before and after adding the particles was monitored and used for calculating static adsorption coefficient. The procedure was repeated for other concentrations of uric acid starting from 0.01 to 0.095 mmol. This concentration range is about ten folds lower for both normal and elevated levels of uric acid in human blood respectively. The purpose of selecting such low concentration range for uric acid was to minimize the matrix effects of blood serum, i.e. non-specific sorption. The non-imprinted titania nanoparticles and other interfering analytes were also tested in the same manner.

Results and discussion

Morphological characterization

Microscopic characterization of molecularly imprinted titania nanoparticles was performed via Atomic force microscopy (AFM) and SEM to study their morphology. The two- and three-dimensional AFM images of as-prepared or imprinted titania nanoparticles are shown in Fig. 1. AFM reveals the morphology, i.e. size and shape of titania nanoparticles. The AFM image of powdered titania sample suggests that imprinted titania nanoparticles are spherical in shape with an average particle size or diameter in the range of 100–150 nm. Moreover, these nanoparticles are not coalesced to form bigger aggregates and size distribution is homogeneous. It may be due to the presence of uric acid—the template attached to the surface of imprinted titania nanoparticles—that may prevent coalescence of these particles.

Figure 2 shows the SEM image and 3D surface construct of a thick film of imprinted titania nanoparticles formed by coating a dense paste of powdered titania on quartz crystal. The microstructure and surface morphology of imprinted nanoparticles is obvious. These images demonstrate that thick films of titania nanoparticles can be fabricated easily leading to the formation of a very fine network of imprinted titania nanoparticles. Since, each of these imprinted titania particles has template-specific interaction sites, the dispersion of titania nanoparticles in thick film ensures that these interaction sites are homogeneously distributed. The high number of interaction centers is complementary for high sensitivity and shorter adsorption pathways.

Spectroscopic characterization

Fourier transformation infrared spectroscopy measurements of uric acid imprinted titania nanoparticles, i.e. as-prepared, and molecularly imprinted but thoroughly washed titania nanoparticles as well as non-imprinted particles were performed to examine the imprinting characteristics and any pronounced differences in chemistry and functionality of these nanoparticles. The FTIR spectra of different types of titania nanoparticles are also shown in Fig. 3. The predominant functional groups (peaks) observed in different types of titania nanoparticles are recorded in Table 1.

As shown in Fig. 3, the characteristic amide's C=O stretching vibrations and secondary amine's C–N stretching vibrations appear at 1,673 and 1,290 cm^{-1}, respectively (Mohan 2004). It is obvious that these groups are present only in case of as-prepared uric acid imprinted titania nanoparticles (before washing), since these functional groups originate from uric acid molecules. Thus, repeated washing of these nanoparticles successfully removes the template molecules leading to the absence of carbonyl and amine peaks in imprinted and thoroughly washed titania nanoparticles. Expectedly, non-imprinted titania particles do not indicate any such absorption for amide and amine groups.

However, the presence of Ti–O group can be confirmed in all three types of nanoparticles from the broad absorption in the range of 600–750 cm^{-1} (Rubab et al. 2014). The IR absorption of alcoholic group (TiO–H) indicates the formation of titania networks with sufficient number of surface hydroxyl groups, which may lead to weak interactions, e.g. hydrogen bonding, between the particles and the template molecules. Moreover, TiO–H absorption of as-prepared titania is different from the washed imprinted and non-imprinted particles due to the presence of amine (N–H) groups in molecularly imprinted titania nanoparticles.

Fig. 1 Two-dimensional (**a**), and three-dimensional (**b**) atomic force micrographs of as-prepared uric acid imprinted titania nanoparticles

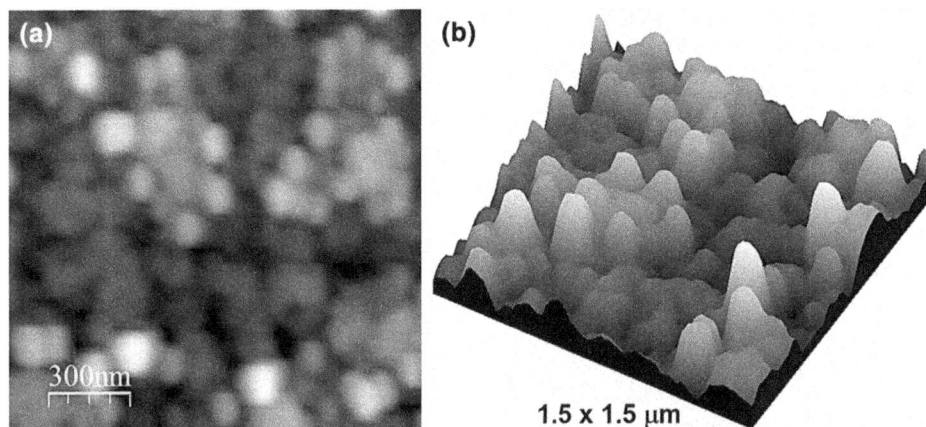

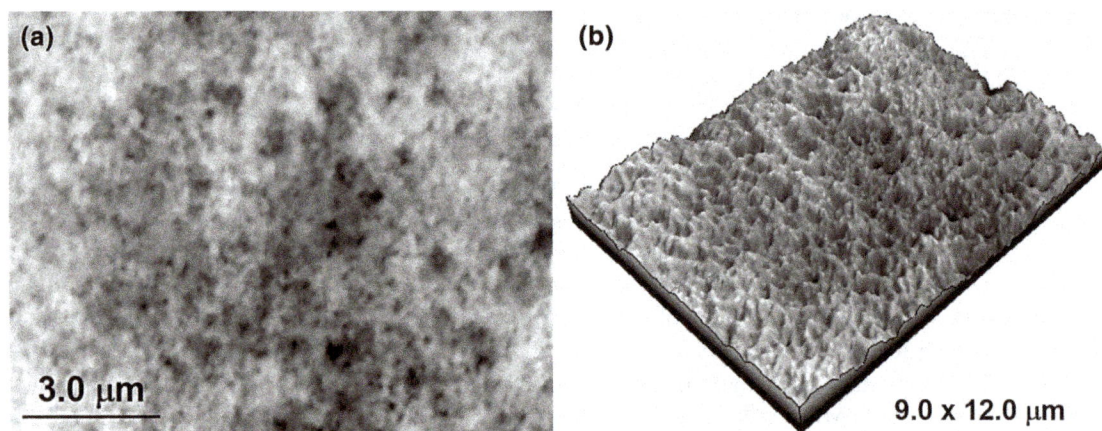

Fig. 2 The SEM image of molecularly imprinted titania nanoparticles (**a**), and corresponding 3D surface construct (**b**) to shown surface morphology of titania nanoparticles

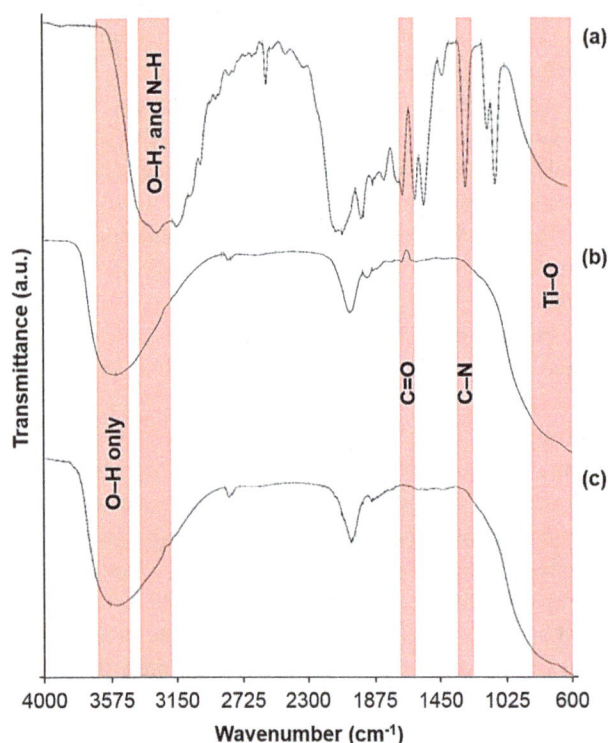

Fig. 3 FTIR spectra of as-prepared uric acid imprinted titania nanoparticles (*a*), imprinted titania nanoparticles after washing and removal of template (*b*), and non-imprinted titania nanoparticles (*c*)

Table 1 FTIR results of uric acid imprinted, washed, and non-imprinted titania nanoparticles

Sample nanoparticles	Amine (C–N) (cm^{-1})	Amide (C=O) (cm^{-1})	Alcohols (Ti–OH) (cm^{-1})
Uric acid imprinted TiO$_2$	1,290 (sh)	1,673 (sh)	3,290 (s, b)
Imprinted TiO$_2$ after washing	–	1,673 (w)	3,555 (b)
Non-imprinted TiO$_2$	–	–	3,560 (b)

FTIR absorption intensity is denoted as *sh* sharp, *s* strong, *b* broad; and *w* weak

Subsequently, the rebinding of uric acid molecules in (and with) imprinted titania particles was studied through re-inclusion of analyte (uric acid) molecules. For this purpose, 0.02 g of imprinted titania nanoparticles was immersed in uric acid solutions of varying concentrations (ranging from 0.01 to 0.095 mmol) for 1 h. After which, the mixture was centrifuged and the supernatant was analyzed on a UV/vis spectrophotometer. The re-inclusion of uric acid in imprinted titania nanoparticles was monitored by recording the differences in the absorbance of uric acid solutions before and after immersing titania nanoparticles. A calibration curve can be plotted from the differences in the absorbance of uric acid solution at various concentrations before and after rebinding (i.e. immersing titania nanoparticles for adsorption and re-inclusion of uric acid), as shown in Fig. 5.

The exact concentrations of uric acid solutions before and after its re-inclusion in imprinted titania nanoparticles were determined to calculate adsorption capacity of titania particles as described below. The quantity of uric acid adsorbed by the imprinted titania nanoparticles could also be calculated from the following equation.

$$Q = (C_i - C_f) V/m$$

Uric acid reinclusion/rebinding studies

The characteristics of uric acid imprinted titania nanoparticles such as selectivity and their binding capacity after complete template removal in the process of washing were studied by UV–vis spectrophotometric analysis. Thorough washing of molecularly imprinted titania nanoparticles removes the template, i.e. uric acid from the particulate matrix. Consequently, the absorbance is reduced after each washing step, as shown in Fig. 4.

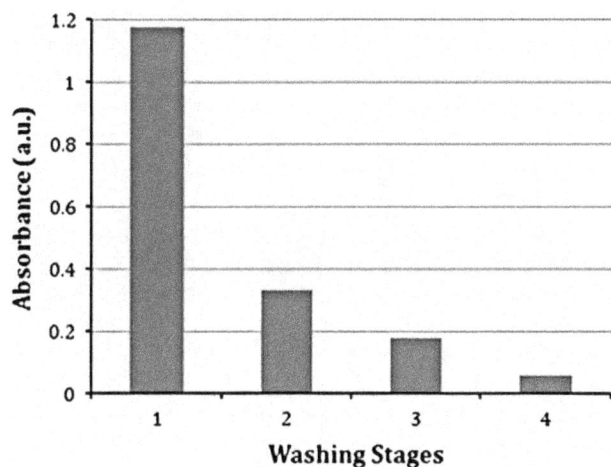

Fig. 4 Gradual decrease in absorbance of uric acid after each washing step: titania nanoparticles were washed repeatedly until no signal of uric acid was observed

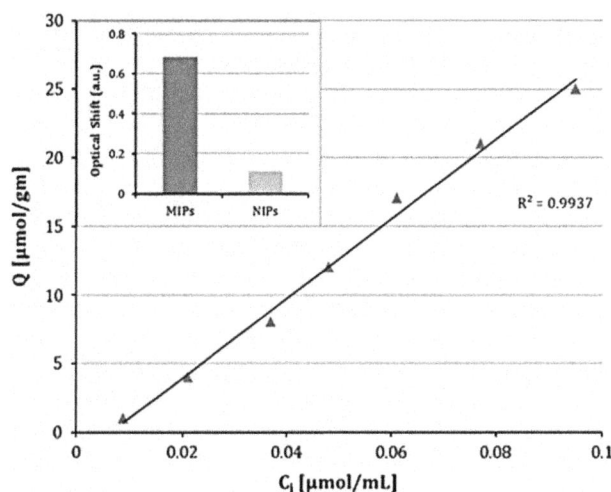

Fig. 6 Response of imprinted titania nanoparticles for uric acid (template), ascorbic acid (interfering compound), and guanine (structurally related) at equimolar concentrations

MINPs and NINPs for the same concentration of uric acid. It is obvious that the response of imprinted titania nanoparticles is about six times higher than non-imprinted ones. This suggests that NINPs can be taken as reference thus, to compensate all of the non-specific binding interactions.

Selectivity studies

In addition, we carried out selectivity experiments by immersing imprinted titania nanoparticles in ascorbic acid and guanine solutions of equimolar concentration, and the results are depicted in Fig. 6. It can be noticed that molecularly imprinted titania nanoparticles show higher binding affinity toward uric acid as compared to ascorbic acid and guanine. The ascorbic acid is considered as interfering compound in uric acid analysis, whereas the guanine is a purine and has somewhat similar structure to uric acid. The high optical shift is attributed to the successful imprinting of uric acid, i.e. the presence of uric acid-specific binding sites within imprinted titania nanoparticles' network. These results suggest that uric acid imprinted titania nanoparticles can perform efficiently in complex mixtures, and their binding is preferential toward uric acid even if closely related substances or interfering species are present in these mixtures.

Comparison

Table 2 gives a comparison of the results obtained in this study with reports published previously during the last few years (Patel et al. 2009; Sun et al. 2011; Khasanah et al. 2012; Wang et al. 2013). It is evident that uric acid imprinted titania nanoparticles, which can be produced rapidly using one-step synthetic procedure reported herein, offer the advantages of high sensitivity and selectivity, and

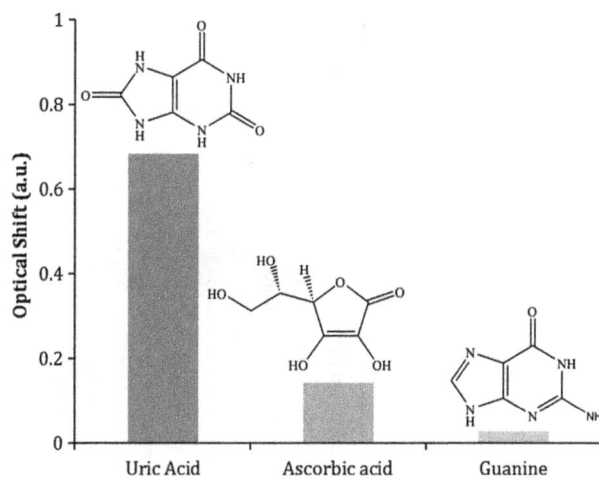

Fig. 5 Representation of adsorption isotherm for uric acid. *Inset* shows the bar graph for uric acid absorption by molecularly imprinted and non-imprinted titania nanoparticles

where, C_i and C_f are initial and final concentrations of uric acid solution expressed in $\mu mol\ mL^{-1}$, V is the volume in mL, and m is the mass of imprinted particles taken in g. So, the unit of Q is $\mu mol\ g^{-1}$. A graph was plotted between C_i and Q to interpret the binding affinity of the imprinted titania nanoparticles with the increasing concentration of uric acid, as shown in Fig. 5. In this way, it is clearly demonstrated that molecular recognition ability of imprinted titania nanoparticles increases linearly with the increasing concentration of uric acid.

In order to determine the non-specific interactions, both imprinted nanoparticles (MINPs) and non-imprinted nanoparticles (NINPs) were separately immersed in 1 mmol solution of uric acid in binding experiments. The inset bar graph in Fig. 5 shows the relative optical shifts of

Table 2 Comparison of results with already published reports

S. no.	Sensing interface	Synthesis/electrode preparation time (h)	Uric acid interactions	Interferents analyzed	Detection limit (μmol)	Remarks	References
1.	Imprinted titania nanoparticles	1–2	Fully reversible non-covalent interactions	Ascorbic acid, guanine	10	One-step rapid synthesis Meet desired sensitivity and specificity Measurements do not require buffer Reversible interactions, reusable, recyclable, and thus economical	This work
2.	Graphene/chitosan/Pd nanoparticles	>24	Electrochemical oxidation	Ascorbic acid, dopamine	0.17	Highly sensitive and specific Lengthy multistep synthesis and electrode modification procedure Require special/phosphate buffer medium for measurements	Wang et al. (2013)
3.	Imprinted poly(methacrylic acid)	–	Irreversible oxidation	–	0.0006	Highly sensitive Unknown selectivity Irreversible interactions	Khasanah et al. (2012)
4.	Graphene/Pt nanoparticles	>24	Electrochemical oxidation	Ascorbic acid, dopamine	0.05	Highly sensitive and specific Lengthy multistep synthesis and electrode modification procedure Require special/phosphate buffer medium for measurements	Sun et al. (2011)
5.	Imprinted poly(melamine-co-chloranil)	>24	Electrostatic and multiple hydrogen bonding	Ascorbic acid and others	~22 (3.71–4.10 μg mL^{-1})	Exhibit sufficient sensitivity and specificity Lengthy multistep synthesis and electrode modification procedure Measurements do not require buffer Not reusable	Patel et al. (2009)

reversible non-covalent interactions with the analyte (uric acid). The latter makes them more useful as sensing interface due to their reusability and recyclability with minimum loss in their sensing properties, thus providing a low-cost method for repeated uric acid detection and assay measurements.

Conclusions

In this work, molecularly imprinted titania nanoparticles (MINPs) are developed for the sensitive and selective recognition of uric acid. Preliminary characterization of titania by AFM, SEM and FTIR indicated the formation of uniformly distributed nanoparticles with effective imprinting characteristics. The results of static adsorption capacity suggest that the binding capability of MINPs increases linearly with the increasing concentration of uric acid and the tested concentration range is comparable with the normal to elevated levels of uric acid in blood serum. The non-imprinted particles (NINPs) are considered as control (reference) as they showed negligible effect in comparison to imprinted titania nanoparticles. In addition, the interfering substances and structurally related compounds such as ascorbic acid and guanine showed significantly lower binding with MINPs that confirms the adequate selectivity of imprinted titania nanoparticles. Thus, imprinted titania nanoparticles could be integrated with a suitable transducer system in future for developing cost effective biomedical diagnostics for real-time monitoring of uric acid at trace levels in complex biological fluids.

Acknowledgments Authors thank Higher Education Commission (HEC) of Pakistan for providing the startup Grant No. PM-IPFP/HRD/HEC/2011/0583.

References

Andersson LI (2000) Molecular imprinting: developments and applications in the analytical chemistry field. J Chromatogr B Biomed Sci App 745:3–13

Ansell RJ, Mosbach K (1998) Magnetic molecularly imprinted polymer beads for drug radioligand binding assay. Analyst 123:1611–1616

Chen Z, Zhao R, Shangguan D, Liu G (2005) Preparation and evaluation of uniform-sized molecularly imprinted polymer beads used for the separation of sulfamethazine. Biomed Chromatogr 19:533–538

Chen L, Xu S, Li J (2011) Recent advances in molecular imprinting technology: current status, challenges and highlighted applications. Chem Soc Rev 40:2922–2942

Dickert FL, Hayden O (1999) Molecular imprinting in chemical sensing. TrAC Trends Anal Chem 18:192–199

Hilt JZ, Byrne ME (2004) Configurational biomimesis in drug delivery: molecular imprinting of biologically significant molecules. Adv Drug Deliv Rev 56:1599–1620

Horcas I, Fernández R, Gómez-Rodríguez JM et al (2007) WSXM: a software for scanning probe microscopy and a tool for nanotechnology. Rev Sci Instr 78:013705.

Iqbal N, Afzal A (2013) Imprinted polyurethane–gold nanoparticle composite films for rapid mass-sensitive detection of organic vapors. Sci Adv Mater 5:939–946.

Kaur H, Halliwell B (1990) Action of biologically-relevant oxidizing species upon uric acid. Identification of uric acid oxidation products. Chem Biol Interact 73:235–247

Kempe M, Mosbach K (1995) Molecular imprinting used for chiral separations. J Chromatogr A 694:3–13

Khasanah M, Mudasir AK, Sugiharto E (2012) Development of uric acid sensor based on molecularly imprinted polymethacrylic acid-modified hanging mercury drop electrode. J Chem Chem Eng 6:209–214

Kryscio DR, Peppas NA (2009) Mimicking biological delivery through feedback-controlled drug release systems based on molecular imprinting. AIChE J 55:1311–1324

Latif U, Mujahid A, Afzal A et al (2011) Dual and tetraelectrode QCMs using imprinted polymers as receptors for ions and neutral analytes. Anal Bioanal Chem 400:2507–2515

Lieberzeit PA, Afzal A, Glanzing G, Dickert FL (2007) Molecularly imprinted sol–gel nanoparticles for mass-sensitive engine oil degradation sensing. Anal Bioanal Chem 389:441–446

Maxwell SRJ, Thomason H, Sandler D et al (1997) Antioxidant status in patients with uncomplicated insulin-dependent and non-insulin-dependent diabetes mellitus. Eur J Clin Invest 27:484–490

Mohan J (2004) Organic spectroscopy: principles and applications. CRC Press, USA

Mosbach K (1994) Molecular imprinting. Trends Biochem Sci 19:9–14

Mujahid A, Afzal A, Glanzing G et al (2010a) Imprinted sol–gel materials for monitoring degradation products in automotive oils by shear transverse wave. Anal Chim Acta 675:53–57.

Mujahid A, Lieberzeit PA, Dickert FL (2010b) Chemical sensors based on molecularly imprinted sol–gel materials. Materials 3:2196–2217

Mujahid A, Iqbal N, Afzal A (2013) Bioimprinting strategies: from soft lithography to biomimetic sensors and beyond. Biotechnol Adv 31:1435–1447.

Patel AK, Sharma PS, Prasad BB (2009) Electrochemical sensor for uric acid based on a molecularly imprinted polymer brush grafted to tetraethoxysilane derived sol–gel thin film graphite electrode. Mater Sci Eng C 29:1545–1553.

Pérez N, Whitcombe MJ, Vulfson EN (2001) Surface imprinting of cholesterol on submicrometer core-shell emulsion particles. Macromolecules 34:830–836

Piletsky SA, Turner NW, Laitenberger P (2006) Molecularly imprinted polymers in clinical diagnostics—Future potential and existing problems. Med Eng Phys 28:971–977

Poma A, Turner AP, Piletsky SA (2010) Advances in the manufacture of MIP nanoparticles. Trends Biotechnol 28:629–637

Ramström O, Mosbach K (1999) Synthesis and catalysis by molecularly imprinted materials. Curr Opin Chem Biol 3:759–764

Rubab Z, Afzal A, Siddiqi HM, Saeed S (2014) Augmenting thermal and mechanical properties of epoxy thermosets: the role of thermally-treated versus surface-modified TiO$_2$ nanoparticles. Mater Express 4:54–64.

Sellergren B (2000) Molecularly imprinted polymers: man-made mimics of antibodies and their application in analytical chemistry. Elsevier, Amsterdam

Sellergren B, Allender CJ (2005) Molecularly imprinted polymers: a bridge to advanced drug delivery. Adv Drug Deliv Rev 57:1733–1741

Sun C-L, Lee H-H, Yang J-M, Wu C-C (2011) The simultaneous electrochemical detection of ascorbic acid, dopamine, and uric acid using graphene/size-selected Pt nanocomposites. Biosens Bioelectron 26:3450–3455.

Vasapollo G, Sole RD, Mergola L et al (2011) Molecularly imprinted polymers: present and future prospective. Int J Mol Sci 12:5908–5945

Wang X, Wu M, Tang W et al (2013) Simultaneous electrochemical determination of ascorbic acid, dopamine and uric acid using a palladium nanoparticle/graphene/chitosan modified electrode. J Electroanal Chem 695:10–16.

Wulff G (2002) Enzyme-like catalysis by molecularly imprinted polymers. Chem Rev 102:1–28

Ye L, Weiss R, Mosbach K (2000) Synthesis and characterization of molecularly imprinted microspheres. Macromolecules 33:8239–8245

Zhang L, Cheng G, Fu C (2003) Synthesis and characteristics of tyrosine imprinted beads via suspension polymerization. React Funct Polym 56:167–173

Zhang H, Ye L, Mosbach K (2006) Non-covalent molecular imprinting with emphasis on its application in separation and drug development. J Mol Recognit 19:248–259

Green synthesis of silver nanoparticles using marine algae *Caulerpa racemosa* and their antibacterial activity against some human pathogens

T. Kathiraven · A. Sundaramanickam ·
N. Shanmugam · T. Balasubramanian

Abstract We present the synthesis and antibacterial activity of silver nanoparticles using *Caulerpa racemosa*, a marine algae. Fresh *C. racemosa* was collected from the Gulf of Mannar, Southeast coast of India. The seaweed extract was used for the synthesis of $AgNO_3$ at room temperature. UV–visible spectrometry study revealed surface plasmon resonance at 413 nm. The characterization of silver nanoparticle was carried out using Fourier transform infrared spectroscopy (FT-IR), X-ray diffraction (XRD) and transmission electron microscope (TEM). FT-IR measurements revealed the possible functional groups responsible for reduction and stabilization of the nanoparticles. X-ray diffraction analysis showed that the particles were crystalline in nature with face-centered cubic geometry.TEM micrograph has shown the formation of silver nanoparticles with the size in the range of 5–25 nm. The synthesized AgNPs have shown the best antibacterial activity against human pathogens such as *Staphylococcus aureus* and *Proteus mirabilis*. The above eco-friendly synthesis procedure of AgNPs could be easily scaled up in future for the industrial and therapeutic needs.

Keywords Silver nanoparticles · Green synthesis · *Caulerpa racemosa* · Antibacterial activity

T. Kathiraven · A. Sundaramanickam (✉) ·
T. Balasubramanian
Centre of Advance Study, Marine Biology, Faculty of Marine Sciences, Annamalai University, Parangipettai 608 502, Tamilnadu, India
e-mail: fish_lar@yahoo.com

N. Shanmugam
Department of Physics, Annamalai University, Annamalai Nagar 608 002, Tamilnadu, India

Introduction

Pathogenic bacteria are playing an important role in the creation of unknown diseases and the development of antibiotic resistance which are the major problems in the current scenario. The applications of nanoparticles are gaining an important function in the current scenario as they possess well-defined chemical, visual and mechanical attributes. Nanoparticles of metals are the most potential agents as they show excellent antibacterial activities due to their large surface area-to-volume ratio, which is getting up as the current interest in the researchers due to the growing microbial resistance against metal ions, antibiotics and the growth of resistant strains (Gong et al. 2007).

Antimicrobial nanoparticles offer various distinctive advantages in reducing acute toxicity, overcoming resistance, and lowering cost, when compared to conventional antibiotics (Pal et al. 2007; Weir et al. 2008). Antibiotics in the NPs form may sustain for long run than in tiny molecules (Nisizawa 1988). Physical and chemical synthesis methods, aimed at controlling the physical properties of the particles are mostly employed for the production of metal nanoparticles. Most of the methods are yet in the developmental phase and various troubles are often experienced with the stableness of the nanoparticles preparations, control of the crystals growth and aggregation of the particles (Brust 2002; Kowshik et al. 2003). Consequently, researchers working in the field of nanoparticles preparation turned their attention towards biological systems (Shiv Shankar et al. 2004). In the biosynthesis of nanoparticles, biological organisms like bacteria, fungi, actinomycetes, yeast, algae and plants were utilized as reducing agent or protective agents (Kaushik et al. 2010; Huh 2011). Biosynthetic method of nanoparticles has emerged as a simple and viable alternative to more complex chemical synthetic procedures to obtain nanomaterials. The rate of reduction of metal ions using biological agents is observed to be much quicker with an ambient temperature and pressure conditions (Kaushik et al. 2010).

Of all the different types of metal nanoparticles, the silver nanoparticles are playing a major role in the field of nanotechnology and nanomedicine. A number of living organisms are already well known to elaborate silver nanostructured compound such as *cyanobacteria*, *bacteria*, *fungi*, *actinomycetes* and plants such as *Cinnamomum camphora* (Huh and Kwon 2011), *Medicago sativa* (Tolaymat et al. 2010; Retchkiman-Schabesy et al. 2006), *Pelargonium graveolens* (Lukman et al. 2011), *Avena sativa* (Shankar et al. 2003), *Azardirachta indica* (Armendariz et al. 2004), *Tamarindus indica* (Shanker et al. 2004), *Emblica offcinalis* (Ankamwar et al. 2005), *Aloe vera* (Chandran et al. 2006), *Coriandrum sativum* (Badrinarayanan 2008), *Carica papaya* (Mude et al. 2009), *Parthenium hysterophorus* (Parashar et al. 2009), *Tritium vulgare* (Armendariz et al. 2009), *Acanthella elongata* (Inbakandan et al. 2010) and *Sesuvivm potulacastrum* (Nabikhan et al. 2010). Biosynthesis of silver NPs using the marine seaweed *Sargassum wightii* was carried out by Shanmugam et al., and they have shown that the sizes of the particles are in the range of 20 nm (Shanmugam et al. 2013). In our present study, we report the synthesis of AgNPs with sizes in the range of 10 nm using (Green algae) *Caulerpa racemosa* extract and also assessed their antagonistic effect against gram-positive and gram-negative bacteria.

Materials and methods

Sample collection

Green seaweed (*C. racemosa*) was collected from the Gulf of Mannar, Southeast coast of India. To maintain the freshness, the seaweed samples were instantly kept in a polythene bag with natural seawater.

Preparation of seaweeds extract

The samples were thoroughly washed with Milli Q water, chopped into fine pieces and then it was shade dried. Dried seaweed was ground well and made into fine powder. 1 g of biomass was kept in a 250-ml conical flask with 100 ml of Milli Q water for 24 h. Finally, the extract was filtered with Whatman No. 1 filter paper and stored it in a refrigerated temperature for further analysis.

Biosynthesis of AgNPs

For the biosynthesis of Ag nanoparticles 10 ml seaweed filtrate was added in 90 ml of 10^{-3} M aqueous $AgNO_3$

solutions at room temperature (Govindaraju et al. 2009). The bio-reduction of silver nitrate into silver nanoparticles can be confirmed by visual observation.

Source of microorganisms

The bacterial strains *Staphylococcus aureus* (ATCC 29123) and *Proteus mirabilis* (ATCC 25933) were obtained from American Type of Culture Collection Centre (ATCC) and were maintained in nutrient agar and LB agar medium procured from Himedia, Mumbai.

Characterization of silver nanoparticles

The reduction of metal ions was periodically monitored by visual inspection as well as by measuring the UV–Vis spectra of the solution by periodic sampling of aliquots of the aqueous component in 10 mm optical-path length quartz cuvette and periodically measured by Perkin Elmer double-beam spectrometer (Model LAMDA 25) operated between 200 and 800 nm. XRD analysis was conducted with Rigaku DMAX 2200 diffractometer using monochromatic CuKα radiation ($\lambda = 0.154056$ Å) running at 30 kV and 30 mA. The scanning was done in the region of 2θ from 30° to 80° at 0.02°min^{-1}. The crystalline size of the nanoparticles was calculated through the Scherrer's equation. The Fourier transform infrared (FT-IR) measurements were carried out to identify the existence of the functional groups in the synthesized silver nanoparticles. Dry powders of the biomass and Ag nanoparticles solutions were centrifuged at 10,000 rpm for 15 min and the resulting suspensions were redispersed in sterile distilled water. The purified pellets were dried and ground with KBr and analyzed on an avatar 330 FT-IR instrument mode at a resolution of 4 cm^{-1}. The morphology of the synthesized AgNPs was determined by high-resolution transmission electron microscopy (TEM). For TEM studies, the solution containing the nanoparticles was placed on copper grid and allowed to dry in a vacuum. The transmission electron micrographs were taken using TEM operated at an accelerating voltage of 90 keV.

Antibacterial assays (well diffusion method)

Antibacterial activity was assayed by using the agar well diffusion test technique. Muller Hinton agar medium (MHA) was prepared, the pH of the medium was maintained at 7.4 and then it was sterilized by autoclaving at 121 °C and 15 lbs pressure for 15 min. 20 ml of the sterilized media was poured into sterilized petri dishes and allowed to solidify at room temperature. A sterile cotton swab is used for spreading each test microorganism from the 24 h inoculated broth evenly on the MHA plates and

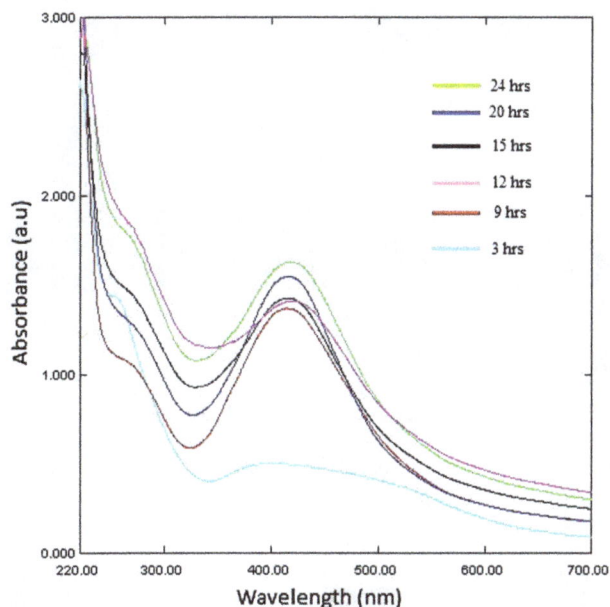

Fig. 1 Shows the UV–Vis spectra of the silver nitrate solutions incubated with marine green algae as a function of time of reaction

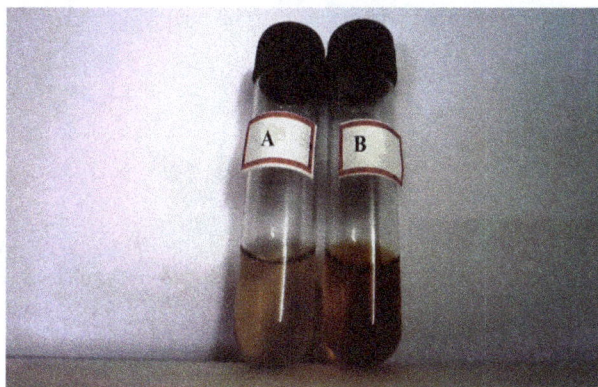

Fig. 2 Tube (A) having seaweed extract and Ag⁺ ions at the initial time (B) reaction mixture after 3 h

left for a few minutes to allow complete absorption of the inoculums. In each of these plates 5-mm diameter wells were made at the centre using an appropriate size sterilized cork borer. Different concentrations of each algal extract were added to the respective wells on the MH agar plates. Concentration ranges from 5, 10 and 15 µl, respectively, were placed in the wells and allowed to diffuse at room temperature for 30 min. No AgNPs was added in the control plate. The AgNPs loaded plates were kept in incubation at 37 °C for 24 h. After incubation, a clear inhibition zone around the wells indicated the presence of antimicrobial activity. All data on antimicrobial activity are the average of triplicate analyses (Nathan 1978).

Results and discussion

UV–visible absorption spectrometer

The absorption spectra of the as-prepared nanosized silver samples were characterized by UV–visible spectroscopy. The biosynthetic nanotechnology is an environmental friendly technology for the synthesis of nanoparticles. In this aspect, C. racemosa has proved to be an important biological component for the extracellular biosynthesis of stable AgNps. It is well known that Ag nanoparticles exhibit light yellowish to brown color. The biosynthesis of silver nanoparticles was measured by UV–Vis spectroscopy. UV–Vis spectra of the silver nitrate solutions incubated with marine green algae as a function of time of reaction.

The surface plasmon resonance (SPR) band of nanosilver occurs initially at 440 nm (3 h). This increases in intensity as a function of time of reaction. It is observed that the nanosilver SPR band is centered at about 413 nm (Fig. 1). From the spectra, it is clear that when the function of reaction time increased, the SPR band is shifted towards shorter wavelength region which shows a decrease in particle size as a result of increased band gap from the formula $E = hc/\lambda$. At lower concentrations, the SPR band is broad and it is due to large anisotropic particles. A smooth and narrow absorption band at 413 nm is observed which is characteristic of almost spherical nanoparticles. The position of SPR band in UV–Vis spectra is sensitive to particle shape, size, its interaction with the medium, local refractive index and the extent of charge transfer between medium and the particles (Figs. 1, 2).

Meanwhile similar studies were carried out with marine alga S. wightii (Govindaraju 2009) and plant extracts were previously obtained (Krishnaraj et al. 2010; Shrivastava 2009).

Fourier transform infrared spectroscopy (FT-IR) measurements

FT-IR spectra were recorded for C. racemosa extract and synthesized silver nanoparticles to identify the possible biomolecules responsible for the reduction of AgNO₃ into AgNPs. FT-IR spectrum of C. racemosa shows different major peaks positioned at 3416, 2924, 2854, 1631, 1389, 1061, 1019 and 660 cm⁻¹ (Fig. 3). The presence of peak at 3416 cm⁻¹ could be ascribed to O–H group in polyphenols or proteins/enzymes or polysaccharide (Song et al. 2009; Susanto et al. 2009). A small peak positioned at 2924 cm⁻¹ may be due to CH-stretching of alkanes. A sharp intense band observed at 1631 cm⁻¹ can be due to the stretching vibration of the (NH)=O group. The observed band at 660 cm⁻¹ is due to α-glucopyranose rings deformation of carbohydrates (Feng 2000). The bands positioned at 1061

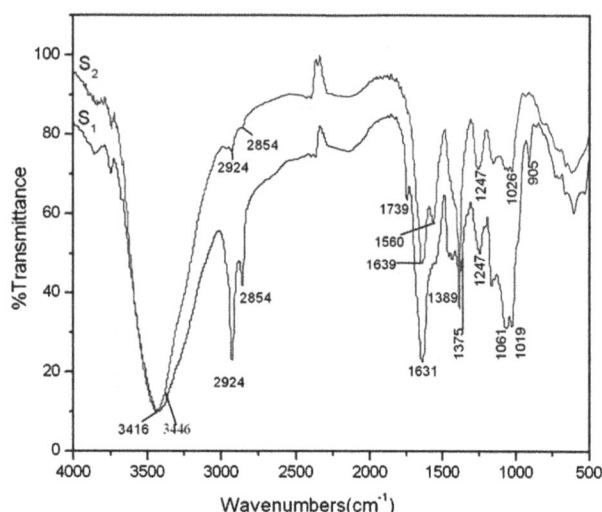

Fig. 3 Shows FT-IR spectra of (S1) *C. racemosa* extract (S2) biologically synthesized silver nanoparticles using *Caulerpa racemosa*

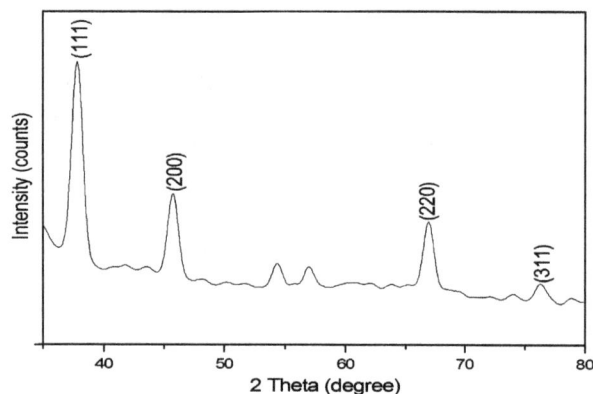

Fig. 4 Shows XRD pattern analysis of silver nanoparticles synthesized by treating *C. racemosa* extract with silver nitrate aqueous solution

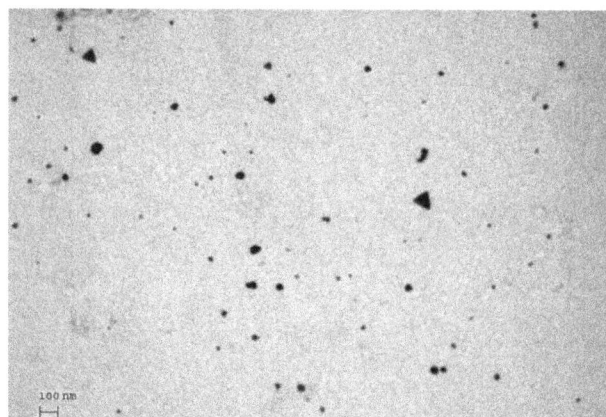

Fig. 5 TEM image silver nanoparticles (10 mL of seaweed solution in 10^{-3} M of AgNO$_3$ in 90 mL of water)

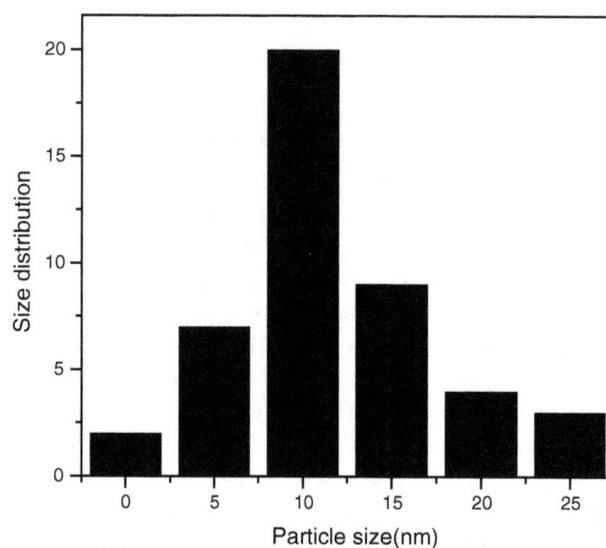

Fig. 6 Paticle size histogram of AgNPs

and 1019 cm^{-1} are due to C–N stretching vibration of aliphatic amines. On the other hand, FT-IR spectrum of the synthesized AgNPs shows the presence of major peaks at 3440 and 1639 cm^{-1} which are associated with OH–stretching vibrations and stretching vibration of the (NH)=O group, respectively. The shifting of the band from 1631 to 1639 cm^{-1} may be due to the binding of (NH)C=O group with the nanoparticles. The (NH)C=O groups within the case of cyclic peptides are involved in stabilizing the nanoparticles. Thus, the peptides may play an important role in the reduction of AgNO$_3$ into Ag nanoparticles.

XRD analysis

The development of single-phase compound was confirmed by X-ray diffraction (XRD) method. The XRD pattern of synthesized AgNPs was observed and compared with the standard powder diffraction card of Joint Committee on Powder Diffraction Standards (JCPDS). Intense diffraction peaks due to AgNPs are clearly observed at 38.24°, and 44.42°, 64.44° and 77.40° are pertaining to the (111) (200), (220) and (311) planes of Bragg's reflection based on the FCC (JCPDS, file No. 04-0783) structure of silver nanoparticles. No reflection peaks related to nitrate ions and other impurities were observed in this pattern, which indicating the high purity of the end product. In addition, the acquired reflections are sharp with good intensity which confirms that the structures of synthesized nanoparticles are well crystalline (Fig. 4). Our findings match with the reports suggest by Govindaraju et al. (2009).

Morphology and size

Transmission electron microscopy (TEM) has been used to identify the size, shape and morphology of nanoparticles.

Table 1 Antimicrobial activity of silver nanoparticles against *Staphylococcus aureus* and *Proteus mirabilis*

S. No	Pathogen name	Zone of inhibition (mm)			
		Control	5 μl	10 μl	15 μl
1	*Staphylococcus aureus*	0	7	9	12
2	*Proteus mirabilis*	0	8	11	14

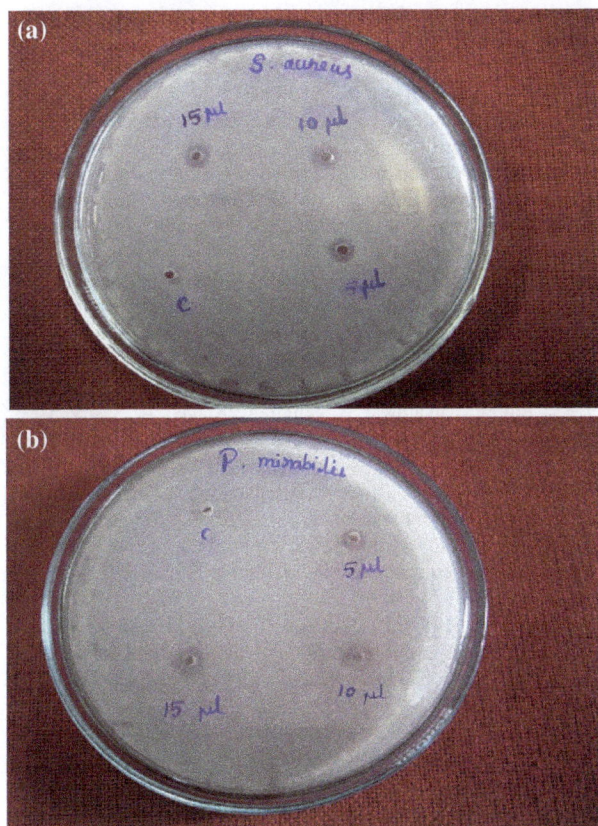

Fig. 7 Shows antibacterial activity of silver nanoparticles assayed by the agar well diffusion method in petri plates. Silver nanoparticles poured in the wells show the zone of inhibition against **a** *Staphylococcus aureus* and **b** *Proteus mirabilis*

From the image (Fig. 5), it is clear that the morphology of silver nanoparticles is almost spherical with few triangular nanoparticles. From the histogram analysis, it is noted that the particles with the size of 10 nm was more pronounced (Fig. 6).

Antibacterial studies

In the present study, the antibacterial activity of green synthesized silver nanoparticles were tested against *P. mirabilis* and *S. aureus* with various concentrations (5, 10 and 15 μl) and the results are shown in Table 1 and Fig. 7. The results of antibacterial activity with a zone of inhibition maximum was found in *P. mirabilis* (14 mm for 15 μl)

and minimum level antibacterial activity present in *S. aureus* (7 mm for 5 μl). This enormous difference may be due to the susceptibility of the organism used in the current study. The nanoparticles get attached to the cell membrane and also penetrate inside the bacteria. When silver nanoparticles enter the bacterial cell, it forms a low molecular weight region in the center of the bacteria to which the bacteria conglomerates thus protecting the DNA from the silver ions. The nanoparticles preferably attack the respiratory chain cell division finally leading to cell death. The nanoparticles release silver ions in the bacterial cells, which enhance their bactericidal activity (Morones et al. 2005; Kvitek et al. 2008). Several studies propose that AgNPs may attach to the surface of the cell membrane disturbing permeability and respiration functions of the cell (Morones et al. 2005). It is also possible that AgNPs not only interact with the surface of membrane, but can also penetrate inside the bacteria (Sondi 2007).

Conclusion

It has been concluded that the extract of marine seaweed *C. racemosa* is capable of producing Ag nanoparticles extracellularly and these nanoparticles are quite stable in solution due to capping likely by the proteins present in the extract. This is an efficient, eco-friendly and simple process. The AgNPs showed potential antibacterial activity against human pathogens like *P. mirabilis* and *S. aureus*. Therefore, nanoparticles of silver in combination with commercially available antibiotics could be used as an antimicrobial agent after further trials on experimental animals.

Acknowledgments We thank the Ministry of Earth Sciences (MoES), New Delhi, for financial support through a scheme/ICMAM-PD/SWQM/CASMB/35/2012. We also thank the authorities of Annamalai University for providing the necessary facilities during the entire course of this work.

References

Ankamwar B, Damle C, Ahmad A, Sastry M (2005) Biosynthesis of gold and silver nanoparticles using *Emblica Officinalis* fruit extract, their phase transfer and transmetallation in an organic solution. J Nanosci Nanotechnol 5:1665–1671

Armendariz V, Isaac H, Jose R, Peralta V, Yacaman MJ, Troiani H, Santiago P, Jorge L, Gardea T (2004) Size controlled gold nanoparticles formation by Avena sativa biomass: use of plants in Nanobiotechnology. J Nanopart Res 6:377–379

Armendariz V, Parsons JG, Lopez ML, Peralta-Videa JR, Jose-Yacaman M, Gardea-Torresdey JL (2009) Extraction of gold nanoparticles from oat and wheat biomass using sodium citrate and cetyltrimethylammonium bromide studied using XAS, HRTEM and UV-Vis. Nanotechnology 20:105607

Badrinarayanan K, Sakthivel N (2008) A simple and green method for the synthesis of silver nanoparticles using *Ricinus communis* leaf extract. Mater Lett 62:4588–4591

Brust M, Kiely CJ (2002) Some recent advances in nanostructure preparation from gold and silver particles: a short topical review. Colloids Surf A Physicochem Eng Aspects 202:175–186

Chandran SP, Chaudhary M, Pasricha R, Ahmad A, Sastry M (2006) Synthesis of gold nanotriangles and silver nanoparticles using *Aloe vera* plant extract. Biotechnol Prog 22:577–579

Feng QL, Wu J, chen GQ, cui FZ, Kim TN, Kim JO (2000) A mechanistic study of the antibacterial effect of silver ions on *Escherichia coli* and *Staphylococcus aureus*. J Biomed Mater 52(4):662–668

Gardea-Torresdey JL, Tiemann KJ, Gamez G, Dokken K, Tehuacanero S, Yacaman MJ (1999) Gold nanoparticles obtained by bio-precipitation from gold(III) solutions. J Nanopart Res 1(3):397–402

Gardea-Torresdey JL, Gomez E, Perlata-Videa JR, Parsons JG, Troiani H, Yacamen MJ (2003) Alfalfa sprouts: a natural source for the synthesis of silver nanoparticles. Langumir 19:1357–1362

Gong P, Li H, He X, Wang K, Hu J, Tan W, Zhang S, Yang X (2007) Preparation and antibacterial activity of Fe_3O_4@Ag nanoparticles. Nanotechnology 18:604–611

Govindaraju K, Kiruthiga V, Ganesh Kumar V, Singaravelu G (2009) Extracellular synthesis of silver nanoparticles by a marine alga *sargassum wightii grevilli* and their antibacterial effects. J Nanosci Nanotechnol 9:1–5

Huh AJ, Kwon YJ (2011) Nanoantibiotics: a new paradigm for treating infectious diseases using nanomaterials in the antibiotics resistant era. J Control Release 156(2):128–145

Inbakandan D, Venkatesan R, Ajmal Khan S (2010) Biosynthesis of gold nanoparticles utilizing marine sponge *Acanthella elongate* (Dendy, 1905). Colloids Surf B 81:634–639

Kaushik N, Thakkar MS, Snehit S, Mhatre MS, Rasesh Y (2010) Biological synthesis of metallic nanoparticles. Nanomed Nanotechnol Biol Med 2:257–262

Kowshik M, Ashtaputre S, Kharrazi S, Vogel W, Urban J, Kulkarni SK (2003) Extracellular synthesis of silver nanoparticles by a silver-tolerant yeast strain. MKY3. Paknikar. Nanotechnology 14:95

Krishnaraj C, Jagan EG, Rajasekar S, Selvakumar P, Kalaichelvan PT, Mohan N (2010) Synthesis of silver nanoparticles using *Acalypha indica* leaf extracts and its antibacterial activity against water borne pathogens. Colloids Surf B Biointerfaces 76:50–56

Kvitek L, Panacek A, Soukupova J, Kolar M, Vecerova R, Prucek R, Holecova M, Zboril R (2008) Effect of surfactants and polymers on stability and antibacterial activity of silver nanoparticles (NPs). J Phys Chem 112:5825–5834

Li S, Shen Y, Xie A, Yu X, Qiu L, Zhang L, Zhang Q (2007) Green synthesis of silver nanoparticles using *Capsicum annuum* L. extract. Green Chem 9:852–858

Lukman AI, Gong B, Marjo CE, Roessner U, Harris AT (2011) Facile synthesis, stabilization, and anti-bacterial performance of discrete Ag nanoparticles using Medicago sativa seed exudates. J Colloid Interface Sci 353:433–444

Morones JR, Elechiguerra JL, Camacho A, Ramirez JT (2005) The bactericidal effect of silver nanopartilces. Nanotechnology 16:2346–2353

Mude N, Avinash I, Aniket G, Mahendra R (2009) Synthesis of silver nanoparticles using callus extract of *Carica papaya*. J Plant Biochem Biotechnol 18:0971–0978

Nabikhan A, Kandasamy K, Raj A, Alikunhi NM (2010) Synthesis of antimicrobial silver nanoparticles by callus and leaf extracts from salt marsh plant, *Sesuvium portulacastrum* L. Colloids Surf B 79:488–493

Nathan P, Law EJ, Murphy DF, MacMillan BG (1978) A laboratory method for selection of topical antimicrobial agents to treat infected burn wounds. Burns 4:177–178

Nisizawa K, Mchaugh DJ (1988) Production and utilization of products from commercial seaweeds. FAO, Rome

Noginov MA, Zhu G, Bahoura M, Adegoke J, Small C, Ritzo BA, Drachev VP, Shalaev VM (2006) The effect of gain and absorption on surface plasmon in metal nanoparticles. Appl Phys B 86:455–460

Pal S, Tak YK, Song JM (2007) Dose the antibacterial activity of silver nanoparticles depend on the shape of the nanoparticle? A study of the gram-negative bacterium *Escherichia coli*. Appl Environ Microbiol 27(6):1712–1720

Parashar V, Parashar R, Sharma B, Pandey AC (2009) Parthenium leaf extract mediated synthesis of silver nanoparticles: a novel approach towards weed utilization Digest. J Nanomater Biostruct 4:723–727

Retchkiman-Schabes PS, Canizal G, Becerra-Herrera R, Zorrilla C, Liu HB, Ascencio J (2006) Biosynthesis and characterization of Ti/Ni bimetallic nanoparticles. Opt Mater 29(1):95–98

Shankar S, Ahmad A, Sastry M (2003) Geranium leaf assisted biosynthesis of silver nanoparticles. Biotechnol Prog 19:1627–1631

Shanker SS, Rai A, Ankamwar B, Singh A, Ahmed A, Sastry M (2004) Biological synthesis of triangular gold nanoprisms. Nat Mater 3:482–488

Shanmugam N, Rajkamal P, Cholan S, Kannadasan N, Sathishkumar K, Viruthagiri G, Sundaramanickam A (2013) Biosynthesis of silver nanoparticles from the marine seaweed *Sargassum wightii* and their antibacterial activity against some human pathogens. Appl Nanosci 4:13204-013-0271

Shiv Shankar S, Akhilesh Rai A, Ahmad A, Sastry M (2004) Rapid synthesis of Au, Ag, and bimetallic Au core–Ag shell nanoparticles using Neem (*Azadirachta indica*) leaf broth. J Colloid Interface Sci 275:496–502

Shrivastava S, Dash D (2009) Applying nanotechnology to human health. J Nanotechnol 12:240–243

Sondi I, Salopek-Sondi B (2007) Silver nanoparticles antimicrobial agent: a case study on *E.Coli* as a model for gram negative bacteria. J Colloid Interface 275:177–182

Song HY, Ko KK, Oh LH, Lee BT (2006) Fabrication of silver nanoparticles and their antimicrobial mechanisms. Eur Cells Mater 11:58

Song JY, Jang HK, Kim BS (2009) Biological synthesis of gold nanoparticles using Magnolia kobus and Diopyros kaki leaf extracts. Process Biochem 44:1133–1138

Susanto H, Feng Y, Ulbricht M (2009) Fouling behavior of aqueous solutions of polyphenolic compounds during ultrafiltration. J Food Eng 91:333–340

Tolaymat TM, El Badawy AM, Genaidy A, Scheckel KG, Luxton TP, Suidan M (2010) An evidence-based environmental perspective of manufactured silver nanoparticle in syntheses and applications: A systematic review and critical appraisal of peer-reviewed scientific papers. Sci Total Environ 408:999–1006

Weir E, Lawlor A, Whelan A, Regan F (2008) The use of nanoparticles in anti-microbial materials and their characterization. Analyst 133:835–845

Yang W, Yang C, Sun M, Yang F, Ma Y, Zhang Z, Yang X (2009) Green synthesis of nanowire-like Pt nanostructures and their catalytic properties. Talanta 78:557–564

21

Synthesis, characterization and magnetic properties of hematite (α-Fe$_2$O$_3$) nanoparticles on polysaccharide templates and their antibacterial activity

M. Mohamed Rafi · K. Syed Zameer Ahmed ·
K. Prem Nazeer · D. Siva Kumar · M. Thamilselvan

Abstract The present study is to synthesize iron oxide nanoparticles on different polysaccharide templates calcined at controlled temperature, characterizing them for spectroscopic and magnetic studies leading to evaluate their antibacterial property. The synthesized iron oxide nanoparticles were characterized by X-ray diffractometer (XRD), Fourier transform infrared spectroscopy, high resolution scanning electron microscopy (HRSEM), high resolution transmission electron microscopy (HRTEM) and vibrating sample magnetometer. The iron oxide nanoparticles were tested for antibacterial activity against gram-positive and gram-negative bacterial species. The XRD confirms the crystalline nature of iron oxide nanoparticles with the mean crystallite size of 10 nm. The functional groups of the synthesized iron oxide nanoparticles were 547, 543 and 544 cm^{-1} characterizing the Fe–O and the broad bands at 3,398, 3,439 and 3,427 cm^{-1} were attributed to the stretching vibrations of hydroxyl group absorbed by iron oxide nanoparticles. HRTEM analyses revealed that the average particle size of the hematite nanoparticles are about 85, 92 and 77 nm for AF, DF and GF, respectively, which was a coincident with the results obtained from the HRSEM analysis. Magnetic measurement exhibited ferromagnetic behavior of the α-Fe$_2$O$_3$ at the room temperature with higher coercivity of $H_C = 2,303$, 2,333 and 1,019 Oe for AF, DF and GF, respectively. Antibacterial test showed the inhibition against *Aeromonas hydrophila* and *Escherichia coli* with significant antagonistic activity.

Keywords Iron oxide nanoparticles · Polysaccharides · Ferromagnetism · Antibacterial activity

M. M. Rafi
Department of Physics, C. Abdul Hakeem College,
Melvisharam 632509, Tamilnadu, India

K. S. Z. Ahmed
Department of Biochemistry, C. Abdul Hakeem College,
Melvisharam 632509, Tamilnadu, India

M. M. Rafi · K. P. Nazeer (✉) · D. Siva Kumar
PG and Research Department of Physics, Islamiah College,
Vaniyambadi 635752, Tamilnadu, India
e-mail: mohamedrafi947@gmail.com;
nazeerprem13@gmail.com

M. Thamilselvan
Department of Physics, TPGIT, Vellore 632002,
Tamilnadu, India

Introduction

Templating is one of the most frequently used methods of synthesizing materials with structural units ranging from nanometers to micrometers. Aqueous gel-like lyotropic liquid crystal with extensive hydrogen bonding and nanoscale hydrophilic compartments can be employed for direct templating of nanoscale features (Braun et al. 2005). The use of biological materials as templates is gaining momentum and biotemplating takes advantage of the structural stability and specificity of biological systems to create novel materials (Sotiropoulou et al. 2008). Synthesis of well-structured, monodisperse nanostructures of iron oxide has tremendous interest. Such nanostructures are extensively used in magnetic materials, as photocatalyst, in sensors, medical applications such as hyperthermia, targeted drug delivery, magnetic resonance imaging, etc. (Zhao et al. 2005). Choosing the right material for synthesis of iron oxide nanoparticles is crucial by varying the processing conditions with various forms of iron oxides such as Fe$_3$O$_4$, γ-Fe$_2$O$_3$ and α-Fe$_2$O$_3$. Precipitation of the aqueous

Fe^{2+}/Fe^{3+} solution (Laurent et al. 2008) or in situ generated Fe^{3+} solution results in Fe_3O_4 and γ-Fe_2O_3, while high temperature reaction results in α-Fe_2O_3 (Zhou et al. 2008). Ferromagnetic particles can also be synthesized through the hematite route owing to the higher stability of hematite and then subsequently reduced to magnetite (Zhao et al. 2008). Recently, the use of iron oxide nanoparticles in photonic applications has also been reported (Ge et al. 2008). Agarose is a hydrophilic polymer and is widely used in biomedical applications and bioengineering (Rochas and Lahaye 1989). The basic disaccharide repeating units of agarose consists of (1,3) linked β-D-galactose (G) and (1,4) linked α-L-(3,6)-anhydrogalactose. Dextran is a water-soluble polysaccharide mainly composed of α-D-(1$\rightarrow$6) linked glucose units and some α-D-(1$\rightarrow$3) linked to glucose branch units. In strongly alkaline solution, dextran interacts with hydroxyl groups present in the iron oxide particles. Gelatin is derived from collagen, and it is commonly used to immobilize drugs and genes to produce controlled released products for pharmaceutical and medical applications (Gupta and Gupta 2008). Gelatin, being water soluble, biodegradable and biocompatible, can also be a promising candidate for the surface modification of iron oxide.

In this communication, we report a simple thermal decomposition route to synthesize α-Fe_2O_3 nanoparticles using only ferrous sulfate hydrate ($FeSO_4 \cdot 7H_2O$) as raw material. The iron oxide bound agarose, dextran and gelatin templates can be used directly by removal of unbound ions, owing to their remarkable stability, biocompatibility and biodegradability. The iron oxide nanoparticles thus synthesized have been characterized for the morphology and magnetic properties. Further, the in vitro antibacterial activities were evaluated against six bacteria including three Gram-positive *Staphylococcus aureus* (*S. aureus*), *Aeromonas hydrophila* (*A. hydrophila*), *Streptococcus pyogenes* (*S. pyogenes*) and three Gram-negative *Pseudomonas aeruginosa* (*P. aeruginosa*), *Enterococcus faecalis* (*E. faecalis*), *Escherichia coli* (*E. coli*) bacteria. To the best of our knowledge, the synergistic impact of magnetic nanoparticles of iron oxide as antibacterial agent is not yet reported so far in the earlier literatures and hence this study is conducted to find the antibacterial activity of magnetic nanoparticles.

Experimental

Materials

All chemicals were procured from M/s Fishur Scientific, India and used without any further purification. Ferrous sulfate ($FeSO_4 \cdot 7H_2O$) was used as source for iron(II) and agarose, dextran and gelatin were the template components of the polysaccharide. High alumina crucibles were employed for the calcination reactions. *Aeromonas hydrophila* (MTCC-1739), *E. coli* (MTCC-1677), *S. aureus* (MTCC-3160), *P. aeruginosa* (MTCC-4030), *E. faecalis* (MTCC-3159) and *S. pyogenes* (MTCC-1928) were obtained from the Institute of Microbial Technology (IMTECH), Chandigarh, India. The lyophilized culture sample was resuspended in nutrient broth with 1.0 % NaCl at 37 °C for 24 h into the viable culture source. All these strains were grown in Tryptic soy broth (TSB) except for *E. coli* strains. *Escherichia coli* strains were grown in Luria–Bertani (LB) medium. The strains were grown aerobically at 37 °C, with 10 mL of medium in 18–150 mm borosilicate glass culture tubes with shaking at 200 rpm under normal laboratory lighting conditions unless specified.

Preparation of agarose-Fe_2O_3 (AF), dextran-Fe_2O_3 (DF), gelatin-Fe_2O_3 (GF)

A quantity of 2.5 g of polysaccharide (agarose, dextran and gelatin) was added to 100 mL distilled water, then the mixture was heated at 90 °C for about 10 min with constant stirring for dissolution of the polysaccharide. 10 g of ferrous sulfate hydrate ($FeSO_4 \cdot 7H_2O$) was added to the solution, the resultant solution was stirred for 30 min. The template–iron mixed solution was treated at 800 °C (heating rate of 5 °C/min) and maintained at that temperature for 120 min, after which it was cooled to room temperature at a rate of 10 °C/min. The so-obtained powder was collected, washed four times in deionized water and ethanol, and dried at 80 °C for 5 h. Thus the obtained sample was characterized.

Characterization techniques

The X-ray diffractometer (Bruker Model: D8 Advance) employing Cu Kα ($\lambda = 1.5406$ Å, $2\theta = 10°$–$80°$) radiation was used to characterize the crystal structure of the sample. IR spectra were recorded with an FTIR spectrometer (Perkin Elmer IR Spectrometer) in the range of 400–4,000 cm^{-1} on pressed disks using KBr as binding material. The surface morphology of the samples was studied by high resolution scanning electron microscopy (HRSEM) (F E I Quanta FEG 200) and high resolution transmission electron microscopy (HRTEM) (JEOL JEM 2100). The magnetic properties of the samples were investigated by using vibrating sample magnetometer (EC&G Princeton Applied Research VSM MODEL 155).

Results and discussion

FTIR spectroscopic studies

The FTIR spectra of agarose, AF, dextran, DF, gelatin and GF are shown in Fig. 1. The dominant bands of the three

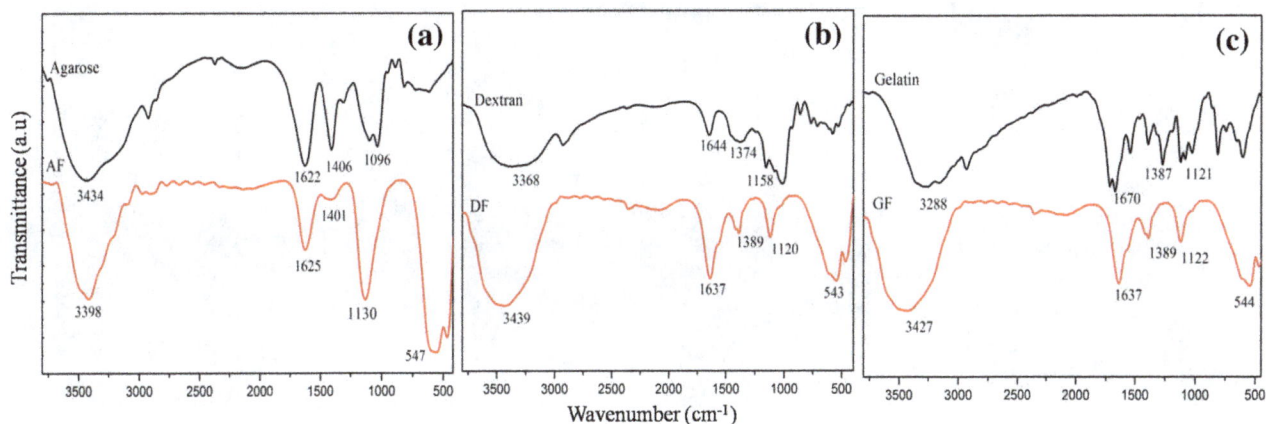

Fig. 1 FTIR spectra of **a** agarose, AF, **b** dextran, DF **c** gelatin, GF

products namely AF at 547 cm^{-1}, DF at 543 cm^{-1} and GF at 544 cm^{-1} are the characteristics of α-Fe$_2$O$_3$ (Barron and Torrent 1996). The presence of C–O bond stretching of the C–O–C group in the anhydroglucose ring was observed at 1,096, 1,120 and 1,121 cm^{-1}, respectively (Ma et al. 2009). The bands at 1,406, 1,374 and 1,387 cm^{-1} are due to C–O bond stretching of the C–O–H group. The symmetric stretching vibration of C–O–C groups and bending vibration of water molecules are observed for agarose, dextran and gelatin at 1,622, 1,644 and 1,670 cm^{-1}, respectively (Kormann et al. 1989). The peaks exhibited at 3,398, 3,439 and 3,427 cm^{-1} are attributed to the stretching vibrations of OH, which is assigned to OH-absorbed by iron oxide nanoparticles.

X-Ray diffraction analyses

The powder XRD patterns of AF, DF and GF are shown in Fig. 2. All samples could be indexed to the hexagonal α-Fe$_2$O$_3$ phase (JCPDS 33-0664), with no characteristic peaks of other impurities, thus suggesting the high phase purity of the as-synthesized products. The strong and sharp diffraction peaks indicate the high crystallinity of these samples (Liu et al. 2005). The XRD data also revealed that the α-Fe$_2$O$_3$ obtained had rhombohedral structure. A hematite crystal has a rhombohedrally centered hexagonal structure of corundum type with a close packed lattice in which two-third of the octahedral sites are occupied by Fe^{3+} ions (Sreeram et al. 2009). In a typical crystal unit, each Fe atom is surrounded by six oxygen atom, where each oxygen atom is bound to four Fe atoms. Due to these characteristics of the corundum structure, the surface hydroxyl configuration of the various crystal face of hematite is quite different (Jitianu et al. 2002). The major diffractions were observed from (012), (104), (110), (024) and (116) planes of α-Fe$_2$O$_3$ for all the cases. The average crystallite sizes of the prepared α-Fe$_2$O$_3$ were calculated

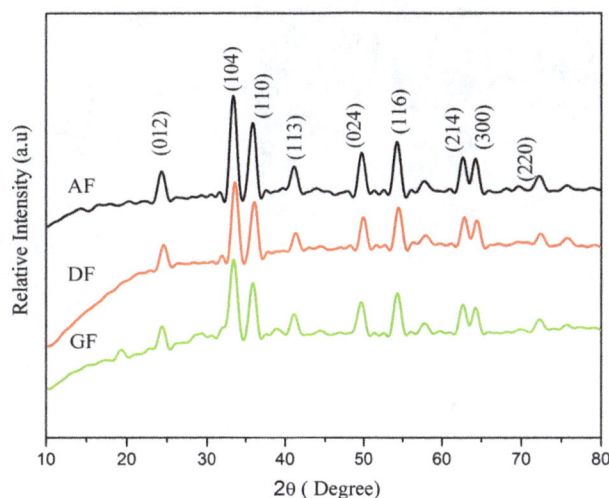

Fig. 2 Powder X-ray diffraction pattern of AF, DF and GF

using Debye–Scherrer equation ($D = 0.9\lambda/(\beta \; Cos\theta)$), (where, D is the crystallite size (diameter), λ is the wave length of X-ray, i.e., 1.540598 Å, β is the value of FWHM, which is expressed in radians and θ is the Bragg's angle) and found to be around 10 nm.

Morphological studies

The surface morphologies of the prepared AF, DF and GF were studied using high resolution scanning electron microscope, as shown in Fig. 3. It is evident from the HRSEM images that the particles are dumbbell nature and the average size of iron oxide particles are about 90, 97 and 80 nm for AF, DF and GF, respectively. The difference in the crystallite size calculated using X-ray diffraction and particle size obtained by SEM is due to the fact that the particles composed of several crystallization domains are observed by X-rays while whole particle is observed with SEM (Vazquez et al. 1998). The increase in grain size

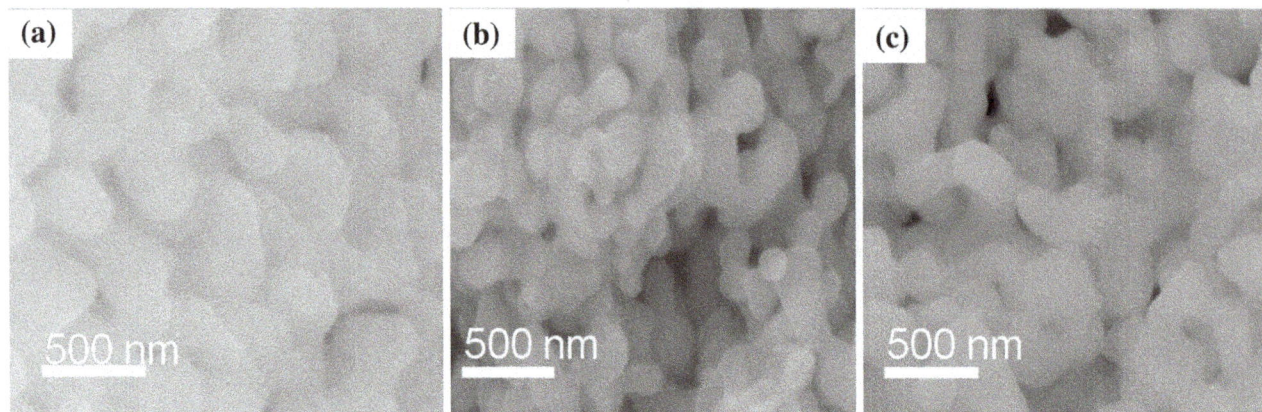

Fig. 3 HRSEM micrographs for **a** AF, **b** DF and **c** GF

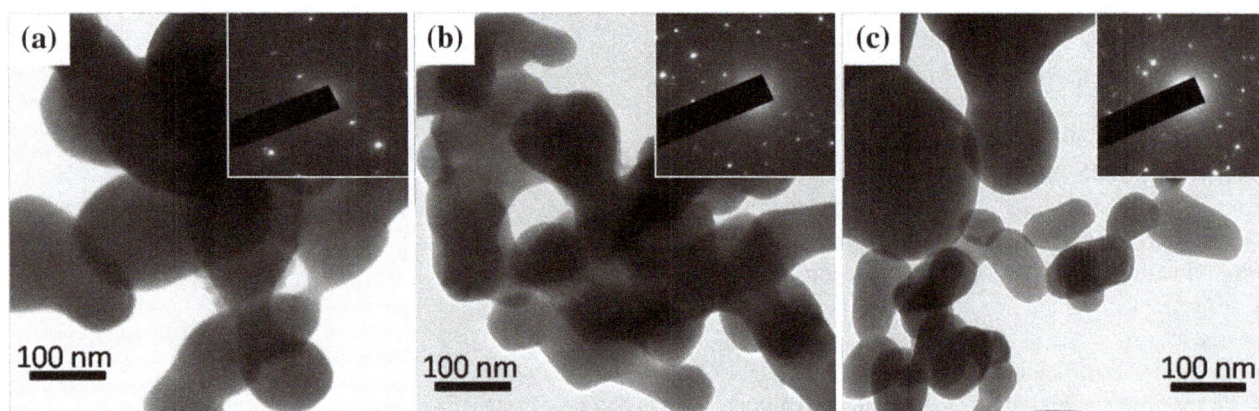

Fig. 4 HRTEM micrographs of **a** AF, **b** DF, **c** GF (*inset* shows SAED patterns)

occurred due to the further agglomeration of primary particles, when the sample was annealed at higher temperature (800 °C).

Further analyses on the hematite nanoparticles are carried out by selected area electron diffraction (SAED) and high resolution TEM (HRTEM). Figure 4a, b, c displays the high magnification TEM image of an AF, DF and GF. The morphology of the nano-sized hematite particles showed a dumbbell shape nature. These dumbbell-shaped nanoparticles are joined together to form bundles of aggregates. The mean size of particles was about 85, 92 and 77 nm for AF, DF and GF which was coincident with the results obtained from the HRSEM analysis. The corresponding selected area electron diffraction (SAED) pattern is shown in inset of Fig. 4. The ED pattern consists of concentric rings along with spots over the rings. This feature indicates the polycrystallinity of the sample (Pawaskar et al. 2002).

Magnetic studies

Generally, the magnetic properties of materials have been dependent on factors, such as the morphology and crystal

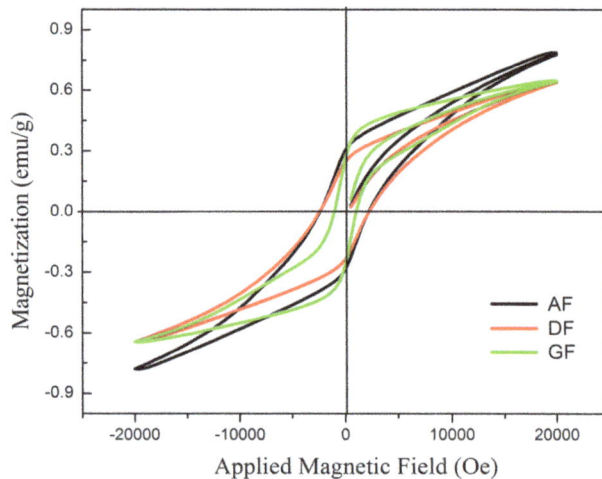

Fig. 5 Magnetization curves of AF, DF and GF at 300 K

structure (including impurities or substitutions) of the samples (Cudennec and Lecerf 2006; Subarna et al. 2007). As the size of the hematite particle decreases to the micro/nanometer scale, materials can exhibit unusual magnetic

Table 1 Antibacterial activity of iron oxide nanoparticles

Samples	Gram positive									Gram negative								
	S. aureus (mg/mL)			*A. hydrophila* (mg/mL)			*S. pyogenes* (mg/mL)			*P. aeruginosa* (mg/mL)			*E. faecalis* (mg/mL)			*E. coli* (mg/mL)		
	12.5	25	50	12.5	25	50	12.5	25	50	12.5	25	50	12.5	25	50	12.5	25	50
AF	–	–	+	–	–	++	–	–	+	–	–	+	–	–	–	–	–	++
DF	–	+	++	–	+	+++	–	+	++	–	+	++	–	–	–	–	+	+++
GF	–	–	++	–	+	++	–	–	++	–	–	++	–	–	–	–	+	++

"–" No significant change (or) inhibition

"+" 1 % increase in primary inhibition

"++" 2 % increase in primary inhibition

"+++" 3 % increase in primary inhibition

behaviors that are quite different from those of conventional bulk materials (Sorescu et al. 1999). Figure 5 shows the magnetic hysteresis loops of AF, DF and GF at 300 K in the applied magnetic field sweeping from −20 to 20 kOe. It can be seen that no saturation of the magnetization for all the three samples as a function of the field is observed up to the maximum applied magnetic field, which is similar to the reported literature (Liu et al. 2005; Niederberger et al. 2002). The hysteresis loop indicates that these hematite nanoparticles have ferromagnetic behavior, and the magnetization at the maximum applied magnetic field (M_{max}) are 0.804, 0.650 and 0.640 emu/g for samples AF, DF and GF, respectively. All the three samples possess wide open M-H loops with the remanent magnetization (M_r), the coercivity (H_c) of 0.306, 0.269, 0.267 emu/g, 2,340, 2,333 and 1,019 Oe, respectively. It is clearly seen that the GF exhibited an extremely small hysteresis loop and low coercivity compared to both AF and DF. In the case of AF and DF, both exhibited larger hysteresis loops and coercivity, further confirming that the hematite particle size of GF was smaller than those of AF and DF (Liu et al. 2008). This was also consistent with the findings from HRSEM and HRTEM. The high coercive values may be partly attributed to a surfactant-free surface because no surfactant is used in this study (Cao et al. 2006; Zeng et al. 2008).

Antimicrobial activity

Antibacterial activity of iron oxide nanoparticles was analyzed using well diffusion method against selected bacterial species. Bacterial inoculums were prepared by growing a single colony overnight in nutrient broth and adjusting the turbidity to 0.5 McFarland standards. Mueller–Hinton agar (MHA) plates were inoculated with this bacterial suspension and various volumes of iron oxide nanoparticles in 12.5, 25 and 50 mg/mL were added at the center well with a diameter of 8 mm. These plates were incubated at 37 °C for 24 h. The tetracyclin was used as a positive control for bacterial species as commercial control for the antibacterial assay. The antibacterial agents were used in their commercial presentation to prepare a stock solution adjusted to the concentration of 1.25 mg/mL^{-1} (30 μg per well) (Magaldi et al. 2004). All experiments were done in triplicates to obtain concurrent results and the standard deviations were tabulated (Table 1). The results showed a inhibition on *A. Hydrophila* and *E. coli* with iron oxide nanoparticles. Based upon the antibacterial parameters, the iron oxide nanoparticles did not show any significant changes in dose of 12.5 mg/mL, but there was a considerable increase in the inhibition with the increase in dosage. Similar phenomenon was reported for the antibacterial effect of iron oxide nanoparticles (Tran et al. 2010).

Conclusion

The Iron oxide nanoparticles were synthesized on to novel biopolymer template using the simple thermal decomposition method. As evidenced from FTIR, good interaction between hematite and polysaccharide functional groups controls the hematite crystal growth. The XRD measurements revealed a pure phase α-Fe$_2$O$_3$. HRSEM and HRTEM confirmed dumbbell nature of the iron oxide nanoparticles obtained at 800 °C and the lowest particle size was found to be 77 nm for GF. The magnetization measurements of the iron oxide nanoparticles prepared at 800 °C exhibit ferromagnetic behavior at room temperature for all the samples. The hematite nanoparticles have outstanding antimicrobial efficiency against some bacterial pathogens. It is concluded that further exploration on this field is needed to develop eco-friendly bionanomaterials for biomedicines.

References

Barron V, Torrent J (1996) Surface hydroxyl configuration of various crystal faces of hematite and goethite. J Colloid Interf Sci 177:407–410

Braun PV, Osenar P, Twardowski M, Tew GN, Stupp SI (2005) Macroscopic nanotemplating of semiconductor films with hydrogen-bonded lyotropic liquid crystals. Adv Funct Mater 15:1745–1750

Cao H, Wang G, Zhang L, Zhang YS, Zhang X (2006) Shape and magnetic properties of single-crystalline hematite (α-Fe$_2$O$_3$) nanocrystals. Chem Phys Chem 7:1897–1901

Cudennec Y, Lecerf A (2006) The transformation of ferrihydrite into goethite or hematite, revisited. J Solid State Chem 179:716–722

Ge J, Yin Y (2008) Magnetically responsive colloidal photonic crystals. J Mater Chem 18:5041–5045

Gupta AK, Gupta M (2008) Synthesis and surface engineering of iron oxide nanoparticles for biomedical applications. Biomaterials 26:3995–4021

Jitianu A, Crisan M, Meghea A, Rau I, Zaharescu M (2002) Influence of the silica based matrix on the formation of iron oxide nanoparticles in the Fe$_2$O$_3$–SiO$_2$ system, obtained by sol–gel method. J Mater Chem 12:1401–1407

Kormann C, Bahnemann DW, Hoffmann MR (1989) Environmental photochemistry: is iron oxide (hematite) an active photocatalyst? A comparative study: α-Fe$_2$O$_3$, ZnO, TiO$_2$. J Photo chem Photo biol A Chem 48:161–169

Laurent S, Forge D, Port M, Roch A, Robic C, Elst LV, Muller RN (2008) Magnetic iron oxide nanoparticles: synthesis, stabilization, vectorization, physicochemical characterizations, and biological applications. Chem Rev 108:2064–2110

Liu XM, Fu SY, Xiao HM, Huang CJ (2005) Preparation and characterization of shuttle-like α Fe$_2$O$_3$ nanoparticles by supermolecular template. J Solid State Chem 178:2798–2803

Liu SL, Zhang LN, Zhou JP, Xiang JF, Sun JT, Guan JG (2008) Fiber like Fe$_2$O$_3$ macroporous nanomaterials fabricated by calcinating regenerate cellulose composite fibers. Chem Mater 20:3623–3628

Ma XF, Chang PR, Yang JW, Yu JG (2009) Preparation and properties of glycerol plasticized-pea starch/zinc oxide-starch bionanocomposites. Carbohydr Polym 75:472–478

Magaldi S, Mata-Essayag S, Hartung de Capriles C, Perez C, Colella MT, Olaizola C, Ontiveros Y (2004) Well diffusion for antifungal susceptibility testing. Int J Infect Dis 8:39–45

Niederberger M, Krumeich F, Hegetschweiler K, Nesper R (2002) An iron polyolate complex as a precursor for the controlled synthesis of monodispersed iron oxide colloids. Chem Mater 14:78–82

Pawaskar NR, Sathaye SD, Bhadbhade MM, Patil KR (2002) Applicability of liquid–liquid interface reaction technique for the preparation of zinc sulfide nano particulate thin films. Mater Res Bull 37:1539–1545

Rochas C, Lahaye M (1989) Average molecular-weight and molecular-weight distribution of agarose and agarose-type polysaccharides. Carbohydr Polym 10:289–298

Sorescu M, Brand RA, Tarabasanu DM, Diamandescu L (1999) The crucial role of particle morphology in the magnetic properties of haematite. J Appl Phys 85:5546–5548

Sotiropoulou S, Sierrasastre Y, Mark SS, Batt CA (2008) Biotemplated nanostructured materials. Chem Mater 20:821–834

Sreeram KJ, Nidhin M, Nair BU (2009) Synthesis of aligned hematite nanoparticles on chitosan–alginate film. Colloids Surf B 71:260–267

Subarna M, Soumen D, Kalyan M, Subhadra C (2007) Synthesis of a α-Fe$_2$O$_3$ nanocrystal in its different morphological attributes: growth mechanism, optical and magnetic properties. Nanotechnology 18:275608

Tran N, Mir A, Mallik D, Sinha A, Nayar S, Webster TJ (2010) Bactericidal effect of iron oxide nanoparticles on *Staphylococcus aureus*. Int J Nano med 5:277–283

Vazquez CV, Blanco MC, Quintela MA, Sanchez RD, Rivas J, Oseroff SB (1998) Characterization of La$_{0.67}$Ca$_{0.33}$MnO$_{3\pm\delta}$ particles prepared by the sol–gel route. J Mater Chem 8:991–1000

Zeng SY, Tang KB, Li TW, Liang ZH, Wang D, Wang YK, Qi YX, Zhou WW (2008) Facile route for the fabrication of porous hematite nanoflowers: its synthesis, growth mechanism, application in the lithium ion battery, and magnetic and photocatalytic properties. J Phys Chem C 112:4836–4843

Zhao W, Gu J, Zhang L, Chen H, Shi J (2005) Fabrication of uniform magnetic nanocomposite spheres with a magnetic core/mesoporous silica shell structure. J Am Chem Soc 127:8916–8917

Zhao W, Chen H, Li Y, Li L, Lang M, Shi J (2008) Uniform rattle-type hollow magnetic mesoporous spheres as drug delivery carriers and their sustained-release property. Adv Funct Mater 18:2780–2788

Zhou H, Wong SS (2008) A facile and mild synthesis of 1-D ZnO, CuO and α-Fe$_2$O$_3$ nanostructures and nanostructured arrays. ACS Nano 2:944–958

22

Mixed convection flow along an inclined permeable plate: effect of magnetic field, nanolayer conductivity and nanoparticle diameter

Puneet Rana · O. Anwar Bég

Abstract Steady, two-dimensional mixed convection boundary layer flow of an incompressible Al_2O_3–water nanofluid along an inclined permeable plate in the presence of transverse magnetic field has been examined numerically. The governing equations (Boussinesq approximation) with associated boundary condition are solved using FEM for nanofluid containing spherical-shaped nanoparticles having volume fraction ranging from 1 to 4 %. Static-based model for calculating the effective thermal conductivity at 300 K, proposed by Leong et al. (J Nanopart Res 8:245–254, 2006) and Murshed et al. (Int J Therm Sci 47:560–568, 2008) has been implemented. Effect of various pertinent parameters with different classical and experimental models for effective dynamic viscosity is discussed.

Keywords Nanofluid · Inclined plate · Mixed convection · Static mechanism · FEM

Introduction

Nowadays, the heat transfer enhancement is one of the most challenging problems in different industrial applications and engineering systems. Choi (1995) was the first person to introduce fluids composed of nanometer-sized particles dispersed in a base fluid which are called as

P. Rana (✉)
Department of Mathematics, Jaypee Institute of Information Technology, Noida, Uttar Pradesh, India
e-mail: puneetranaiitr@gmail.com

O. A. Bég
Gort Engovation-Propulsion, Nanomechanics and Biophysics, Southmere Avenue, Bradford BD73NU, UK

nanofluids and disclosed the various advantages of the application of nanofluids, such as improved heat transfer, minimal clogging, size reduction of the heat transfer system, microchannel cooling, and miniaturization of systems in that study.

The classical theoretical approach for the conductivity measurement solid–fluid suspensions (Maxwell 1873; Hamilton and Crosser 1962) could not justify the experimental results in which the large enhancement of the thermal conductivity in case of low concentrations of nanoparticles. To overcome this limitations of classical theory, the new mechanisms have been first proposed by Wang et al. (2003) for the enhanced thermal transport in case of nanofluids, such as particle motion, surface action, and electro-kinetic effects. They suggested that nanoparticle size and nanolayer thermal conductivity are important parameters for enhancing the thermal conductivity of nanofluids. Later on, Xuan and Li (2000) recommended several possible mechanisms for enhanced thermal conductivity of nanofluids, such as the increased surface area of nanoparticles, particle–particle collisions and the dispersion of nanoparticles. Based on static mechanics, several attempts have been made to formulate appropriate effective thermal conductivity (Xue and Xu 2005; Leong et al. 2006; Tillman and Hill 2007; Murshed et al. 2008). Koo and Kleinstreuer (2004) stated that the effective thermal conductivity is composed of the particle's conventional static part and a Brownian motion part. Moreover, the Brownian motion effect was found to become more effective at higher temperature as observed experimentally.

The electrically conducting fluid flow past a heated surface under magnetic effects has attracted lot of researchers due to its many engineering applications such as in cooling of nuclear reactors, petroleum industries,

MHD power generators, plasma studies etc. Abdelkhalek (2006) investigated the effects of mass transfer on steady two-dimensional laminar MHD mixed convection. Aydin and Kaya (2009a) discussed the magnetic field effect on about a permeable vertical flat plate. The combined effect of inclination and magnetic field in mixed convection flow has been discussed by Aydin and Kaya (2009b).

In case of nanofluid, Hamad (2011) investigated the effect of magnetic field effect on natural convection flow of a nanofluid over stretching sheet analytically. Later on, many researchers discussed the effect of magnetic field on different geometries with classical assumptions in the years 2011–2012. Recently, effect of nanofluid properties on magnetohydrodynamic pump (MHD) has been investigated by Shahidian et al. (2012). In their study, effect of thermal conductivity based on nanolayer concept has been taken into consideration for the case of Al_2O_3 nanofluid.

In the present chapter, we have simulated mixed convection flow of nanofluid along an inclined plate, using the static model introduced by Leong et al. (2006) and Murshed et al. (2008) for thermal conductivity of nanofluids at 300 K. The effect of Brownian motion is assumed to be neglected at this moderate temperature. Both classical and experimental correlations for the viscosity of Al_2O_3–water nanofluid are implemented. The objective of the present chapter is to study the effect of magnetic field, nanoparticle diameter, nanolayer conductivity to base fluid conductivity ratio, inclination angle and nanoparticle volume fraction on the steady boundary layer nanofluid flow and heat transfer characteristics.

Mathematical model

In the present analysis, the steady, two-dimensional, incompressible laminar boundary layer flow of a Al_2O_3–water nanofluid past an inclined semi-infinite permeable flat plate with an angle δ has been taken. The flow is considered in the direction of x axis along the inclined flat plate whereas the y axis is taken normal to the plate. The surface of the flat plate is maintained at a constant temperature (T_w) higher than the constant temperature (T_∞) of the ambient nanofluid. A magnetic field having uniform strength B_0 is applied in the y direction, perpendicular to the plate and viscous dissipation term is ignored. Under the assumption of small magnetic Reynolds number, the induced magnetic field is considered to be zero. The thermal equilibrium and no slip have also been assumed between nanoparticles and base fluid. The geometry of the flow configuration is shown in Fig. 1. The boundary layer and Boussinesq approximations are assumed to be valid. The basic equations for nanofluids can be written as:

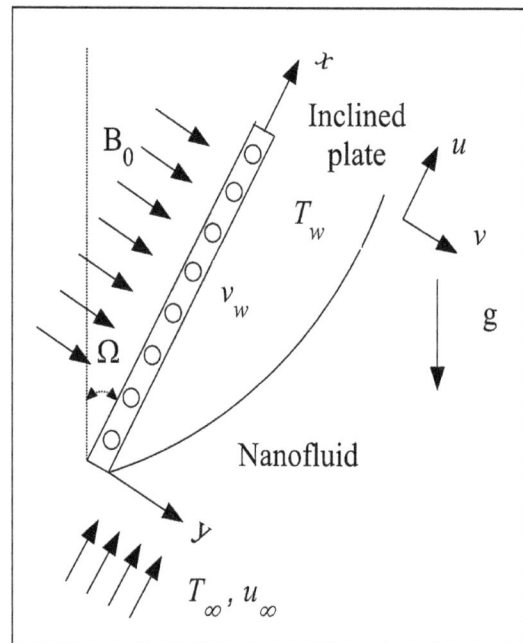

Fig. 1 Physical model and coordinate system

$$\frac{\partial u}{\partial x} + \frac{\partial v}{\partial y} = 0, \tag{1}$$

$$u\frac{\partial u}{\partial x} + v\frac{\partial u}{\partial y} = \frac{1}{\rho_{nf}}\left[\mu_{nf}\frac{\partial^2 u}{\partial y^2} + (\rho\beta)_{nf}g(T - T_\infty)\cos(\delta)\right.$$

$$\left. - \sigma_{nf}B_0^2(u - u_\infty)\right] \tag{2}$$

$$u\frac{\partial T}{\partial x} + v\frac{\partial T}{\partial y} = \frac{1}{(\rho C_p)_{nf}}\left[k_{nf}\frac{\partial^2 T}{\partial y^2}\right], \tag{3}$$

subject to the boundary conditions

$$\begin{aligned} x > 0 \ \ y = 0 \ \ &T = T_w \ \ u = 0 \ \ v = \pm V_w, \\ x = 0 \ \ y > 0 \ \ &T = T_\infty \ \ u = u_\infty, \\ y \to \infty \ \ &T = T_\infty \ \ u = u_\infty. \end{aligned} \tag{4}$$

Effective viscosity models

The three different viscosity models for nanofluids containing alumina particles (Al_2O_3) are given below:

Model I (classical)

Model (I) attributed to Brinkman (1952) is as follows:

$$\frac{\mu_{nf}}{\mu_f} = \frac{1}{(1 - \phi)^{2.5}}. \tag{5}$$

Wang et al. (2003) experimentally observed that the Al_2O_3–water mixture shows an increase of viscosity between 20 and 30 % for 3 %vol. Al_2O_3 solutions as compared to water alone.

Model II

The experimental correlation for Al_2O_3–water (Model II) is as follows (Wang et al. 2003):

$$\frac{\mu_{nf}}{\mu_f} = 1 + 7.3\phi + 123\phi^2. \tag{6}$$

The Al_2O_3–water viscosity observed by Pak and Cho (1998) was higher than that of water as compared to Wang et al. (2003).

Model III

Pak and Cho (1998) correlation for the viscosity of Al_2O_3–water nanofluid is as follows:

$$\frac{\mu_{nf}}{\mu_f} = 1 + 39.11\phi + 533.9\phi^2. \tag{7}$$

The physical properties of the nanofluid are as Oztop and Abu-Nada (2008) and Tiwari and Das (2007):

$$\rho_{nf} = (1-\phi)\rho_f + \phi\rho_p,$$
$$(\rho\beta)_{nf} = (1-\phi)(\rho\beta)_f + \phi(\rho\beta)_p, \tag{8}$$
$$(\rho C_p)_{nf} = (1-\phi)(\rho C_p)_f + \phi(\rho C_p)_p.$$

Effective thermal conductivity model

The existing classical models (Maxwell 1873; Hamilton and Crosser 1962) are found to be incapable to anticipate the anomalously high thermal conductivity of nanofluids. This is due to reason that classical models do not incorporate the effects of particle size, the particle/liquid interfacial layer and distribution which are considered as significant mechanisms for enhanced thermal conductivity of nanofluids.

Effective thermal conductivity for spherical-shaped nanoparticles can be incorporated from the following expression (Leong et al. 2006; Murshed et al. 2008):

$$k_{nf} =$$
$$\frac{k_{lr}\phi(k_p - k_{lr})[2\gamma_2^3 - \gamma_1^3 + 1] + (k_p + 2k_{lr})\gamma_2^3[\phi\gamma_1^3(k_{lr} - k_f) + k_f]}{\gamma_2^3(k_p + 2k_{lr}) - (k_p - k_{lr})\phi(\gamma_2^3 + \gamma_1^3 - 1)}, \tag{9}$$

where, $\gamma_1 = 1 + (2 * \hat{h}/d_p)$, $\gamma_2 = 1 + (\hat{h}/d_p)$, $\hat{h}$ is the interfacial thickness at the surface. The interfacial layer thermal conductivity can be represented as $k_{lr} = \omega k_f$, where $\omega > 1$ is an empirical parameter (Leong et al. 2006; Murshed et al. 2008). Thus, we get,

$$\frac{k_{nf}}{k_f} =$$
$$\frac{\omega\phi(k_p - \omega k_f)[2\gamma_2^3 - \gamma_1^3 + 1] + (k_p + 2\omega k_f)\gamma_2^3[\phi\gamma_1^3(\omega - 1) + 1]}{\gamma_2^3(k_p + 2\omega k_f) - (k_p - \omega k_f)\phi(\gamma_2^3 + \gamma_1^3 - 1)}, \tag{10}$$

Hashimoto et al. (1980) demonstrated the interfacial layer thickness on the basis of electron density at interface, which is given as $\hat{h} = \sqrt{2\pi}\hat{\sigma}$, where $\hat{\sigma}$ is a parameter which quantifies interfacial boundary diffusioness and for spherical particle its value is 1 nm.

Introducing the following non-dimensional variables:

$$\xi = \frac{x}{L}, \quad \psi(x,y) = (v_f u_\infty x)^{1/2} f(\xi, \eta),$$
$$\eta = y\left(\frac{u_\infty}{v_f x}\right)^{1/2}, \theta(\eta) = \frac{T - T_\infty}{T_w - T_\infty}, \tag{11}$$

where $\psi(x,y)$ represents free stream function that satisfies the Cauchy–Riemann Equation with $u = \frac{\partial\psi}{\partial y}$ and $v = -\frac{\partial\psi}{\partial x}$, the velocity components can be reduced as:

$$u = u_\infty f', v = -\left(\frac{v_f u_\infty}{x}\right)^{1/2}\left\{\frac{1}{2}f + \xi\frac{\partial f}{\partial\xi} - \frac{\eta}{2}f'\right\}. \tag{12}$$

The transformed momentum and energy equations (2) and (3) can be written as:

$$\frac{1}{(1-\phi+\phi\rho_p/\rho_f)}\left\{ \begin{array}{c} \left(\frac{\mu_{nf}}{\mu_f}\right)f''' + [1 - \phi + \phi(\rho\beta)_p/(\rho\beta)_f]Ri\xi\theta\cos(\delta) \\ -M\xi(f' - 1) \end{array} \right\}$$
$$+ \frac{1}{2}ff'' = \xi\left(f'\frac{\partial f'}{\partial\xi} - f''\frac{\partial f}{\partial\xi}\right), \tag{13}$$

$$\frac{1}{(1-\phi+\phi(\rho C_p)_p/(\rho C_p)_f)}\left[\frac{1}{Pr}\left(\frac{k_{nf}}{k_f}\right)\theta''\right] + f\theta' = \xi\left(f'\frac{\partial\theta}{\partial\xi} - \theta'\frac{\partial f}{\partial\xi}\right), \tag{14}$$

The transformed boundary conditions are:

$$f(\xi,0) + 2\xi\frac{\partial f}{\partial\xi} = f_w\xi^{1/2}, \quad f'(\xi,0) = 0, \quad \theta(\xi,0) = 1,$$
$$f'(\xi,\infty) = 1, \quad \theta(\xi,\infty) = 0. \tag{15}$$

where prime denotes the differentiation w.r.t. η.

The important parameters, dictating the flow dynamics are defined by:

$$Pr = \frac{\mu C_p}{k} = \frac{v_f}{\alpha}, \quad Ri = \frac{Gr}{Re^2}, \quad Gr = \frac{g_e\beta(T_w - T_\infty)L^3}{v_f^2},$$
$$Re = \frac{u_\infty L}{v_f}, \quad Ha = \frac{\sigma B_0^2 L^2}{\mu}, \quad M = Ha/Re, f_w = -2\frac{L}{v_f}V_w Re^{-1/2}. \tag{16}$$

The local skin friction coefficient and the local Nusselt number are given as:

$$Cf_x = \mu_{nf}\left(\frac{\partial u}{\partial y}\right)_{y=0}/\rho_f u_\infty^2 = (Re_x)^{-1/2}\frac{\mu_{nf}}{\mu_f}f''(0),$$
$$C_f = (Re_x)^{1/2}Cf_x = \frac{\mu_{nf}}{\mu_f}f''(0), \tag{17}$$

$$Nu_x = -\left(\frac{k_{nf}}{k_f}\right)\left(\frac{\partial T}{\partial y}\right)_{y=0}/(T_w - T_\infty) = -(Re_x)^{1/2}\left(\frac{k_{nf}}{k_f}\right)\theta'(0),$$
$$Nu = (Re_x)^{-1/2}Nu_x = -\left(\frac{k_{nf}}{k_f}\right)\theta'(0), \tag{18}$$

where C_f and Nu are modified skin friction and Nusselt number, respectively.

Finally, the average Nusselt number is determined from:

$$\mathrm{Nu}_{\mathrm{avg}} = \left(\frac{1}{L}\right) \int_0^L \mathrm{Nu}(\xi)\,\mathrm{d}\xi, \tag{19}$$

where, L = characteristic length of plate. Simpson's 1/3 rule of integration is implemented.

Finite element solution

Finite element method (FEM) was basically developed in reference to structural problems, but now it has been used as a mathematical tool for solving the linear and non-linear ordinary or partial differential equations as well as integral equations. The finite element method not only overcomes the shortcoming of the traditional variational methods, it is also endowed with the features of an effective computational technique. This method is so general that it can be applied to a wide variety of engineering problems, including heat transfer (Bhargava and Rana 2011; Rana et al. 2012), fluid mechanics (Rana and Bhargava 2011; Rana et al. 2013), rigid body dynamics (Dettmer 2006), solid mechanics (Hansbo and Hansbo 2004) and many other fields.

Finite element formulation

Let the domain be divided into quadratic rectangular elements Ω_e. The finite element model may be obtained by substituting finite element approximations of the form:

$$f = \sum_{j=1}^{9} f_j N_j(\xi,\eta), \quad h = \sum_{j=1}^{9} h_j N_j(\xi,\eta), \quad \theta = \sum_{j=1}^{9} \theta_j N_j(\xi,\eta) \tag{20}$$

with $w_1 = w_2 = w_3 = N_j$ $(j = 1, 2, 3)$ where $N_j(\xi,\eta)$ are the quadratic interpolation functions for a rectangular element Ω_e as follows: The finite element model of the equations thus formed is given by

$$\begin{bmatrix} [K^{11}] & [K^{12}] & [K^{13}] \\ [K^{21}] & [K^{22}] & [K^{23}] \\ [K^{31}] & [K^{32}] & [K^{33}] \end{bmatrix} \begin{bmatrix} \{f\} \\ \{h\} \\ \{\theta\} \end{bmatrix} = \begin{bmatrix} \{b^1\} \\ \{b^2\} \\ \{b^3\} \end{bmatrix} \tag{21}$$

where $[K^{mn}]$ and $[b^m]$ $(m, n = 1, 2, 3)$ are the matrices of order 3×3 and 3×1, respectively, and therefore each element matrix is of the order 9×9. All the matrices are defined as follows:

$$K_{ij}^{11} = \int_{\Omega_e} N_i \frac{\partial N_j}{\partial \eta} \,\mathrm{d}\Omega_e, \quad K_{ij}^{12} = -\int_{\Omega_e} N_i N_j \,\mathrm{d}\Omega_e, K_{ij}^{13} = 0,$$

$$K_{ij}^{21} = \int_{\Omega_e} \left\{ N_i \xi \frac{\partial \bar{h}}{\partial \eta} \frac{\partial N_j}{\partial \xi} + \left(\frac{1}{2}\right) N_i N_j \frac{\partial \bar{h}}{\partial \eta} \right\} \mathrm{d}\Omega_e,$$

$$K_{ij}^{22} = -\frac{1}{(1 - \phi + \phi \rho_s/\rho_f)} \left(\frac{\mu_{nf}}{\mu_f}\right) \int_{\Omega_e} \frac{\partial N_i}{\partial \eta} \frac{\partial N_j}{\partial \eta} \,\mathrm{d}\Omega_e$$
$$- \int_{\Omega_e} N_i \xi N_j \frac{\partial \bar{h}}{\partial \xi} \,\mathrm{d}\Omega_e - M \int_{\Omega_e} N_i \xi N_j \,\mathrm{d}\Omega_e$$

$$K_{ij}^{23} = \left(\frac{1 - \phi + \phi(\rho\beta)_s/(\rho\beta)_f}{1 - \phi + \phi\rho_s/\rho_f}\right) \mathrm{Ri} \int_{\Omega_e} N_i \xi N_j \,\mathrm{d}\Omega_e,$$

$$K_{ij}^{31} = \int_{\Omega_e} N_i \frac{\partial \bar{\theta}}{\partial \eta} N_j \,\mathrm{d}\Omega_e + \int_{\Omega_e} \xi N_i \frac{\partial \bar{\theta}}{\partial \eta} \frac{\partial N_j}{\partial \xi} \,\mathrm{d}\Omega_e$$

$$K_{ij}^{32} = -\int_{\Omega_e} N_i \xi N_j \frac{\partial \bar{\theta}}{\partial \xi} \,\mathrm{d}\Omega_e$$

$$K_{ij}^{33} = \frac{1}{\left(1 - \phi + \phi(\rho C_p)_s/(\rho C_p)_f\right)} \left(\frac{1}{\mathrm{Pr}} \left(\frac{k_{nf}}{k_f}\right) \int_{\Omega_e} \frac{\partial N_i}{\partial \eta} \frac{\partial N_j}{\partial \eta} \,\mathrm{d}\Omega_e\right)$$

$$b_i^1 = 0, b_i^2 = -\frac{1}{(1 - \phi + \phi\rho_s/\rho_f)}$$
$$\times \left[\left(\frac{\mu_{nf}}{\mu_f}\right) \oint_{\Gamma_e} N_i q_{n_2} \,\mathrm{d}s + M \oint_{\Gamma_e} \xi N_i q_{n_2} \,\mathrm{d}s\right]$$

$$b_i^3 = -\frac{1}{\mathrm{Pr}} \frac{1}{\left(1 - \phi + \phi(\rho C_p)_s/(\rho C_p)_f\right)} \left(\frac{k_{nf}}{k_f}\right) \oint_{\Gamma_e} N_i q_{n_3} \,\mathrm{d}s,$$

where

$$\bar{h} = \sum_{i=1}^{9} \bar{h}_i N_i, \quad \bar{f} = \sum_{i=1}^{9} \bar{f}_i N_i, \quad \bar{\theta} = \sum_{i=1}^{9} \bar{\theta}_i N_i,$$

$$\frac{\partial \bar{h}}{\partial \eta} = \sum_{i=1}^{9} \bar{h}_i \frac{\partial N_i}{\partial \eta}, \quad \frac{\partial \bar{h}}{\partial \xi} = \sum_{i=1}^{9} \bar{h}_i \frac{\partial N_i}{\partial \xi}, \quad \frac{\partial \bar{\theta}}{\partial \xi} = \sum_{i=1}^{9} \bar{\theta}_i \frac{\partial N_i}{\partial \xi}$$

Since the interpolation functions are easily derivable for a rectangular elements. Thus, we transform the finite element integral statements defined over quadrilaterals to a rectangle. Similar procedure has been followed for the rectangular element. For example, a nine-node quadratic elements with each element mapped using isoparametric mapping (Reddy 1985; Rana et al. 2013) from $\xi - \eta$ domain (Ω_e) to a unit square $\xi' - \eta'$ domain (Ω^0) is shown in Fig. 2. Thus, the integral is transferred from problem coordinate system (ξ, η) to specific coordinate system (ξ', η') (master element).

$$\xi = \sum_{i=1}^{9} \xi_i N_i(\xi', \eta'), \quad \eta = \sum_{i=1}^{9} \eta_i N_i(\xi', \eta')$$

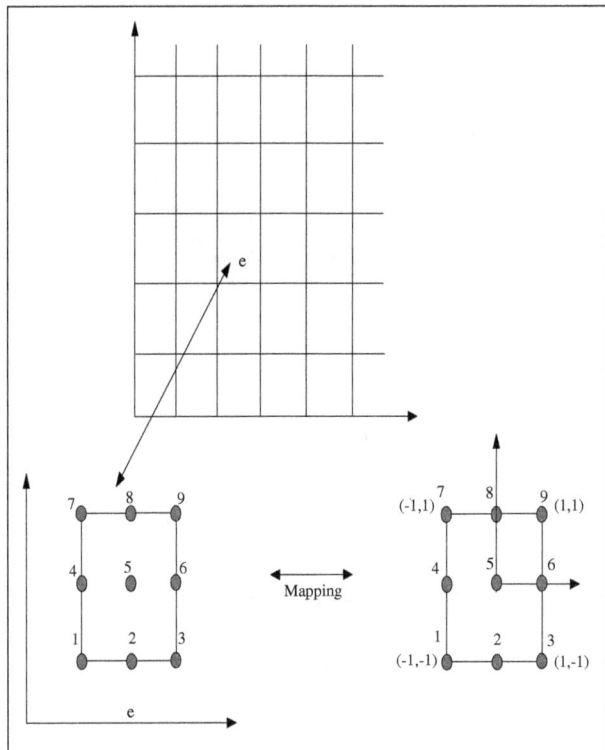

Fig. 2 Mapping of master rectangular element to a rectangular element of a finite element mesh

where $N_i(\xi', \eta')$ are the local biquadratic basis functions on the (ξ', η') domain. The integrals can be evaluated in (ξ', η') domain using following relationships:

$$\left\{ \begin{matrix} \frac{\partial N_i}{\partial \xi'} \\ \frac{\partial N_i}{\partial \eta'} \end{matrix} \right\} = [J] \left\{ \begin{matrix} \frac{\partial \psi_i}{\partial \xi} \\ \frac{\partial \psi_i}{\partial \eta} \end{matrix} \right\} \text{ where}$$

$$J = \begin{bmatrix} \frac{\partial \xi}{\partial \xi'} & \frac{\partial \eta}{\partial \xi'} \\ \frac{\partial \xi}{\partial \eta'} & \frac{\partial \eta}{\partial \eta'} \end{bmatrix} \text{ is the Jacobian matrix}$$

$$\left\{ \begin{matrix} \frac{\partial N_i}{\partial \xi} \\ \frac{\partial N_i}{\partial \eta} \end{matrix} \right\} = [J]^{-1} \left\{ \begin{matrix} \frac{\partial N_i}{\partial \xi'} \\ \frac{\partial N_i}{\partial \eta'} \end{matrix} \right\}, \ dA = J d\xi d\eta$$

The necessary condition for $[J]^{-1}$ is that the determinant J, called Jacobian J should be non-zero at every point (ξ', η').

Results and discussion

Comprehensive numerical computations are conducted for various values of the parameters that describe the flow characteristics and the results are illustrated graphically. A numerical study is made to see the effect of different controlling parameter on Al_2O_3–water nanofluid with thermophysical properties shown in Table 1. Moreover,

Table 1 Thermophysical properties of water and nanoparticles

	ρ (kg/m^3)	C_p (J/kg K)	k (W/m K)	$\beta \times 10^{-5}$ (/K)
Pure water	997.1	4,179	0.613	21
Alumina (Al_2O_3)	3,970	765	40	0.85

Table 2 Grid independence study for different models of viscosity keeping Ri = 1, $M = 1$, $d_p = 5$ nm, Pr = 6.2, $\omega = 10$, $\phi = 0.04$

Biquadratic elements	Nu_{avg}			
	Pure water	Model I	Model II	Model III
10×10	0.7910	1.3820	1.3314	1.2115
10×20	0.8066	1.4096	1.3581	1.2355
20×20	0.8379	1.4653	1.4117	1.2837
20×40	0.8442	1.4765	1.4225	1.2934
30×40	0.8487	1.4839	1.4297	1.2999
40×50	0.8533	1.4933	1.4407	1.3109
40×80	0.8569	1.4989	1.4442	1.3128
50×80	0.8574	1.4993	1.4449	1.3135

Table 3 Comparison of results for Nusselt number $(-\theta'(0,0))$ for various Pr keeping $Ri = 0$ and $\phi = 0$

Pr	Nield and Kuznetsov (2003)	Chamkha et al. (2003)	Aydin and Kaya (2009a)	Present results
0.01	–	0.051830	0.051437	0.051301
0.1	0.1580	0.142003	0.148123	0.147901
1	0.3320	0.332173	0.332000	0.331980
10	0.7300	0.728310	0.727801	0.727800
100	1.5700	1.572180	1.573141	1.573140

Table 4 Comparison of results for reduced Nusselt number $-\theta'(0,0)$ for various Pr at $Ri = 0$, $\phi = 0$

Pr	ξ	Yih (1999)	Chamkha et al. (2003)	Watanabe and Pop (1994)	Aydin and Kaya (2009a)	Present results
0.733	0.0	0.297526	0.29760	0.29755	0.29753	0.297506
	0.5	0.357022	0.35704	0.35699	0.35709	0.356520
	1.0	0.382588	0.38319	0.38336	0.38363	0.389246
	1.5	0.398264	0.39998	0.39959	0.40012	0.400020
	2.0	0.409168	0.40945	0.41091	0.41134	0.411600
1.0	0.0	0.332057	0.33217	0.33206	0.33206	0.332037
	0.5	0.402864	0.40310	0.40280	0.40259	0.402475
	1.0	0.433607	0.43390	0.43446	0.43460	0.434221
	1.5	0.452634	0.45280	0.45413	0.45302	0.454270
	2.0	0.465987	0.46611	0.46798	0.46612	0.466261

velocity and temperature profiles of nanofluid are compared with pure water.

An extensive mesh testing has been conducted to ensure a grid- independence solution of given boundary value problem. The present code has been tested for grid independence by calculating the average Nusselt number on the plate. Different combinations of bi- quadratic elements for both pure water and Al_2O_3–water keeping $Ri = 1$, $M = 1$, $d_p = 5$ nm, $Pr = 6.2$, $\omega = 10$, $\phi = 0.04$ are also explored as shown in Table 2. In the case of Al_2O_3–water, average Nusselt number (Nu_{avg}) of different models has been also shown. Further, the validity of the present numerical code has been assured for a limiting case. Tables 3 and 4 show that an excellent correlation has been achieved with the earlier results of Watanabe and Pop (1994), Yih (1999), Chamkha et al. (2003) and Aydin and Kaya (2007).

For obtaining the numerical solutions, the suitable guess value of ξ_{max} (length of the plate) and η_{max} (length of the domain) has been chosen which satisfy all boundary conditions. It has been observed that for the moderate values of η_{max} (>5.0), there is no appreciable effect on the results. Therefore, for computational purpose infinity has been set as 5.0. However, for ξ_{max}, the results are obtained even for large ξ_{max} (up to 8); but for demonstration purpose, the results are shown only for $0 \leq \xi \leq 2.0$.

The entire flow domain contains 13,041 grid points. At each node, three functions are to be evaluated; hence after assembly of the element equations, we obtain a system of 39,123 equations which are non-linear. Therefore, an iterative scheme has been employed in the solution. After imposing the boundary conditions, a system of 38,818 equations have been solved with an accuracy of 10^{-4}. The iterative process is terminated when the following condition is satisfied:

$$\sum_{i,j} \left| \Theta_{i,j}^m - \Theta_{i,j}^{m-1} \right| \leq 10^{-4} \tag{22}$$

where, Θ stands for either f, h or θ, and m denotes the iterative step.

Gaussian quadrature is implemented for solving the integrations. Excellent convergence has been achieved.

In Fig. 3, the effective dynamic viscosity of Al_2O_3–water nanofluid has been plotted against nanoparticle volume fraction for different models. It has been noted from correlation that viscosity measured from experiments (Wang et al. 1999; Pak and Cho 1998) is far different from classical predictions (Maxwell 1873; Hamilton and Crosser 1962). Figure 4 shows the characteristic of the effective thermal conductivity, which is a function of nanolayer to base fluid conductivity ratio ($\omega = k_{lr}/k_f$) and nanoparticle diameter (d_p). ω has relatively high effect for small

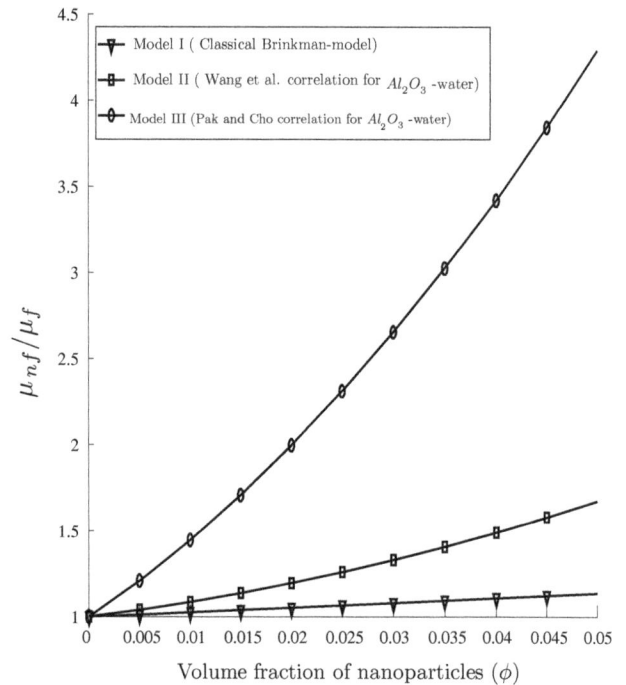

Fig. 3 Comparison of viscosity models

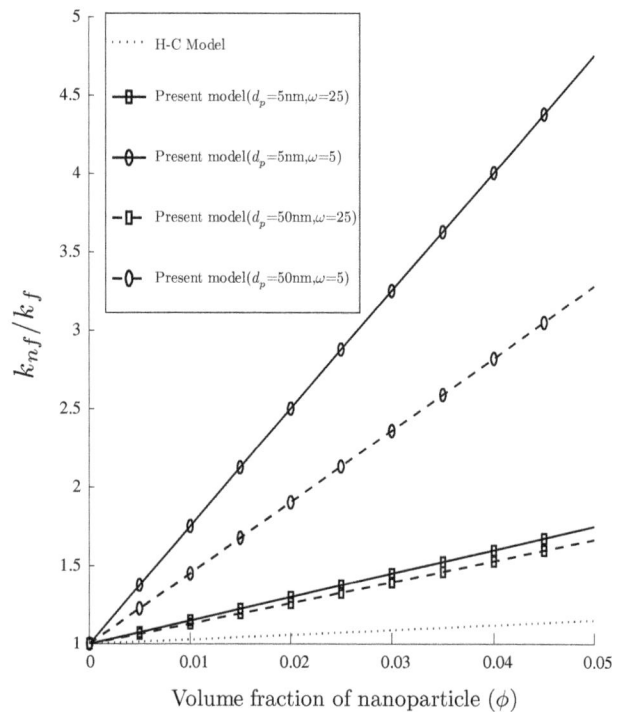

Fig. 4 Comparison of thermal conductivity models

nanoparticle diameters. Consequently, the presence of nanoparticles has a strong effect on thermal conductivity of the nanofluid for small nanoparticle diameters and high nanolayer to base fluid conductivity ratio. Thus, compared with the traditional H–C model, the present model (Leong et al. 2006; Murshed et al. 2008) shows a better flexibility in predicting the heat transfer characteristic.

Selected computations are presented in Figs. 5, 6, 7, 8, 9, 10, 11 and 12. The following ranges of the main parameters are considered: $0.01 \leq \mathrm{Ri} \leq 10.0$, $1.0 \leq M \leq 10.0$, $1 \leq d_p(\mathrm{nm}) \leq 50$, $5.0 \leq \omega(k_{lr}/k_f) \leq 50.0$, Models I–III, $0 \leq \phi \leq 0.04$, $Pr = 6.2$, Al$_2$O$_3$–water. In all cases, default values of the governing parameters are: Ri = 1.0, Pr = 6.2, M = 1.0, d_p = 5 nm, ω = 10, Al$_2$O$_3$–water with Model I, unless otherwise stated. The thermophysical properties of the nanofluid at 300 K are given in Table 2 (Hamad 2011).

Effect of magnetic field parameter

The effect of magnetic field parameter (M) on velocity profile $f'(\xi, \eta)$ and temperature profile $\theta(\xi, \eta)$ is shown in Fig. 5a, b, respectively. The effect of the magnetic field parameter (M) has been explained from Eq. (2), the sign of the last term in the right hand side of Eq. (2). For the forced convection regime, this term $\frac{\sigma B_0^2}{\rho}(u - u_\infty)$ will always be negative since $u_\infty > u$. Thus, a force will aid in the direction of main flow. On the other hand, in case of free convection regime, when $Ri \rightarrow \infty$, buoyancy-driven flow will dominate the external flow regime, hence the above term will be positive because $u_\infty < u$, which will diminish the main flow regime. This term has been resolved into two components: The first component, $\frac{\sigma B_0^2}{\rho} u_\infty$, represents the imposed pressure force, whereas the second component, $\frac{\sigma B_0^2}{\rho} u$, represents the Lorentz force which slows down the fluid motion in the boundary layer region. When the imposed pressure force overcomes the Lorentz force ($u_\infty < u$), the effect of the magnetic parameter increases the velocity. Similarly, when the Lorentz force which slows the fluid motion dominates the imposed pressure force ($u_\infty < u$), the effect of the magnetic parameter (M) decreases velocity flow and hence it decreases both momentum and thermal boundary layer thickness (i.e., increases the velocity and temperature gradient at the wall). Similar results have been observed by Aydin and Kaya (2009a). Moreover, further addition of nanoparticles (increase in volume fraction) increases both velocity and temperature gradient. As magnetic parameter (M) is multiplied by ξ, which leads to increase in velocity and temperature in the boundary layer, as we go away from the leading edge.

The effect of the magnetic parameter (M) on the modified skin friction and Nusselt number is shown in Fig. 5c,

d, respectively. Both the skin friction and the Nusselt number increase with an increase in the magnetic parameter (M). This is associated with an increase in the magnitude of wall velocity and temperature gradients.

Effect of nanoparticle diameter

The effect of the nanoparticle diameter on velocity profile $f'((\xi, \eta))$ and temperature profile $(\theta(\xi, \eta))$ is shown in Fig. 6a, b, respectively. The addition of nanoparticles decreases the momentum boundary layer and increases the thermal boundary layer, however, this behavior is more pronouncing for small size nanoparticles. The small size nanoparticle ($d_p = 1$ nm) increases the temperature which is simply due to the increase of thermal conductivity. Figure 6c, d depicts the effect of the nanoparticle diameter (d_p) on the skin friction and the Nusselt number, respectively. Both the skin friction and the Nusselt number decrease with the increase in nanoparticle diameter. This is due to potential instability of nanofluid with larger nanoparticles. The settling velocity of nanoparticles (V_g) can be calculated from Stokes law (only accounts for gravitational and buoyant forces) is as follows:

$$V_g = \frac{2}{9}\left(\frac{\rho_p - \rho_{nf}}{\mu_{nf}}\right)(d_p/2)^2 g, \tag{23}$$

where g is the acceleration due to gravity. It can be noticed that from Eq. (23), that the stability of a suspension (defined by lower settling rates) improves if: (a) the viscosity of the suspension (μ_{nf}) is high (b) the density of the solid material (ρ_p) is close to that of the fluid (ρ_{nf}), and (c) the particle diameter (d_p) is small.

Effect of nanolayer thermal conductivity and viscosity models

The effect of nanolayer to base fluid conductivity ratio (ω) on velocity profile $(f'(\xi, \eta))$ and temperature profile $(\theta(\xi, \eta))$ is shown in Fig. 7a, b, respectively. As we have already mentioned that ω (> 1) is an empirical parameter which depends on the orderness of fluid molecules in the interface, nature and surface chemistry of nanoparticles. The value of ω should lies between 1 and 65.2529 ($1 < \omega < k_p/k_f$) for Al$_2$O$_3$ nanoparticles. It is observed that both velocity and temperature increase with the increase of nanolayer to base fluid conductivity ratio (ω) for the default parameters. It is due to the increase in thermal conductivity of nanofluid with the increase in ω.

The effect of the nanolayer to base fluid conductivity ratio (ω) on the modified skin friction and the Nusselt number is shown in Fig. 7c, d, respectively. Both the

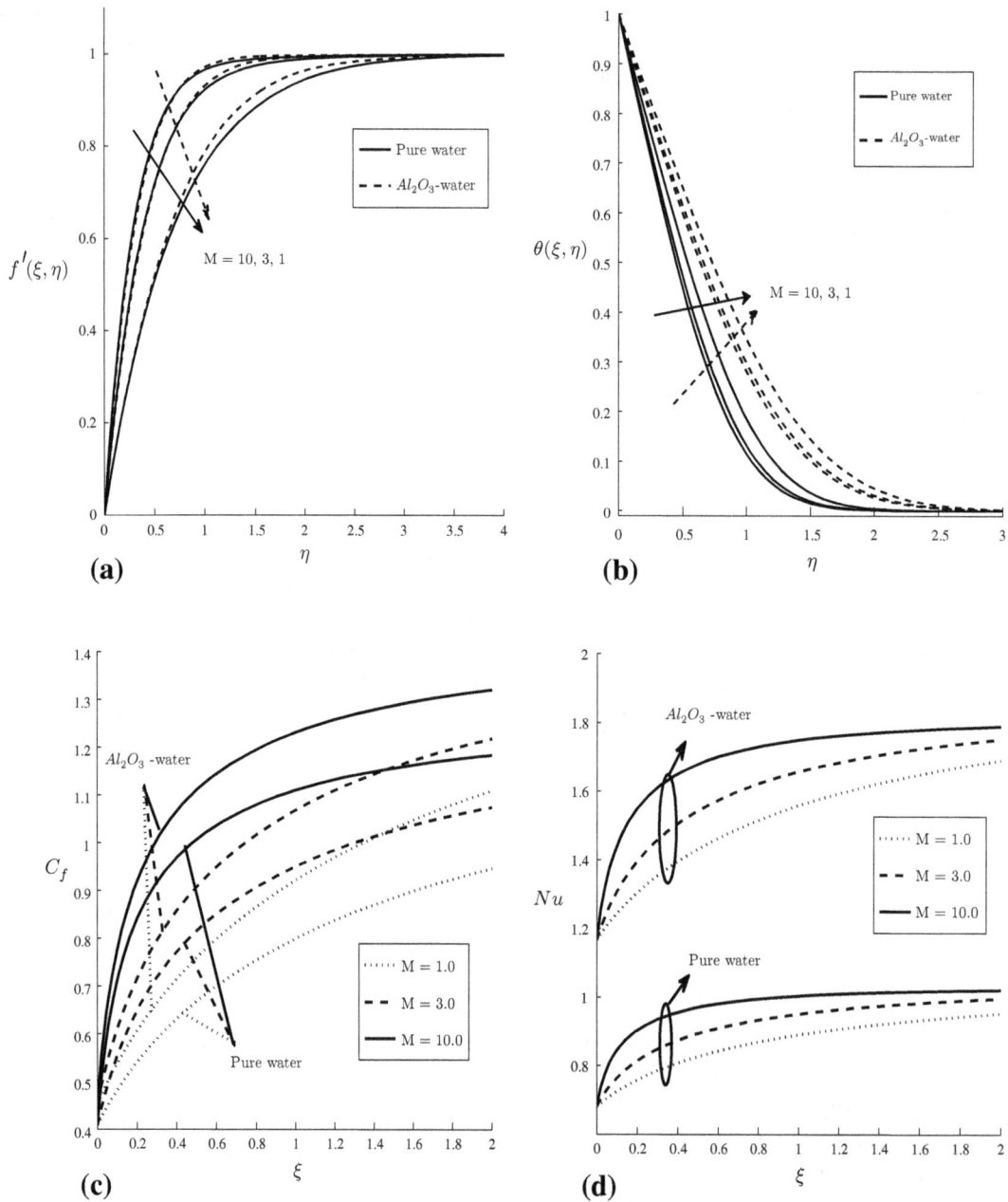

Fig. 5 Effect of magnetic parameter (M) for pure water and Al_2O_3–water ($\phi = 0.04$). **a** Dimensionless velocity profiles. **b** Dimensionless temperature profiles. **c** Modified skin friction against the streamwise distance ξ. **d** Modified Nusselt number against the streamwise distance ξ

modified skin friction and the Nusselt number increase with the increase in this conductivity ratio. Thus, we can say that the value of nanolayer to base fluid conductivity ratio plays an important role. The effect of three different models for viscosity on velocity profile $(f'(\xi,\eta))$ and skin friction $(\theta(\xi,\eta))$ is shown in Fig. 8a, b, respectively. Different models for viscosity have been compared for this problem because of unavailability of most appropriate viscosity model. It is clear that for model I, the velocity is

higher among all other models. This is due to lower viscosity in Model I as compared to other which also justifies the trend of skin friction.

Effect of angle of inclination

The effect of the angle of inclination (δ) on the modified skin friction and the Nusselt number is shown in Fig. 9a, b, respectively. With an increase in plate inclination, the

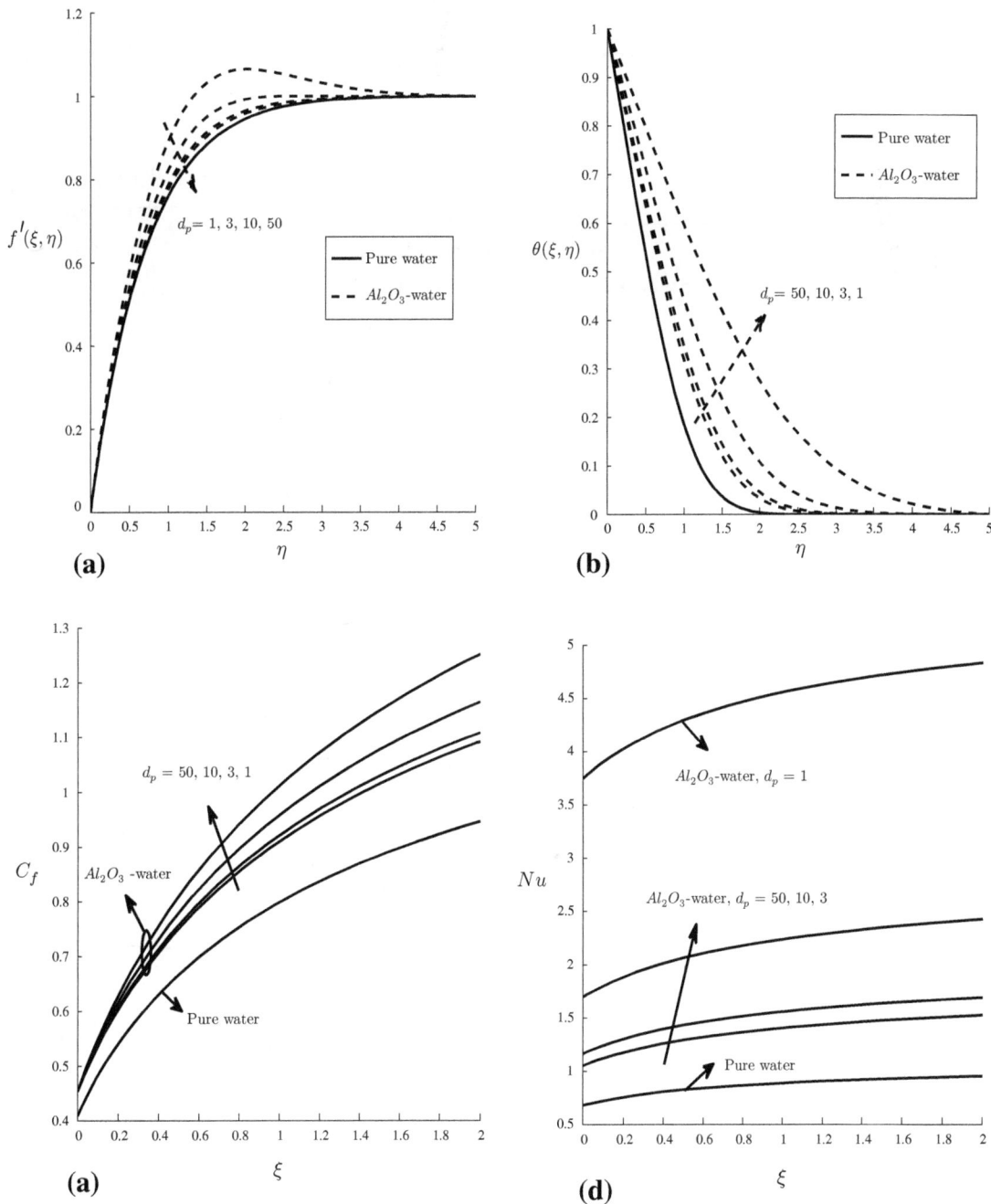

Fig. 6 Effect of nanoparticle diameter (d_p) for pure water and Al$_2$O$_3$–water ($\phi = 0.04$). **a** Dimensionless velocity profiles. **b** Dimensionless temperature profiles. **c** Modified skin friction against the streamwise distance ξ. **d** Modified Nusselt number against the streamwise distance ξ

velocity decreases. The case of $\delta = 0$ corresponds to the vertical plate configuration and for this scenario, the velocity is found to be maximum. For the case $\delta \to \pi/2$, $\cos(\pi/2) \to 0$, i.e., buoyancy effects vanish (horizontal plate scenario, where the gravity field is normal to the plate surface and exerts no effect on the flow). The plate orientation is simulated via the modified buoyancy term,

$\frac{\sigma B_0^2}{\rho}(u - u_\infty)\cos(\delta)$, arising in the momentum boundary layer Eq. (2). As δ increases, $\cos(\delta)$ decreases. This causes the buoyancy effect to be depleted with increasing inclination of the plate. The effects of three lateral mass flux (transpiration) cases at the plate surface, i.e., of suction ($f_w = 0.25$), solid wall ($f_w = 0$) and injection

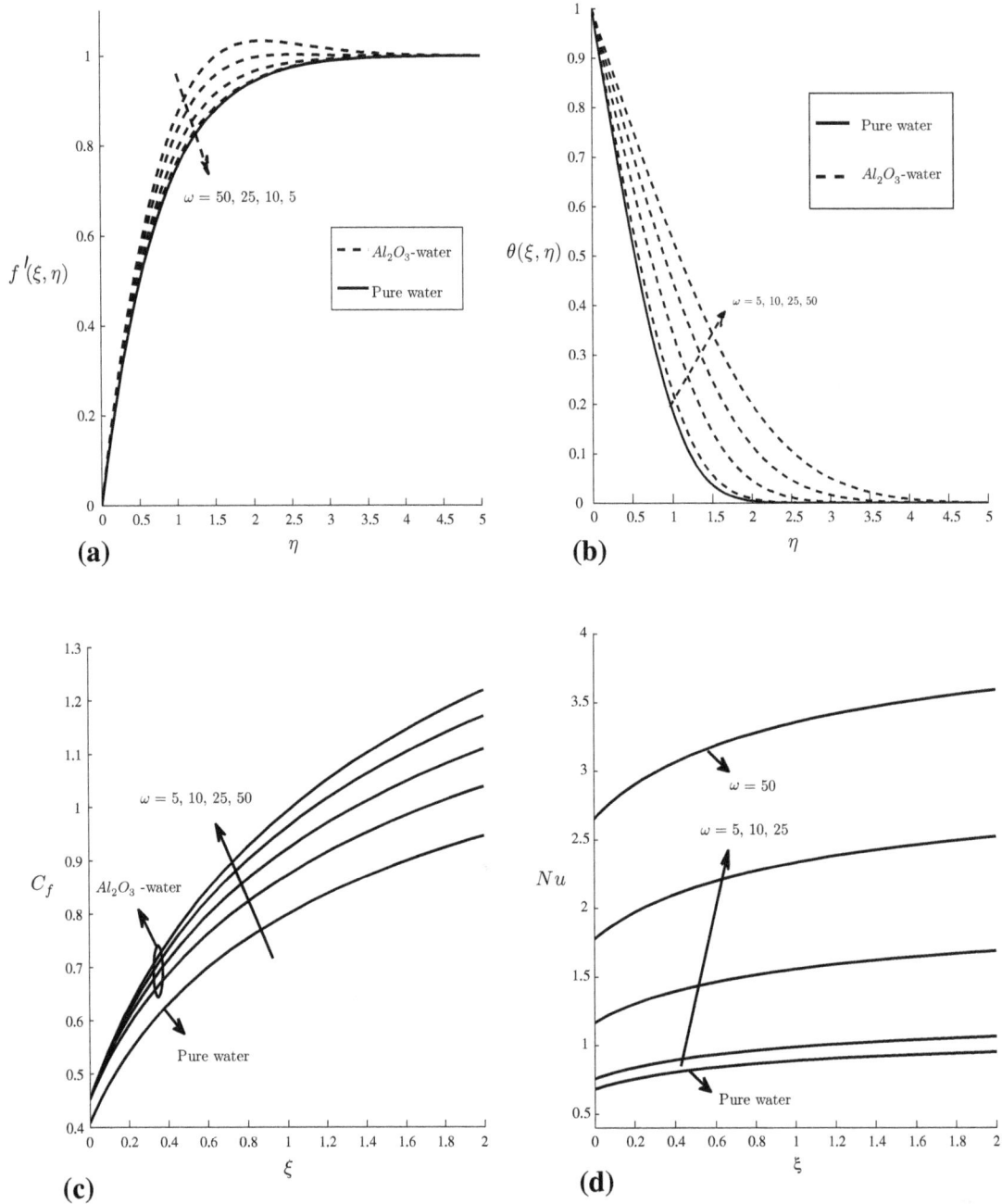

Fig. 7 Effect of nanolayer to base fluid conductivity ratio (ω) for pure water and Al_2O_3–water ($\phi = 0.04$). **a** Dimensionless velocity profiles. **b** Dimensionless temperature profiles. **c** Modified skin friction against the streamwise distance ξ. **d** Modified Nusselt number against the streamwise distance ξ

($f_w = -0.25$) on Nusselt number (Nu) have been shown in Fig. 10.

Figure 11 depicts the variation of average Nusselt number (Nu_{avg}) with d_p for different ω. Average Nusselt number decreases exponentially with the decrease of nanoparticle size for each case. In Fig. 12, the average Nusselt number has

been plotted against the volume fraction (ϕ) ranging from $0 \leq \phi \leq 0.04$ for different models of viscosity. The graph shows that the average Nusselt number increases linearly with the volume fraction. Moreover, average Nusselt number (Nu_{avg}) increases more sharply with the volume fraction for model I as compared to other models.

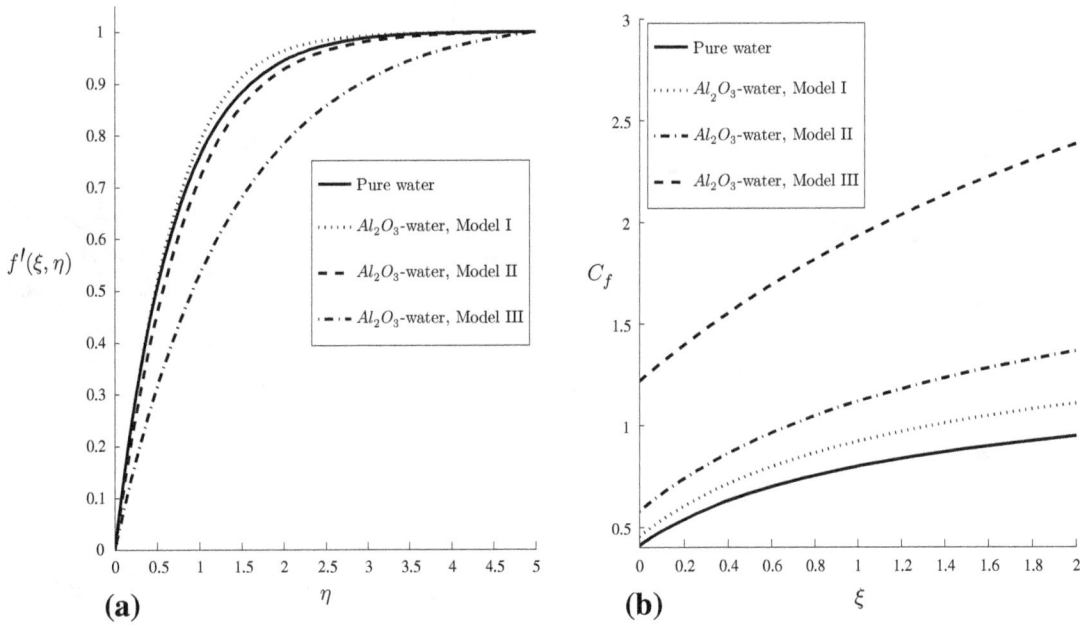

Fig. 8 Effect the different models for pure water and Al_2O_3–water ($\phi = 0.04$). **a** Dimensionless velocity profiles. **b** Modified skin friction against the streamwise distance ξ

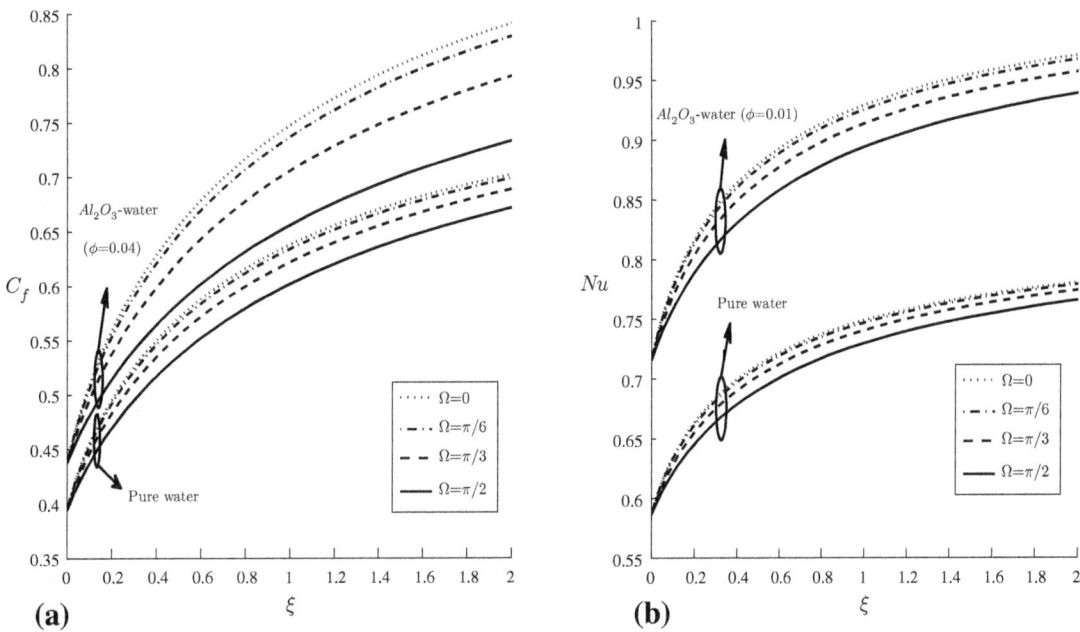

Fig. 9 Effect of angle of inclination (δ) for pure water and Al_2O_3–water ($\phi = 0.04$). **a** Modified skin friction against the streamwise distance ξ. **b** Modified Nusselt number against the streamwise distance ξ

Conclusions

In this present paper, mixed convection Al_2O_3–water nanofluid flows in two-dimensional vertical plate have been investigated to study heat transfer enhancement due to application of the nanoparticles to the base fluid. Numerical results for the local Nusselt number and local skin friction, temperature profile and velocity profile are

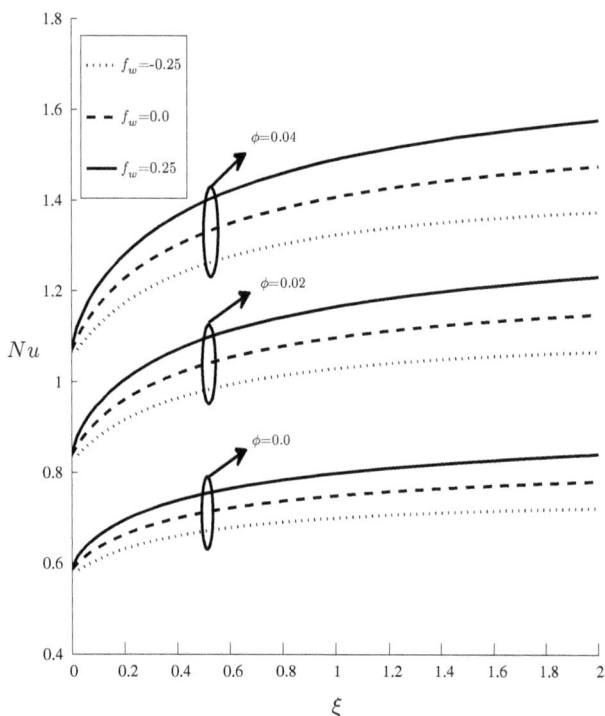

Fig. 10 Effect of suction/injection parameter (f_w) on Nusselt number against the streamwise distance (ξ) for different nanoparticle volume fraction (ϕ)

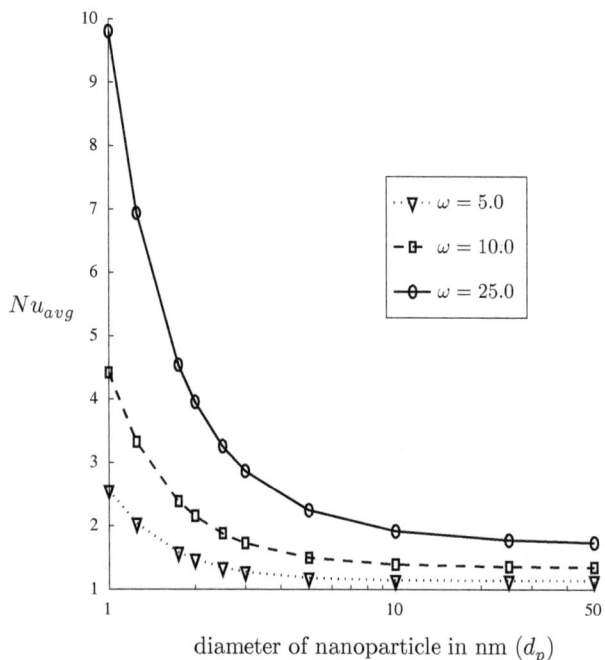

Fig. 11 Variation of average Nusselt number (Nu$_{avg}$) with diameter of nanoparticle in nm (d_p) for different nanolayer and base fluid conductivity ratio (ω)

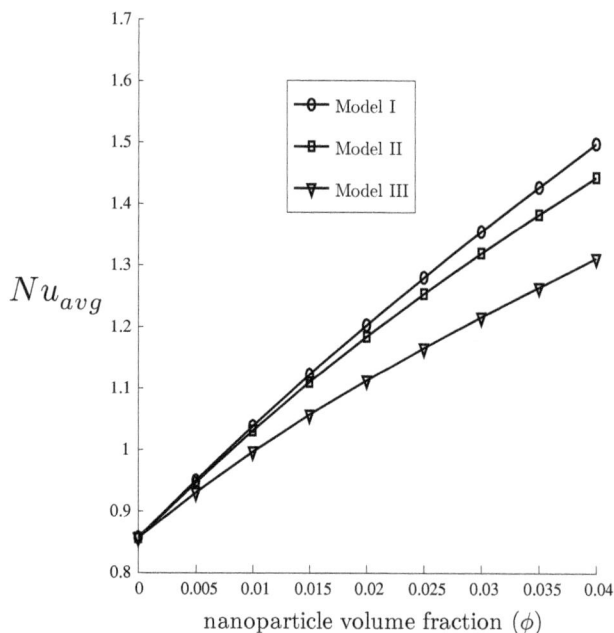

Fig. 12 Variation of average Nusselt number (Nu$_{avg}$) with nanoparticle volume fraction (ϕ) for different models

presented graphically for various governing parameters. The average Nusselt number has also been compared for different nanoparticle diameter and volume fraction. The main conclusions are as follows:

1. At a given Richardson number (Ri) the enhancement in heat transfer has been noted with the increase in magnetic field, both for pure water and nanofluids. But the effect is more pronounced with nanofluids.

2. Increase in nanoparticle diameter (d_p) decreases the surface heat transfer. Moreover, it does not have any significant change on the heat transfer beyond 25 nm at a constant Richardson number (Ri) and magnetic field.

3. Higher the value of nanolayer to base fluid conductivity ratio (ω), more will be the heat transfer enhancement. It has also been shown that the heat transfer increases with the increase of nanoparticle volume fraction (ϕ).

4. Both the angle of inclination of the plate (δ) and transpiration parameter (f_w) have prominent effect in the case of nanofluids.

5. Average Nusselt number (Nu$_{avg}$)is higher for classical viscosity model (model I) assumptions as compared to other.

References

Abdelkhalek MM (2006) The skin friction in the MHD mixed convection stagnation point with mass transfer. Int Commun Heat Mass Transf 33(2):249–258

Aydin O, Kaya A (2007) Mixed convection of a viscous dissipating fluid about a vertical flat plate. Appl Math Model 31(5):843–853

Aydin O, Kaya A (2009a) MHD mixed convection of a viscous dissipating fluid about a permeable vertical flat plate. Appl Math Model 33(11):4086–4096

Aydin O, Kaya A (2009b) MHD mixed convective heat transfer flow about an inclined plate. Heat Mass Transf 46(1):129–136

Bhargava R, Rana P (2011) Finite element solution to mixed convection in mhd flow of micropolar fluid along a moving vertical cylinder with variable conductivity. Int J Appl Math Mech 7:29–51

Brinkman HC (1952) The viscosity of concentrated suspensions and solutions. J Chem Phys 20(4):571–581

Chamkha AJ, Mujtaba M, Quadri A, Issa C (2003) Thermal radiation effects on MHD forced convection flow adjacent to a non-isothermal wedge in the presence of a heat source or sink. Heat Mass Transf 39(4):305–312

Choi SUS (1995) Enhancing thermal conductivity of fluids with nanoparticles in developments and applications of non-Newtonian flows. In: Siginer DA, Wang HP (eds) ASME, FED 231/, vol 66, pp 99–105

Dettmer W, Peric D (2006) A computational framework for fluidrigid body interaction: finite element formulation and applications. Comput Methods Appl Mech Eng 195(13–16):1633–1666

Hamad MAA (2011) Analytical solution of natural convection flow of a nanofluid over a linearly stretching sheet in the presence of magnetic field. Int Commun Heat Mass Transf 38(4):487–492

Hamilton RL, Crosser OK (1962) Thermal conductivity of heterogeneous two-component systems. Ind Eng Chem Fundam 1(3):187–191

Hansbo A, Hansbo P (2004) A finite element method for the simulation of strong and weak discontinuities in solid mechanics. Comput Methods Appl Mech Eng 139(33–35):3523–3540

Hashimoto T, Fujimura M, Kawai H (1980) Domain-boundary structure of styrene-isoprene block copolymer films cast from solutions. 5. Molecular-weight dependence of spherical microdomains. Macromolecules 13(6):1660–1669

Koo J, Kleinstreuer C (2004) A new thermal conductivity model for nanofluids. J Nanopart Res 6(6):577–588

Leong KC, Yang C, Murshed SMS (2006) A model for the thermal conductivity of nanofluids-the effect of interfacial layer. J Nanopart Res 8(2):245–254

Maxwell JC (1873) A treatise on electricity and magnetism, vol 1. Clarendon Press, Oxford

Murshed SMS, Leong KC, Yang C (2008) Investigations of thermal conductivity and viscosity of nanofluids. Int J Therm Sci 47(5):560–568

Nield DA, Kuznetsov AV (2003) Boundary-layer analysis of forced convection with a plate and porous substrate. Acta Mech 166(1):141–148

Oztop HF, Abu-Nada E (2008) Numerical study of natural convection in partially heated rectangular enclosures filled with nanofluids. Int J Heat Fluid Flow 29(5):1326–1336

Pak BC, Cho YI (1998) Hydrodynamic and heat transfer study of dispersed fluids with submicron metallic oxide particles. Exp Heat Transf: Int J 11(2):151–170

Rana P, Bhargava R (2011) Numerical study of heat transfer enhancement in mixed convection flow along a vertical plate with heat source/sink utilizing nanofluids. Commun Nonlinear Sci Numer Simul 16:4318–4334

Rana P, Bhargava R, Bég OA (2012) Numerical solution for mixed convection boundary layer flow of a nanofluid along an inclined plate embedded in a porous medium. Comput Math Appl 64(9):2816–2832

Rana P, Bhargava R, Bég OA (2013) Finite element modeling of conjugate mixed convection flow of Al_2O_3 water nanofluid from an inclined slender hollow cylinder. Phys Scripta 87:055005

Reddy JN (1985) An introduction to the finite element method. McGraw-Hill Book Co, New York

Shahidian A, Ghassemi M, Mohammadi R (2012) Effect of nanofluid properties on magnetohydrodynamic pump (MHD). Adv Mater Res 403:663–669

Tillman P, Hill JM (2007) Determination of nanolayer thickness for a nanofluid. Int Commun Heat Mass Transf 34(4):399–407

Tiwari RK, Das MK (2007) Heat transfer augmentation in a two-sided lid-driven differentially heated square cavity utilizing nanofluids. Int J Heat Mass Transf 50(9):2002–2018

Wang BX, Zhou LP, Peng XF (2003) A fractal model for predicting the effective thermal conductivity of liquid with suspension of nanoparticles. Int J Heat Mass Transf 46(14):2665–2672

Wang X, Xu X, Choi SUS (1999) Thermal conductivity of nanoparticle-fluid mixture. J Thermophys Heat Transf 13(4):474–480

Watanabe T, Pop I (1994) Thermal boundary layers in magnetohydrodynamic flow over a flat plate in the presence of a transverse magnetic field. Acta Mech 105(1):233–238

Xuan Y, Li Q (2000) Heat transfer enhancement of nanofluids. Int J Heat Fluid Flow 21(1):58–64

Xue Q, Xu WM (2005) A model of thermal conductivity of nanofluids with interfacial shells. Mater Chem Phys 90(2):298–301

Yih KA (1999) MHD forced convection flow adjacent to a non-isothermal wedge. Int Commun Heat Mass Transf 26(6):819–827

23

Nickel nanoparticles embedded in carbon foam for improving electromagnetic shielding effectiveness

Rajeev Kumar · Saroj Kumari · Sanjay R. Dhakate

Abstract To improve electromagnetic shielding effectiveness of light weight carbon foam (CF), magnetic nanoparticles were embedded in it during processing. The CF was developed from the coal tar pitch and mixture of coal tar pitch-Nickel (Ni) nanoparticles by sacrificial template technique and heat treated to up 1,000 °C. To ascertain the effect of Ni nanoparticles embedded in CF, it was characterized by scanning electron microscopy, X-ray diffraction, Raman spectroscopy, vector network analyzer and vibration sample magnetometer. It is observed that Ni nanoparticles embedded in the carbon material play an important role for improving the structure and electrical conductivity of CF-Ni by catalytic carbonization. The structural investigation suggests that the Ni nanoparticles embedded in the carbon material in bulk as well on the surface of CF. The CF demonstrates excellent shielding response in the frequency range 8.2–12.4 GHz in which total shielding effectiveness (SE) dominated by absorption losses. The total SE is −25 and −61 dB of CF and CF-Ni, it is governed by absorption losses −48.5 dB in CF-Ni. This increase is due to the increase in dielectric and magnetic losses of ferromagnetic Ni nanoparticles with high surface area. Thus, light weight CF embedded with small amount of magnetic nanoparticles can be useful material for stealth technology.

Keywords Carbon foam · Nickel nanoparticles · Raman spectroscopy · Electromagnetic shielding effectiveness · Magnetization

R. Kumar · S. Kumari · S. R. Dhakate (✉)
Physics and Engineering of Carbon, Division of Material Physics and Engineering, CSIR-National Physical Laboratory, Dr. K. S. Krishnan Marg, 110012 New Delhi, India
e-mail: dhakate@mail.nplindia.org

Introduction

The increasing use of large number of wireless gadgets in contemporary technological world and related electromagnetic (EM) radiation is becoming a serious problem that disturbs stable working conditions of electronic appliances and may harm to human being (Chung 2000). In civil and military aerospace vehicles, protection from EM radiations as well as regulation of thermal heating of electronic power systems is necessary to protect them from any form of damages. In particular, light weight shielding and structural materials are needed to mitigate EM interference (EMI) from electronic systems and to protect human from hazards of space radiation, e.g., aerospace vehicles, especially in 8.2–12.4 GHz frequency region (X-band) (Coleman et al. 2006; Du et al. 2006). The EMI shielding refers to the reflection and absorption of electromagnetic radiation by a material, which thereby acts as shield against the penetration of radiation through the shield. Electrical conductor such as metal and carbon materials mainly shield by reflection of radiation. On the other hand, magnetic materials mainly used for absorption of the radiation. The most materials used for shielding are chosen due to their electrical conductivity rather than their magnetic behavior. While in certain application such as sheath technology, shielding materials have mandatory to absorb maximum EM radiation. The magnetic absorbers depend on the magnetic hysteresis effect, which is attained in magnetic materials such as ferrites. But densities of the magnetic materials are generally high and absorbing bandwidths for magnetic absorbers are usually narrow. On the other hand, dielectric materials are light weight, but do not match up to the absorptivity of magnetic absorbers (Seo et al. 2004; Park et al. 2006; Oh et al. 2004). These two types of materials have different advantages and disadvantages when they are applied as absorbers. They can be used together as a composite, and the magnetic material is usually the base one, but high density of the material is still of great concern. Practically, an effective EMI shielding material being lightweight is an important requirement in aerospace transportation vehicles and space structures (Yang et al. 2005). The shielding effectiveness (SE) value should be > -50 dB in the X-band frequency (Wang et al. 2012). Among the different materials, carbon-based materials are comparatively light weight and structurally strong.

Carbon foam (CF) is a sponge-like rigid high-performance engineering materials in which carbon ligaments are interconnected to each other (Klett et al. 2000, 2004). Recently, it has attracted lot of attention owing to outstanding properties such as low density, large surface area with open cell wall structure, good thermal and mechanical stability (Gallego and Klett 2003; Inagaki et al. 2004).

There are different types of electromagnetic (EM) radiation shielding materials or RAM has been developed since after the advent of radar, but very few reports on the CF are available in literature. As discussed above, generally carbon is conducting materials in which SE is dominated by reflection losses that depend upon the processing temperature of CF and starting precursor.

Yang et al. (2004) reported the development of CF from mesophase pitch by foaming technique which is heated at temperature between 400 and 800 °C and studied its microwave (2–18 GHz) absorption characteristics. It is found that CF heat treated at 600 and 700 °C exhibits better microwave absorption (reflection loss 10 dB). Fang et al. (2006) reported the numerical prediction and experimental validation of CF as microwave (8.2–12.4 GHz) absorber. The CF is fabricated by traditional technique through the polymer foam replication method and it is heat treated at 700, 750 and 800 °C, and characterized them for microwave absorption. The reflection coefficient of 20 mm thick foam is in the order of 8–10 dB. Fang et al. (2007) have also studied the electromagnetic characteristics of CF having different pore size and electrical conductivities which are controlled by the carbonization temperature (700–760 °C). The electromagnetic parameters of these carbon foam and their corresponding pulverized powders are measured by a resonant cavity perturbation technique at a frequency of 2.45 GHz. The CF has dielectric loss several times larger than their corresponding pulverized powder. This suggests that low temperature heat treated foam EM-SE is dominated by absorption. Recently, Moglie et al. (2012) studied the EM-SE of CF (GRAFOAM FPA-20 and FRA-10) in the frequency band 1–4 GHz using the nested reverberation chamber method. It is reported that with increasing thickness of CF total SE increases. Micheli and Marcheti (2012) reported the effect of multi-walled

carbon nanotubes epoxy resin mixture filled in the pores of CF and studied its role on EM radiation absorption properties of CF in frequency range 2–3 GHz. It is observed that absorption properties reaching peak of the reflection coefficient -45 dB for a normally incident plane wave. Blacker et al. (2008) reported that EM-SE greater than about 40 dB in the frequency range of 400 MHz–18 GHz of rigid porous electrically conductive CF is achieved by incorporating electrically conductive carbon nanofiber in a polymer matrix, which is used for the joining the CF enclosures. Blacker et al. (2011) reported that the development of electrically gradated CF material exhibits different electrical resistivity values at or near different surfaces of CF in which the electrical resistivity increases with increasing the thickness of foam and decreases with increasing processing temperature. These electrically gradated CF materials may be used as radar absorbers. But in the case of aerospace and military application, the shield material should have high SE which is mainly dominated by absorption losses as well thermal stability.

The carbon nanomaterials in combination with magnetic particles are used to improve EM radiation absorption and have become more intensive absorber as compared to carbon nanomaterials (Huang and Wu 2000; Liu et al. 2006; Che et al. 2004). Che et al. (2004, 2006) have reported that magnetic nanoparticles/carbon nanotubes (CNTs) exhibited improved microwave absorption properties because of their proper EM matching between the dielectric loss and the magnetic loss.

The purpose of this study is to improve the microwave absorption properties of carbon foam by the incorporation of magnetic nanoparticles. In the present work, CF is developed from a mixture of coal tar pitch and Ni nanoparticles by sacrificial template technique and then heat treated to 1000 °C. The results reveal that SE is dominated by EM radiation absorption mechanism rather than reflection mechanism in the Ni nanoparticle-embedded CF.

Experimental

Synthesis of nickel nanoparticles

The Ni nanoparticles were synthesized by the process reported literature (Saravanan et al. 2001). The 1 g of nickel chloride hexahydrate ($NiCl_2 \cdot 6H_2O$) was dissolved in 100-ml ethylene glycol. The solution was heated to 70–80 °C. Later on, 4.5-ml hydrazine hydrate was added to above solution for complete reduction. After 10 min, 3.0 g of NaOH (1 M) solution was added slowly in continuation of stirring. The complete solution was kept under magnetic stirrer for 1 h at 60 °C. After 1 h, reaction was completed and black nickel nanoparticles were formed. These nanoparticles were collected, washed several times by ethanol and dried in the room temperature. The reduction reaction can be expressed as

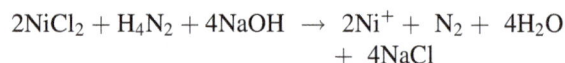

$$2NiCl_2 + H_4N_2 + 4NaOH \rightarrow 2Ni^+ + N_2 + 4H_2O + 4NaCl$$

It is seen that N_2 gas is continuously bubbled during the reaction and as results it automatically creates N_2 atmosphere throughout the reaction. Hence, no extra N_2 gas is required for the synthesis of nickel nanoparticles.

Fabrication of nickel nanoparticles incorporated carbon foam

Initially, coal tar pitch was modified with the nickel nanoparticles 1.0 wt% and heat treated at 400 °C for 25 h in inert atmosphere. The carbon foam was prepared by sacrificial template technique in which the polyurethane (PU) foam (density 0.030 g/cc and average pore size 0.45 mm) was used as template. The Ni nanoparticle-based modified pitch was grounded by ball milling in a tungsten carbide jar for about 5–6 h to get the particle size of mixture less than 30μm, so that these particles can easily penetrate inside the cells of PU foam. The above mixture was converted into water-based slurry (with 3 % polyvinyl alcohol) and it was impregnated inside pore of PU template foam by vacuum infiltration technique (Kumar et al. 2013a, b, c; Yadav et al. 2011). The impregnated PU foam was converted into carbon foam by series of heat treatments in air as well as in an inert atmosphere up to 1,000 °C. Initially, the modified coal tar pitch and Ni nanoparticles mixture impregnated PU foams were heat treated @

1 °C min^{-1} up to 275 °C in the nitrogen atmosphere for 1 h followed by oxidation and stabilization in air atmosphere at temperature 300 °C. The stabilized foam was carbonized in tubular high temperature furnace at 1,000 °C with heating rate 10 °C h^{-1} in inert atmosphere. The schematic of process for synthesis of nickel nanoparticle-based CF is shown in Fig. 1. The CF is designated as CF and Ni nanoparticle-embedded CF as CF-Ni.

Characterization

Raman spectra of the CF and Ni-CF samples were recorded using Renishaw in via Raman spectrometer, UK with laser as an excitation source at 514 nm. The crystal structure of CF samples was studied by X-ray diffraction (XRD, D–8 Advanced Bruker diffractometer) using CuK$_\alpha$ radiation ($\lambda = 1.5418$ Å). The morphology of Ni nanoparticles was observed by transmission electron microscope (TEM, Tecnai, G20-Super-Twin, 300 kV). The TEM sample was prepared by Ni nanoparticle dispersed in solvent and put over the carbon-coated copper grid. The surface morphology of the CF samples was observed by scanning electron microscope (SEM, VP-EVO, MA-10, Carl-Zeiss, UK) operating at an accelerating potential of 10.0 kV. The electrical conductivity CF (size $60 \times 20 \times 4$ mm) was measured as per standard ASTM C611-98. The dc four probe contact method was used in which a Keithley 224 programmable current source was used for providing current. The voltage drop was measured by Keithley 197A auto ranging digital microvoltmeter. The values reported in text are average of six readings of voltage drops at different places of the sample. The compressive strength of CF was

Fig. 1 Fabrication of Ni nanoparticle-embedded coal tar pitch-based carbon foam using sacrificial (PU foam) template technique

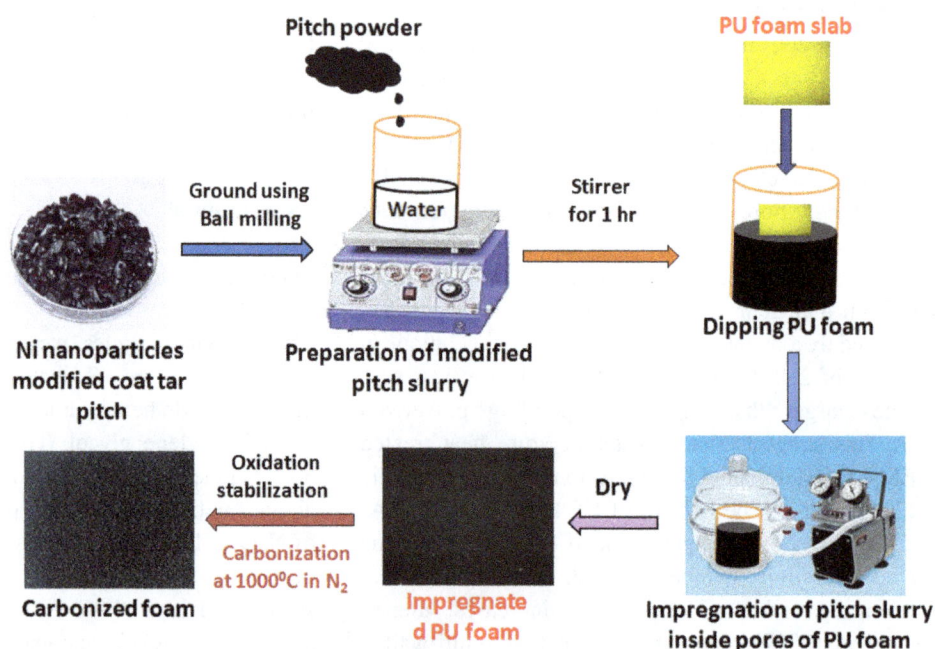

measured by Instron Universal Testing Machine, model 1122 as per ASTM C695-91. EM-SE and EM attributes (complex permittivity and permeability) were measured by waveguide using vector network analyzer (VNA E8263BAgilent Technologies). The rectangular samples (26 × 13 mm) of thickness 2.75 mm were placed inside the cavity of sample holder which matches the internal dimensions of X-band (8.4–12.4 GHz) wave guide. The sample holder was placed between the flanges of the waveguide connected between the two ports of VNA. The magnetic property of the CF samples was measured by vibration sample magnetometer (VSM) model 7304, Lakeshore Cryotronics Inc., USA with a maximum magnetic field of 1.2 T, using Perspex holder vibrating horizontally at frequency of 76 Hz.

Results and discussion

In Fig. 2a, TEM image of Ni nanoparticles shows that the particles are of different shape and size, particularly somewhat star or starfruit shaped and it looks like agglomerated form. The average size of Ni nanoparticles is in between 50 and 100 nm. Figure 2b shows the powder XRD patterns of the Ni nanoparticles. The nanoparticles are single-phase with face-centered cubic (fcc) structure, and no phase of NiO or other impurity is observed. The peaks at $2\theta = 44.45°, 51.73°$ and $76.84°$ correspond to diffraction plane of Ni (111), Ni (200) and Ni (220) reflections, respectively (The international centre for powder diffraction data, powder diffraction files 2001). This result is subsequently confirmed by the TEM examination of the annealed sample where larger metallic Ni nanoparticles are clearly observed.

The bulk density of CF is 0.50 g cc^{-1} and on addition of Ni nanoparticles its density increases to 0.56 g cc^{-1} even at 1 wt% of nickel nanoparticles. The Ni nanoparticles are

in magnetic behavior, but in CF its acts also as an electrical conductivity enhances. The Ni has affinity toward carbon and it forms Ni–C. The CF is heat treated to 1,000 °C, at this temperature graphitic structure does not develop and as a result it is in amorphous nature and it possesses combination of sp^2 and sp^3 hybridized carbon. The Ni can readily react with sp^3 bonded carbon and as a result sp^3 contribution in CF-Ni decreases. This has positive effect on the electrical conductivity. The electrical conductivity of CF is 50 S cm^{-1} which is due to the delocalized π electron in the carbon network while electrical conductivity of CF-Ni increases to 65 S cm^{-1}, this is due to the increases in sp^2 carbon content in it as compared to CF.

The mechanical strength is one of the essential requirements of CF because compressive forces are often encountered during its application as structural component. Therefore, compressive strength of CF should be sufficient to avoid any form of structural damage. The compressive strength of CF depends mainly on two factors namely microstructure and bulk density. It is observed that the compressive strength of CF is 7.0 ± 1 MPa and that of CF-Ni is 8.8 ± 1.2 MPa. This enhancement in strength is related to increase in bulk density and decrease in stress concentration center.

Figure 3 depicts the SEM micrographs of CF and CF-Ni. In case of CF, cells of the different sizes and distributions of cells are not uniform (Fig. 3a). The cell walls (i.e., ligaments) are broken during machining of samples for SEM characterization because of the brittle and porous nature of material (Kumar et al. 2013a, b, c). Figure 3b, SEM image of CF-Ni, in which the brightness of the surface increases due to the increase in electrical conductivity of CF-Ni by catalytic carbonization. The Ni nanoparticles are embedded in bulk carbon as well as on the surface and pore walls. Figure 3c shows magnified view of the pore walls in which

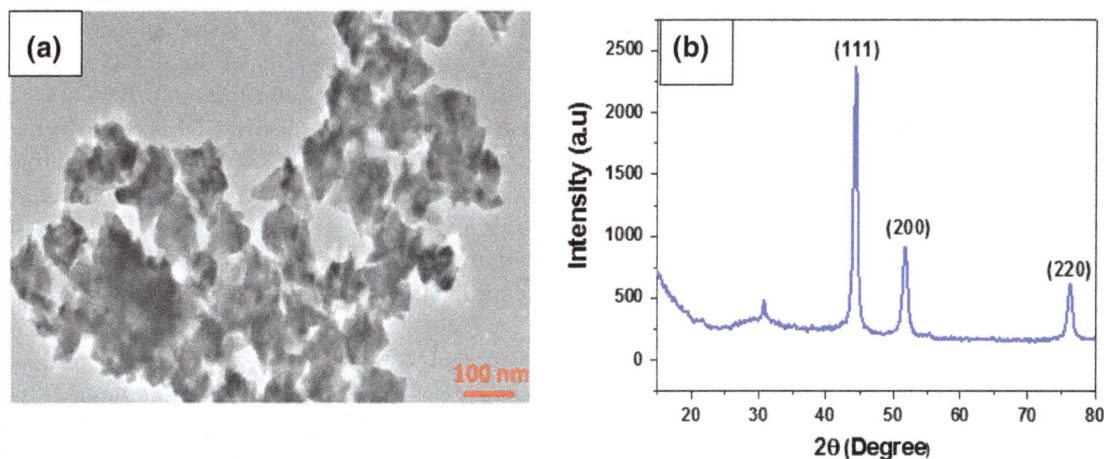

Fig. 2 a TEM micrograph and b XRD spectra of Ni nanoparticles

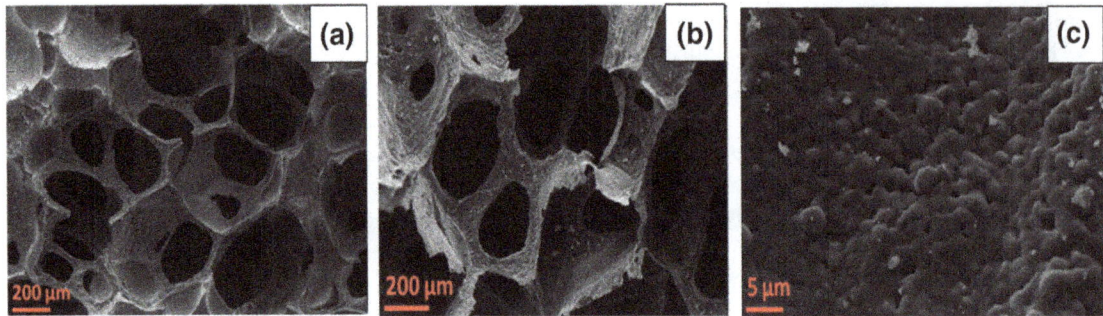

Fig. 3 SEM micrographs of **a** CF, **b** CF-Ni, **c** Ni nanoparticles embedded with carbon on the surface

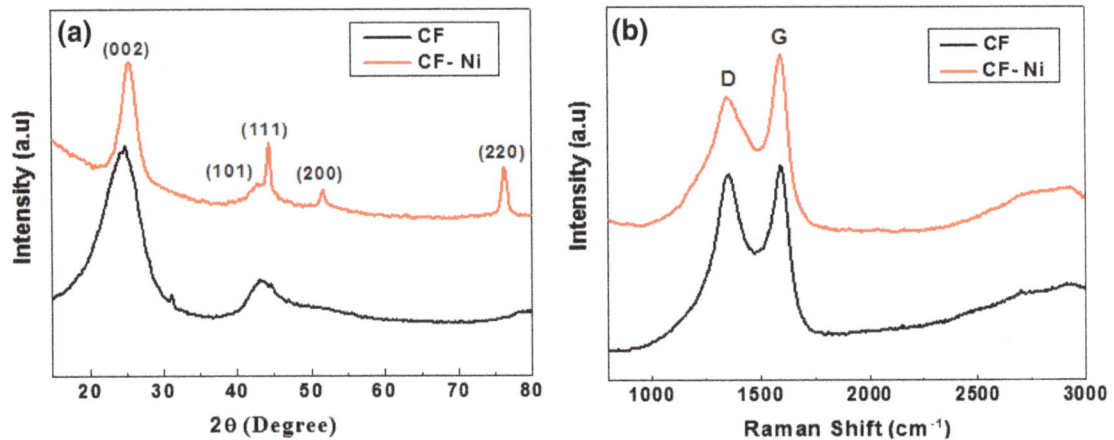

Fig. 4 **a** XRD spectra and **b** Raman spectra of CF and CF-Ni

embedded Ni nanoparticles are visible because the melting point of Ni is well above 1,400 °C.

Figure 4a shows the XRD spectra of CF and CF-Ni. In case of CF, there are two prominent peaks at $2\theta = 24.85$ and 43.15°, which correspond to the carbon of (002) and (101) diffraction plane. The XRD spectra of CF-Ni consist of peaks at $2\theta = 25.55°$, 42.76°, 44.45°, 51.77° and 76.35° corresponding to carbon (002) and (101), Ni crystal lattice (111), (200), and (220) planes. In case of CF-Ni, the 002 peak appears at higher diffraction angle. This suggests, Ni nanoparticle reacts with carbon, it acting as catalyst in improving the crystallite structure of carbon. The catalytic effect of carbon influences the structure in which sp^2 hybridization content increases. This also confirmed by Raman spectroscopy investigation. Figure 4b shows the Raman spectra of CF and CF-Ni and illustrates common features in the Raman shift 1,000–3,000 cm^{-1} region, the G, D and 2D band which lies at around 1,560, 1,360 and 2,700 cm^{-1}, respectively. The G band corresponds to the E 2 g phonon at the Brillion zone center. The D band is due to the breathing modes of sp^2 atoms and requires a defect for its activation and it only gives knowledge to the amount of disorder in the given structure (Ferrari and Robertson 2000; Dhakate et al. 2011). Figure 3b shows the Raman

spectra of CF and CF-Ni, consisting of two bands D and G appearing at Raman shift 1,350 and 1,598 cm^{-1}, respectively. In case of CF, both bands almost appear at same intensity suggest the amorphous-type carbon while in case CF-Ni the G band intensity is higher as compared to D band. This suggests the structure of carbon modified with addition of Ni. The intensity ratio of D and G peak, i.e., (ID/IG) gives idea of defect level in the structure of CF, i.e., sp^3 bonded carbon. The ID/IG ratio in CF is higher this may be due to the presence of lesser amount sp^2 bonded carbon i.e., more defects. The CF has ID/IG ratio 0.9645 while in case of CF-Ni it decreases to 0.7065. This demonstrates that the Ni embedded in CF has positive effect on ID/IG ratio. The results of XRD and Raman spectroscopic are in agreement with each other and support the electrical conductivity data.

Electromagnetic shielding can be explained by measuring SE in terms of reflection and absorption losses. The SE of any shield material is the capacity to attenuate EM radiation that can be expressed in terms of ratio of incoming (incident) and outgoing (transmitted) power (Yong et al. 2005). The EM attenuation offered by a shield may depend on the three mechanisms: reflection of the wave from the front face of shield, absorption of the wave

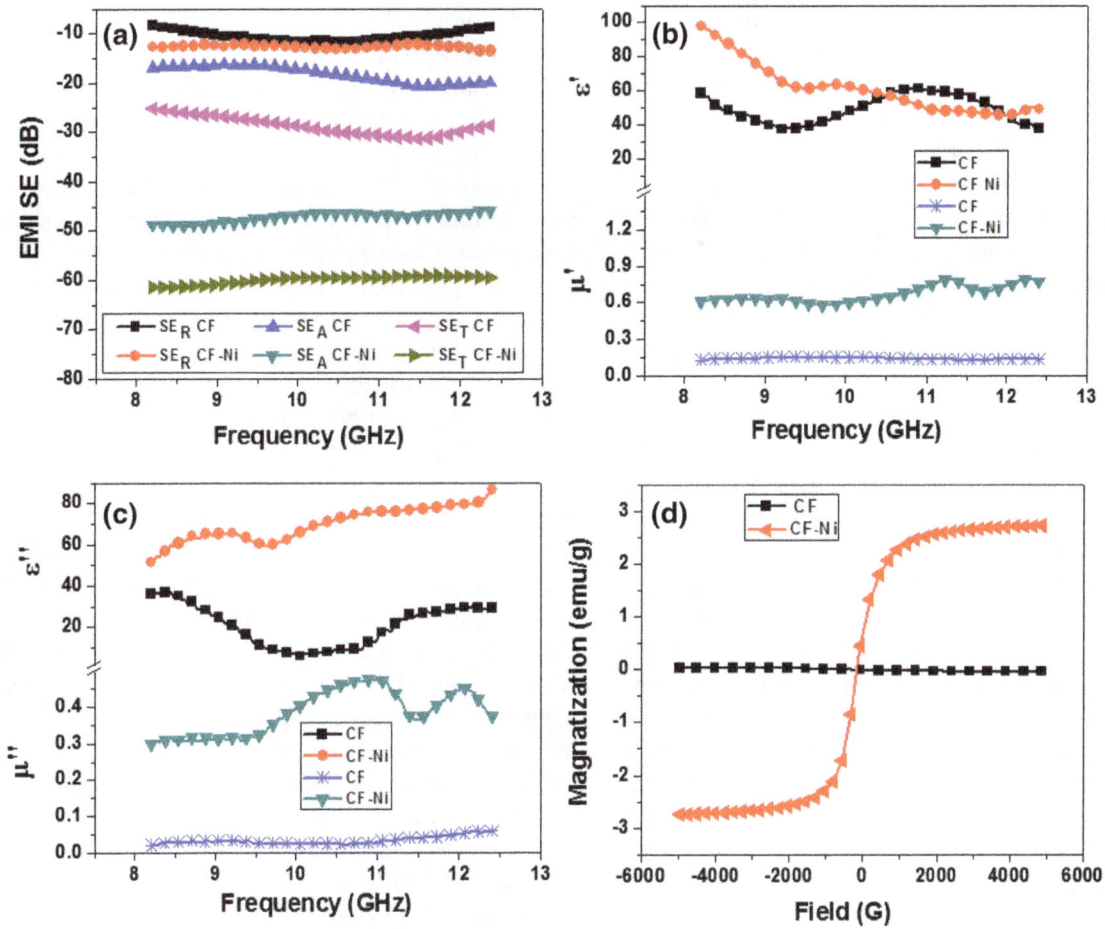

Fig. 5 **a** EMI SE SE_A, SE_R, SE_T of CF and CF-Ni, **b** Real permittivity and permeability (ε', μ'), **c** imaginary permittivity and permeability (ε'', μ'') and **d** room temperature magnetization plot of CF and CF-Ni

as it passes through the shield and multiple reflections of the waves at various interfaces (Han and Wang 2007). Therefore, SE is endorsed to three types of losses, i.e., reflection loss (SE_R), absorption loss (SE_A) and multiple reflection losses (SE_M) and can be expressed (Eq. 1) as:

$$SE_T(dB) = SE_R + SE_A + SE_M = 10\log(P_t/P_i) \quad (1)$$

where P_i and P_t are power of incident and transmitted EM waves, respectively. As, P_t is always less than P_i, therefore, SE is a negative quantity such that a shift towards more negative value means increase in magnitude of SE. It is important to note that the losses associated with multiple reflections can be ignored ($SE_M \sim 0$) in all practical cases when achieved SE_T is more than -10 dB. Therefore, SE_T can be expressed as (Eq. 2)

$$SE_T(dB) = SE_R + SE_A \quad (2)$$

Figure 5a shows the EM-SE of CF and CF-Ni in the X-band (8.2–12.4 GHz) frequency region. It is observed that total shielding effectiveness (SE_T) varies with frequency in both the cases. The value of SE_T for CF is

25 ± 3 dB while that of CF-Ni is 61 ± 5 dB. It is interesting to note that in CF-Ni, SE_T is more than double to that of CF. In case of CF, the SE_T is shared in ratio of reflection:absorption losses::1:2, i.e., reflection losses are -8.5 ± 2 dB and absorption losses are -16.86 ± 1 dB. While in CF-Ni, SE_T is governed by absorption losses (-48.5 ± 4 dB) and partially shared by reflection losses ($SE_R = -12.4 \pm 1$ dB).

The enhancement in the reflection components from -8.5 dB to 12.4 dB in case of CF-Ni is due to the increase in the electrical conductivity of CF-Ni as discussed in earlier section. While absorption losses enhancement from 16.86 to -48.5 dB in CF-Ni is due to the magnetic properties of material. Furthermore, in CF-Ni, Ni particles of size 200–300 nm on the surface of CF and embedded in CF provide higher surface and large interfacial area because of high aspect ratio of nanoparticles, which further enhances the SE due to the absorption.

To probe further, EM parameters which are responsible for EM radiation absorption, relative complex permittivity ($\varepsilon^* = \varepsilon' - i\varepsilon''$) and relative complex permeability

$(\mu^* = \mu' - i\mu'')$ of CF were measured in the frequency region 8.2–12.4 GHz to correlate with shielding properties. These complex parameters have been estimated from experimental scattering parameters (S_{11} and S_{21}) by standard Nicholson–Ross and Weir theoretical calculation (Nicolson and Ross 1970; Singh et al. 2012). The estimated real part of the EM parameters (ε', μ') is directly associated with the amount of polarization occurring in the material which symbolizes the storage ability of the electric and magnetic energy, while imaginary part (ε'', μ'') signifies the dissipated electric and magnetic energy. From Fig. 5b and c, it is clearly demonstrated that both real and imaginary parts of the permittivity (ε', ε'') vary with frequency. The real permittivity in CF is lower as compared to the CF-Ni and it decreases with increasing the frequency in CF-Ni.

The imaginary permittivity in CF-Ni increases with increasing the frequency (Fig. 5c). The decreasing permittivity in CF-Ni with frequency could be ascribed to the decreasing capability of the dipoles to sustain the in-phase movement with speedily pulsating electric vector of the incident radiation. According to EM theory (Colaneri and Shacklette 1992; Das et al. 2000), the ac conductivity (σ_{ac}) and skin depth (δ) are related to the imaginary permittivity (ε''), frequency (ω) and real permeability (μ') as $\sigma_{ac} = \omega\varepsilon_0\varepsilon''$, ($\sigma = \sigma_{ac} + \sigma_{dc}$) and $\delta = \sqrt{2/\sigma_{AC}\omega\mu'}$ which gives reflection (SE$_R$) and absorption losses (SE$_A$) as:

Reflection loss (SE$_R$) as

$$SE_R(dB) = 10\log\{\sigma_{ac}/16\omega\varepsilon_0\mu'\} \quad (3)$$

and absorption loss (SE$_A$) as

$$SE_A(dB) = 20\{t/\delta\}\log e = 20d\sqrt{\mu\omega\sigma_{ac}/2}\log e$$
$$= 8.68\{t/\delta\} \quad (4)$$

$$SE_A(dB) = 8.68t\sqrt{\sigma\omega\mu'/2} \quad (5)$$

From the Eq. (3), it deduces that SE$_R$ is related to frequency (ω), conductivity (σ_{ac}) and real permeability while the ac conductivity ($\sigma_{ac} = \omega\varepsilon_0\varepsilon''$) depends upon the frequency and imaginary permittivity (ε''). As shown in Fig. 5b, real permeability increases with increasing the frequency and it is maximum in case of CF-Ni and minimum in CF. While imaginary permittivity (Fig. 5c) also varies with frequency and it is minimum in case of CF and maximum in CF-Ni. This clearly demonstrates that SE$_R$ is minimum in cases of CF and maximum in CF-Ni due to the higher value of electrical conductivity and imaginary permittivity. The higher value of imaginary permittivity does not store energy, but stored energy will be converted into thermal energy inside the material, which attenuates the EM radiation instead of just storing it. The carbon can easily dissipate the thermal energy due to its high value thermal conductivity.

Equation (5) shows that SE$_A$ depends upon thickness, real permeability, frequency and conductivity of the shield material, from Fig. 5b, maximum real permeability is in case of CF-Ni as compared to CF where thickness is same in both the cases.

In case of CF-Ni, the existence of interfaces between Ni nanoparticles and carbon is responsible for interfacial polarization which further contributes to dielectric losses. Interfacial polarization occurs in heterogeneous materials due to the accumulation of charges at the interfaces and the formation of large dipoles. The magnetic nanoparticles act as tiny dipoles which get polarized in the presence of EM field and as a result better absorption.

The real permeability value of CF-Ni is more as compared to than that of CF (Fig. 5b). This is due to the improvement of magnetic properties along with the parallel reduction of eddy current losses since embedded Ni nanoparticles in magnetic behavior. In the X-band frequency region, the natural resonances can be attributed to the small size Ni particles in CF. Anisotropy energy of the small size materials, especially in the nanoscale, would be higher due to the surface anisotropic field effect of smaller material (Leslie-Pelecky and Rieke 1996). The higher anisotropy energy is also contributed in the enhancement of the EM radiation absorption.

The real and imaginary parts of complex permeability increase with increasing frequency in CF-Ni and maximum at frequency ~11 GHz. While in the CF, it is very small due to the non-magnetic behavior. The magnetic nanoparticles embedded in the CF lead to better matching of input impedance along with the reduction of skin depth. This is attributing further in the increase of absorption losses of CF-Ni. This fact also varied by measuring the saturation magnetization of the CF and CF-Ni. Figure 5d shows the room temperature magnetization plot of CF and CF-Ni. The data of magnetization reveal that CF does not show any magnetization though out the magnetic field because carbon is non-magnetic in nature. However, CF-Ni shows the magnetization that displays narrow hysteresis loop. The CF-Ni possesses saturation magnetization 2.5 emu g^{-1} at 4.9 kg. These results are in agreement with results of electromagnetic attributes. This clearly demonstrates that even small amount of magnetic nanoparticles embedded in conducting CF is very much effective for absorption of EM radiation due to its large surface area and its magnetic properties.

Conclusions

In this investigation, Ni nanoparticle-embedded CF is developed by sacrificial template technique from a mixture coal tar pitch and Ni nanoparticles. It reveals that addition of Ni nanoparticles in the CF influences electrical conductivity, complex permittivity and permeability, mechanical strength and EM-SE. The Ni nanoparticles are magnetic in nature, even though addition of nanoparticles improved the electrical conductivity of CF. This is due to catalytic effect of Ni nanoparticle in the carbon material which is revealed by XRD and Raman studies. The enhancements in electrical conductivity not significantly influence the EM-SE by reflection losses, but it is dominated by absorption of EM radiations due to the magnetic properties and large surface area of nanoparticles. Results of interfacial polarization occur in heterogeneous material (Ni nanoparticle-embedded CF). The shielding effectiveness of Ni nanoparticle-embedded CF increases by 144 % and absorption losses by 170 %. In addition to this, compressive strength of the CF increases from 7 to 8.8 MPa. This clearly demonstrates that the addition of small amount of magnetic nanoparticles in CF can be useful as EM radiation or radar absorbing material which can be used for stealth technology in civil and military applications.

Acknowledgments Authors are highly grateful to Director, NPL and Dr. R.B. Mathur for his kind permission to publish the results. Also thankful to Mr. Jai Tawale and Dr. K.N. Sood for doing SEM. Dr. R.P. pant for XRD, Dr. S.K. Dhawan for Shielding properties measurements and Dr. R.K. Kotnala for VSM.

References

Blacker JM, Plucinski JW (2011) Electrically graded carbon foam, US,7,867,608 B2

Che R, Peng LM, Duan XF, Chen Q, Liang X (2004) Microwave absorption enhancement and complex permittivity and permeability of Fe encapsulated within carbon nanotubes. Adv Mater 16(5):401–405

Che R, Zhi C, Liang C, Zhou X (2006) Fabrication and microwave absorption of carbon nanotubes/CoFe 2 O 4 spinel nanocomposite. Appl Phy Lett 88(3):033105–3

Chung DDL (2000) Materials for electromagnetic shielding. J Mater Eng Perform 9:350–354

Colaneri NF, Shacklette LW (1992) EMI shielding measurements of conductive polymer blends. IEEE Trans Instrum Meas 41:291–297

Coleman JN, Khan U, Blau JW, Gunko YK (2006) Small but strong: a review of the mechanical properties of carbon nanotubes-polymer composites. Carbon 44(9):1624–1652

Das NC, Das D, Khastgir TK, Chakraborthy AC (2000) Electromagnetic interference shielding effectiveness of carbon black and carbon fibre filled EVA and NR based composites. Compos A 31:1069–1081

Dhakate SR, Chauhan N, Sharma S, Tawale J, Singh S, Sahare PD, Mathur RB (2011) An approach to produce single and double

layer graphene from re-exfoliation of expanded graphite. Carbon 49(6):1946–1954

Du F, Ma Y, Lv X, Huang Y, Li F, Chen Y (2006) (2006) The synthesis of single-walled carbon nanotubes with controlled length and bundle size using the electric arc methods. Carbon 44(7):1327–1330

Fang Z, Cao X, Li C, Zhang H, Zhang J, Zhang H (2006) Investigation of carbon foams as microwave absorber: numerical prediction and experimental validation. Carbon 44(15): 3348–3378

Fang Z, Li C, Sun J, Zhang H, Zhang J (2007) The electromagnetic characteristics of carbon foams. Carbon 45(15):2873–2879

Ferrari AC, Robertson J (2000) Interpretation of Raman spectra of disordered and amorphous carbon. Phy Rev B 61:14095–14107

Gallego NC, Klett JW (2003) Carbon foams for thermal management. Carbon 41(7):1461–1466

Han X, Wang YS (2007) Effect of emulsion polymerization conditions on the electromagnetic properties of magnetic and conductive polyaniline nanoparticles. J Funct Mater Devices 13:529–536

Huang CY, Wu CC (2000) The EMI shielding effectiveness of PC/ABS/nickel-coated- carbon-fibre composites. Eur Polymer J 36(12):2729–2737

Inagaki M, Morishita T, Kuno A, Kito T, Hirano M, Suwa T et al (2004) Carbon foams prepared from polyimide using urethane foam template. Carbon 42(3):497–502

Jesse MB, Douglas JM, Carbon foam EMI shield, US 2008/0078576 A1

Klett J, Hardy R, Romine E, Walls C, Burchell T (2000) High-thermal-conductivity, mesophase-pitch-derived carbon foams: effect of precursor on structure and properties. Carbon 38:953–973

Klett JW, McMillan AD, Gallego NG, Burchell TD, Walls CA (2004) Effects of heat treatment conditions on the thermal properties of mesophase pitch-derived graphitic foams. Carbon 42:1849–1852

Kumar R, Dhakate SR, Saini P, Mathur RB (2013a) Improved electromagnetic interference shielding effectiveness of light weight carbon foam by ferrocene accumulation. RSC Adv 3:4145–4151

Kumar R, Dhakate SR, Mathur RB (2013b) The role of ferrocene on the enhancement of the mechanical and electrochemical properties of coal tar pitch-based carbon foams. J Mater Sci 48:7071–7080

Kumar R, Dhakate SR, Gupta T, Saini P, Singh BP, Mathur RB (2013c) Effective improvement of the properties of light weight carbon foam by decoration with multi-wall carbon nanotubes. Mater Chem A 1:5727–5735

Leslie-Pelecky DL, Rieke RD (1996) Magnetic properties of nano-structured materials. Chem Mater 18:1770–1783

Liu JR, Itoh M, MachidaK-I (2006) Magnetic and electromagnetic wave absorption properties of α-Fe/Z-type Ba-ferrite nanocomposites. Appl Phy Lett 88(6):062503–3

Micheli D, Marcheti M (2012) Mitigation of Human exposure to electromagnetic fields using carbon foam and carbon nanotubes. Engineering 4:928–943

Moglie F, Micheli D, Laurenzi S, Marchetti M, Primiani VM (2012) Electromagnetic shielding performance of carbon foams. Carbon 50:1972–1980

Nicolson AM, Ross GF (1970) Measurement of the intrinsic properties of materials by time-domain techniques, IEEE Trans. Instrum Meas 19:377–382

Oh JH, Oh KS, Kim CG, Hong CS (2004) Design of radar absorbing structures using glass/epoxy composites containing carbon black in X-band frequency ranges. Compos Par B Eng 35:49–56

Park KY, Lee SE, Kim CG, Han JH (2006) Fabrication and electromagnetic characteristics of electromagnetic wave absorbing sandwich structures. Compos Sci Technol. 66:576–584

Saravanan P, Jose TA, Thomas PJ, Kulkarni GU (2001) Bull. Mater Sci 24:515–521

Seo S, Chin WS, Lee DG (2004) Characterization of electromagnetic properties of polymeric composite materials with free space method. Compos Struct 66:533–542

Singh AP, Garg P, Alam F, Singh K, Mathur RB, Tandon RP, Chandra A, Dhawan SK (2012) Phenolic resin-based composite sheets filled with mixtures of reduced graphene oxide, -Fe2O3 and carbon fibers for excellent electromagnetic interference shielding in X-band. Carbon 50:3868–3875

The international centre for powder diffraction data, powder diffraction files, 2001, Card Number: JCPDS-040850

Wang H, Wang G, Li W, Wang Q, Wei W, Jiang Z, Zhang S (2012) A material with high electromagnetic radiation shielding effectiveness fabricated using multi-walled carbon nanotubes wrapped with poly(ether sulfone) in a poly(ether ether ketone) matrix. J Mater Chem 22:21232–21237

Yadav A, Kumar R, Bhatia G, Verma GL (2011) Development of mesophase pitch derived high thermal conductivity graphite foam using a template method. Carbon 49:3622–3630

Yang J, Shen ZM, Hao ZB (2004) Microwave characteristics of sandwich composites with mesophase pitch carbon foams as core. Carbon 42:1882–1885

Yang Y, Gupta MC, Dudley KL, Lawrence RC (2005) Novel carbon nanotube-polystyrene foam composites for electromagnetic interference shielding. Nano Lett 5(11):2131–2134

Yong Y, Gupta MC, Dudley KL, Lawrence RW (2005) Conductive carbon nanofiber- polymer foam structures. Adv Mater 17: 1999–2003

24

Evaluation of the antimicrobial activity and cytotoxicity of phytogenic gold nanoparticles

T. V. M. Sreekanth · P. C. Nagajyothi ·
N. Supraja · T. N. V. K. V. Prasad

Abstract Among the nanoscale materials, noble metal nanoparticles have been attracting the scientific community due to their unique properties and selectivity in biological applications. In the present investigation, gold nanoparticles (AuNPs) were synthesized using rhizome extract of *Dioscorea batatas* through a simple, clean, inexpensive and eco-friendly method. Treating 1 mM chloroauric acid ($HAuCl_4$) with the rhizome extract at 50 °C resulted in the formation of AuNPs. The reduction of AuNPs was observed by the color change of the solution from colorless to dark red wine. The synthesized nanoparticles were characterized using the techniques UV–Vis spectrophotometers, Fourier transform infrared spectroscopy, X-ray diffraction, scanning electron microscopy and transmission electron microscopy. Green synthesized AuNPs were found to be toxic against gram-positive and gram-negative bacteria in liquid media. MTT (dimethyl thiazolyl diphenyl tetrazolium salt) assay showed 21.5 % cell inhibition in lower concentration (0.2 mM) and >50 % cell inhibition after 48 h exposure at higher concentrations (0.8–1 mM).

T. V. M. Sreekanth
Department of Life Chemistry, Catholic University of Daegu, Hayang-eup, Gyeongsan-si, Geyongbuk 712-702, Republic of Korea

P. C. Nagajyothi
Department of Nanomaterials Chemistry, Dongguk University, Seokjang-dong 707, Gyeongju 780 714, Republic of Korea

N. Supraja · T. N. V. K. V. Prasad (✉)
Regional Agricultural Research Station, Nanotechnology Laboratory, Institute of Frontier Technology, Acharya N G Ranga Agricultural University, Tirupati, AP 517 502, India
e-mail: tnvkvprasad@gmail.com

Keywords Green synthesis · *Dioscorea batatas* · Antibacterial studies · Cytotoxicity · AuNPs

Introduction

Nanotechnology is a term regularly used in recent years to describe a set of technologies that deal with objects whose measured size is in the range of 1–100 nm in at least one dimension (Prasad and Giridhara Krishna 2012). The gold nanoparticles can be synthesized using the top-down (physical methods such as thermal decomposition, diffusion, irradiation, etc.) and bottom-up (chemical polyol synthesis method, electrochemical synthesis, chemical reduction), and biological entities for fabrication of nanoparticles (Tikariha et al. 2012). The most popular chemical synthetic methods that are used to prepare nanoparticles require high pressure, high temperature and toxic chemicals like, sodium borohydrate, citrate and alcohols as reducing and capping agents. These agents may have associated environmental pollution because of the toxic organic solvent and external reducing agent used that will affect the environment (Sharma et al. 2009; Bar et al. 2009). Gold nanoparticles (AuNPs) have received major attention by the scientific community as well as industry because of their unique physical and chemical properties. It has been reported that there are a number of potential avenues in which AuNPs were applied including photonics, electronics (Ramgopal et al. 2011), chromatographic techniques (Mayer et al. 2011), Catalysis (Li-Na et al. 2011), identification of pathogens in clinical specimens (Singaravelu et al. 2007), biosensing, gene therapy and DNA sequencing (Thirumurugan et al. 2010). Use of plant extract, fungi, yeast, bacteria and algae could be an alternative to chemical and physical methods for the production

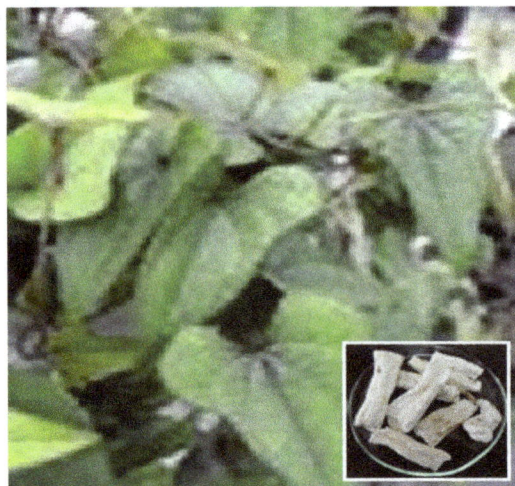

Fig. 1 Typical picture showing the plant *Dioscorea batatas* and rhizome (*inset*)

Fig. 2 Color changing of rhizome extracts containing **a** before synthesis of gold nanoparticles and **b** after synthesis of gold nanoparticles at 50 °C

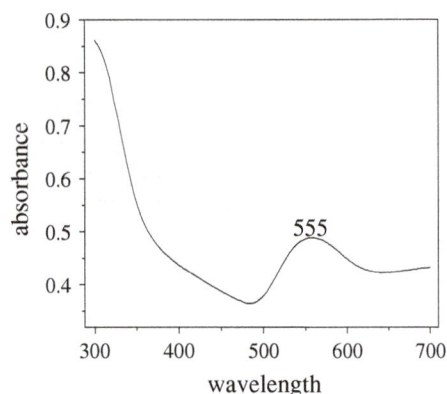

Fig. 3 UV–Vis absorption spectrum showing the localized surface plasma resonance at 555 nm of AuNPs synthesized using the rhizome extract of *Dioscorea batatas*

of nanoparticles in an eco-friendly manner. Several plants such as *Momordica charantia* (Sunil et al. 2012), *Memecylon umbellatum* (Arunachalam et al. 2013), *Mangifera indica* (Phillip 2010), *Trigonella foenum-graecum* (Aswathy Aromal and Philip 2012), *Carthamus tinctorius* (Nagajyothi et al. 2012), *Teraxacum officinale* (Tetty et al. 2012), *Citrus reticulate* (Nagajyothi et al. 2013) and *Lonicera Japonica* (Nagajyothi et al. 2012) have been successfully used for efficient and rapid extracellular synthesis of gold nanoparticles. *Dioscorea* grows extensively on mountains as well as in fields within Korea and Japan. The rhizomes of *Dioscorea batatas* are traditionally used to treat coronary artery disorder caused by blood clotting (Au et al. 2004), and to improve immunity (Phillip 2010; Choi and Hwang 2002). It is also reported the antioxidative reactivity (Choi et al. 2003) and its ability to decrease blood glucose levels (Morrison et al. 2006). In this paper, we present a simple and rapid biosynthesis of gold nanoparticles using *Dioscorea batatas* extract. As prepared, gold nanoparticles were characterized by various methods, such as UV–Vis spectroscopy, FT-IR, SEM–EDS, TEM and XRD. This work provided a potential approach for the production of gold nanoparticles without the involvement of additional chemicals and physical steps.

Materials and methods

Biosynthesis of gold nanoparticles

Dioscorea rhizomes were collected from Gyeongju Oriental Medical College, Gyeongju, South Korea (Fig. 1). The rhizomes were air-dried for 10 days, then kept in a hot air oven at 60 °C for 24 h. Rhizomes were ground to fine powder. 5 g of powder was mixed in 100 ml of water and boiled at 100 °C for 10 min and the solution was filtered using Whatman filter paper (No. 41) and the extract was collected and stored in a plastic bottle. 190 ml of 1 mM chloro auric acid (HAuCl$_4$ purchased from Sigma-Aldrich Chemical Pvt. Ltd.) was added to 10 ml of rhizome extract to make up a final solution 200 ml (10 ml rhizome extract + 190 ml HAuCl$_4$ solution). A change in the color of solution was observed during the heating process at 50 °C.

Characterization of AuNPs

The Localized Surface Plasmon Resonance (LSPR) of AuNPs was recorded using UV–visible spectrophotometer (Cary 4000 UV–Vis spectrophotometer) with the scanning range 200–800 nm. Air-dried, platinum coated samples were performed using an analytical scanning electron microscope (Hitachi s-3500N) equipped with an energy dispersive X-ray spectrum (EDS), TEM (H-7100 Hitachi), XRD Philips (Netherlands) and FT-IR (Bruker model,

Fig. 4 Energy dispersion X-ray spectrum (EDS) micrograph of bio-synthesized AuNPs showing the elemental presence of Au, C and K

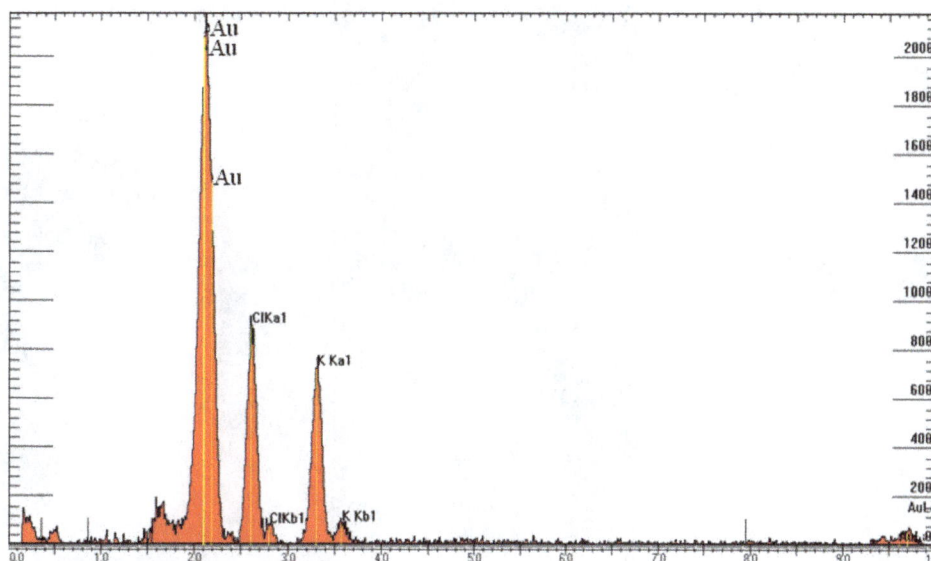

Fig. 5 Transmission electron microscopic micrograph (200 nm *bar*) showing polydispersed gold nanoparticles (size 18.48–56.18 nm) synthesized using the extract of *Dioscorea batatas* rhizome

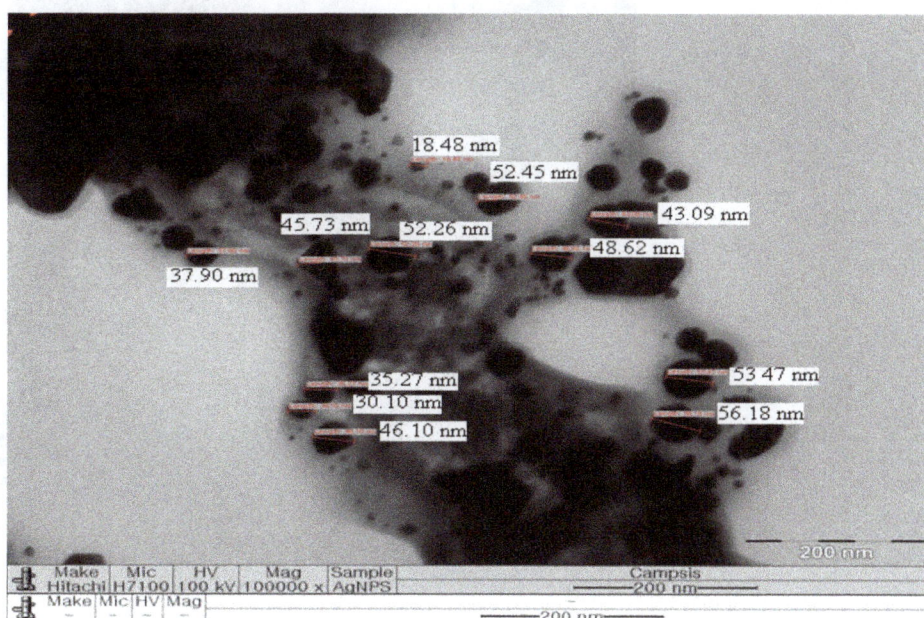

TENSOR 37). Particle sizing experiments were carried out by means of laser diffractometry, using Zeta sizer nano series (Malvern). Measurements were taken in the range of 0.1 and 1×10^8 μm.

Bacterial growth curve

The antibacterial activity of gold nanoparticles against gram-positive and gram-negative bacteria was analyzed by their growth curve. To examine the effect of various concentrations (1, 0.5. 0.25 and 0.125 mM) of AuNPs on bacteria growth, organisms were grown overnight in nutrient broth (NB) at 37 °C. The growth of gram-positive and gram-negative in broth media was indexed by measuring the optical density (at $\lambda = 600$ nm) using ELISA. Ampicillin used as control.

MTT (dimethyl thiazolyl diphenyl tetrazolium salt) assay

This assay is based on the ability of mitochondria succinate dehydrogenase enzymes in living cells to reduce the yellow water soluble substrate dimethyl thiazolyl diphenyl tetrazolium (MTT) into a purple formazan product which is analyzed spectrophotometer. B16/F10 melanoma cell line was used to examine the toxicity of the present green

Fig. 6 a–d Transmission
electron microscopic
micrographs showing
anisotropic structures of AuNPs
synthesized using the rhizome
extract of *Dioscorea batatas*

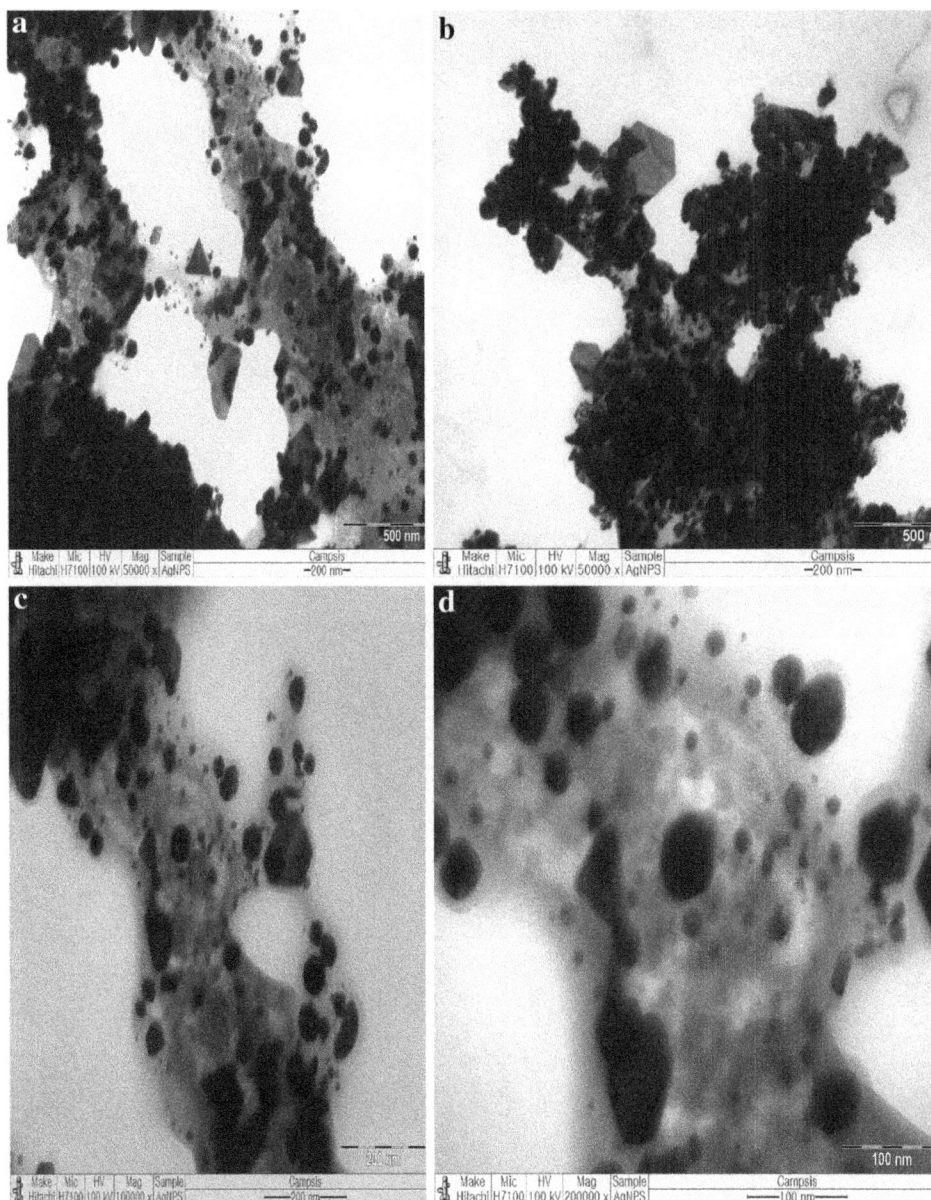

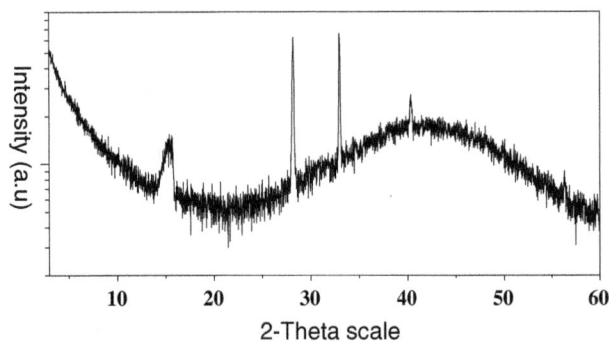

Fig. 7 XRD micrograph showing Bragg's reflections corresponds to
the face-centered cubic (FCC) structure of AuNPs synthesized using
the rhizome extract of *Dioscorea batatas*

synthesized AuNPs. Cells were sub cultured in DMEM
supplemented with 10 % FBS and 1 % penicillin–streptomycin at 5 % CO_2 at 37 °C. At about 90 % confluence,
cells were harvested using 0.25 % trypsin and were seeded
in 24-well plates. To each well, 500 μl of diluted cell
suspension (3.05×10^5 cell/ml) was added. After 24 h,
when the monolayer formed the supernatant was flicked off
and 200 μl of different test samples (1.0, 0.8, 0.6, 0.4 and
0.2 mM) were added to the cells, followed by incubation
for 48 h. The medium was removed by suction and 200 μl
of MTT was added to each well. The plates were gently
shaken and incubated for 4 h. The supernatant was
removed, 200 μl of DMSO was added, and the plates were

a

b

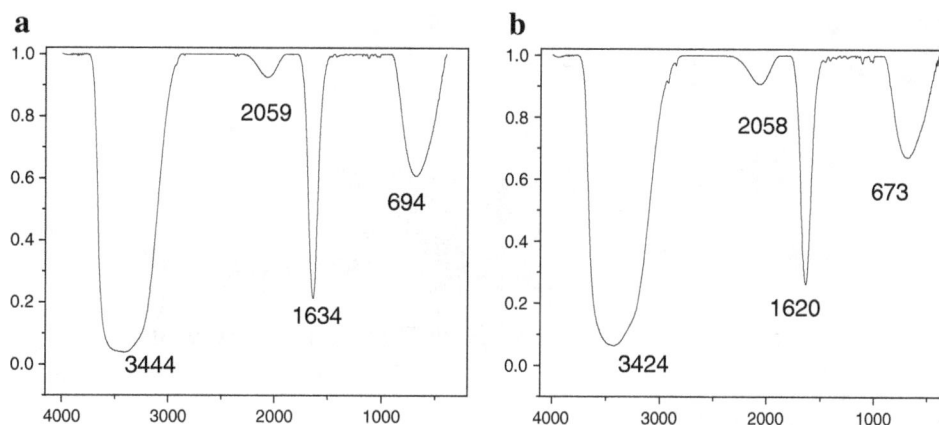

Fig. 8 FT-IR spectrum of **a** *Dioscorea batatas* rhizome extract and **b** the functional groups responsible for the reduction and stabilization of AuNPs

gently shaken to solubilize the formed formazan. The absorbance was measured using a microplate reader at a wavelength of 570 nm. The experiment was done twice and averaged.

Statistical analysis

Each experiment was maintained with three replications and the data were analyzed using two ways ANOVA.

Results and discussion

UV–Vis Spectrophotometric studies: Recording of localized surface plasmon resonance (LSPR)

Bioreduction of aqueous $AuCl_4^-$ ions can easily be followed by UV–Vis spectrophotometer, and one of the most important features in optical absorbance spectra of metal nanoparticles is surface plasmon band, which is due to collective electron oscillation around the surface mode of the particles. Previous studies have shown that gold exhibits red wine color and silver exhibits yellowish-brown color due to the excitations of their surface plasmon response (SPR) (Mulvaney 1996), when dissolved in water. The change in color of the solution was observed (from colorless to dark red wine color) after keeping the solution at 50 °C for 25 min (Fig. 2).

In the case of gold, the reduction started within 5 min after the addition and completed in 30 min. The possible explanation of difference in the reduction time could be due to the difference in their reduction potential for both the metal ions. Metal nanoparticles such as gold have free electrons, which give rise to SPR absorption band (Noginov et al. 2007) at and around 555 nm (Fig. 3). The reduction and stabilization of Au^+ ions could be

done by combinations of biomolecules found in the extracts such as proteins, aminoacids, polysaccharides and vitamins which are evidenced in FT-IR studies (Thirumurugan et al. 2010).

Transmission electron microscopy (TEM) and energy dispersion spectroscopic (EDS) measurements

The EDS spectrum of AuNPs synthesized at 50 °C is shown in Fig. 4. Strong signals from the gold atoms in the nanoparticles were observed, and signals from Si, K and C atoms were also recorded. The presence of C and K, signals were likely due to X-ray emission from carbohydrates/proteins/enzymes present in the cell wall of the biomass. The presence of the elemental gold can be observed in the graph obtained from EDS analysis, which also supports the XRD results. The TEM images of AuNPs are shown in (Figs. 5, 6). The TEM images have shown that the formed AuNPs were polydispersed and were predominately spherical in nature. But it is evident from TEM micrographs that triangular, hexagonal, rod and irregular shaped nanoparticles were also formed.

X-Ray diffraction (XRD): study of crystalline structure of gold nanoparticles

The XRD pattern of the AuNPs is shown in (Fig. 7). Bragg reflections obtained in the micrograph clearly indicated the presence of (111) and (200) sets of lattice planes which is a consequence of crystalline nature of formed AuNPs and indexed as face-centered-cubic (FCC) structure of gold. In addition to the Bragg peaks representative of FCC AuNPs, additional as yet unassigned peaks are also observed suggesting that the crystallization of bio-organic phase occurs on the surface of the nanoparticles.

Fig. 9 Particle size distribution of AuNPs by size, volume and correlation synthesized using *Dioscorea batatas* rhizome extract

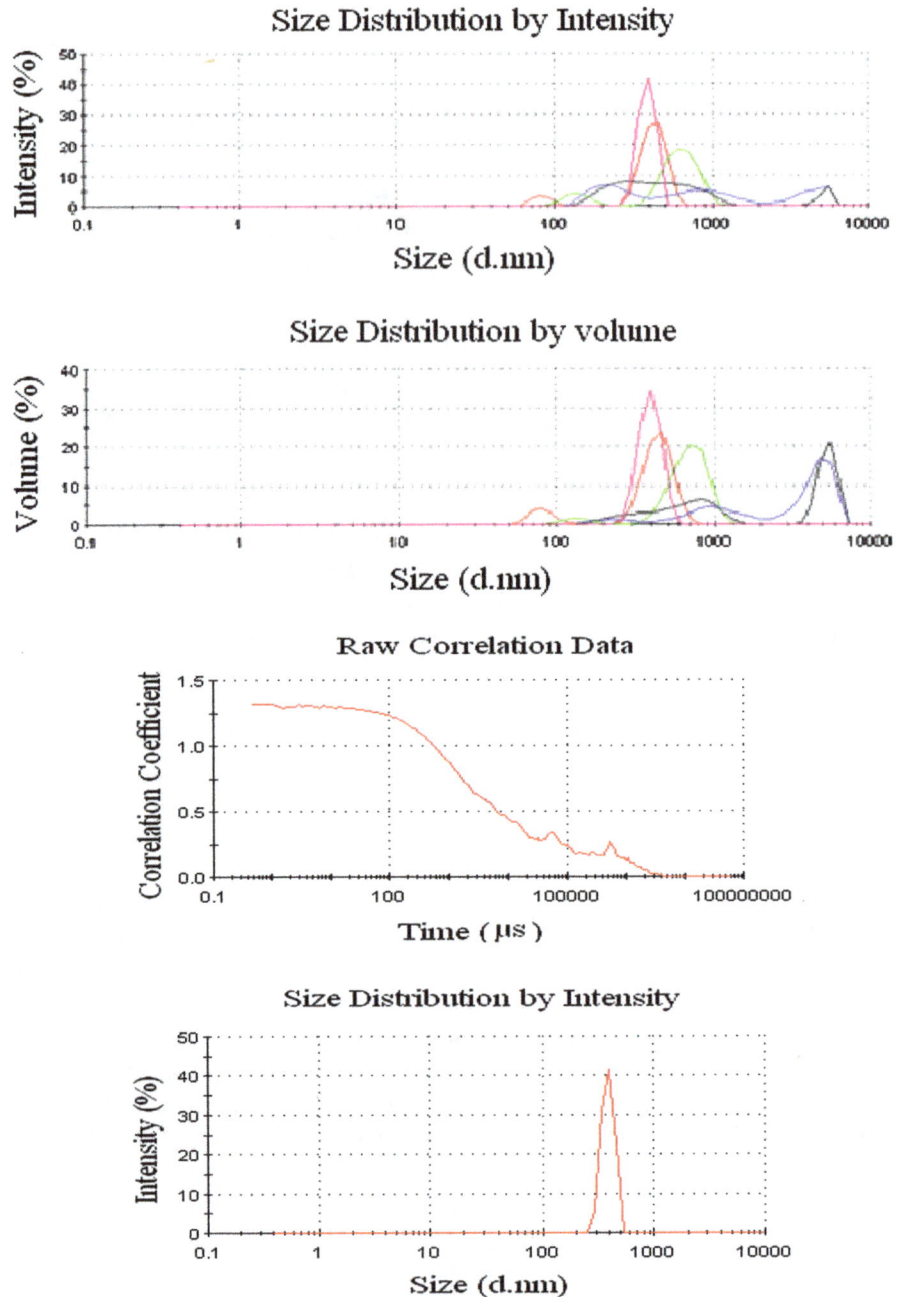

Fourier transform infrared (FT-IR) spectroscopic studies

FT-IR results reveal the absorption bands at 3,444, 1,634 and 694 cm^{-1} (rhizome extraction) (Fig. 8); 3,424, 1,620 and 673 cm^{-1} (AuNPs at 50 °C) (Fig. 9b), respectively. The vibrational bands corresponding to the bonds such as amines (N–H stretch), –C=C (alkane), C–Cl (Halogens) which was in the region range of 694–3,444 cm^{-1}. The most wide spectrum absorption was observed at 3,424 and 3,444 cm^{-1} and it can be attributed to the stretching vibrations of amino (N–H) (Rajasekharreddy et al. 2011), absorption peaks centered at 1,620 and 1,634 cm^{-1} can be

attributed to the stretching vibration of –C=C (alkane) (Zhu 2000). Amines are a particularly attractive class of reducing agents because of their structural or chemical properties (Newman and Blanchard 2006). Thus, the FT-IR micrograph reveals that amides that are present in the extract are responsible for the reduction and stabilization of the gold nanoparticles.

Dynamic light scattering (DLS) technique: particle size measurement of hydrosol

Particle size determination of the formulated AuNPs was shown under different categories like size distribution by

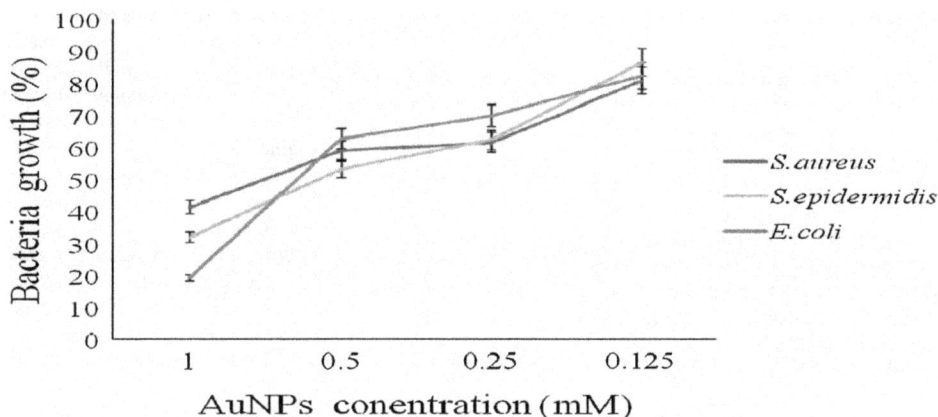

Fig. 10 Growth curves of, *Staphylococcus aureus* (KCTC 1916), *Staphylococcus epidermidis* (KCTC 1917) and *Escherichia coli* (KCTC 2441) in the presence of different AuNPs concentrations

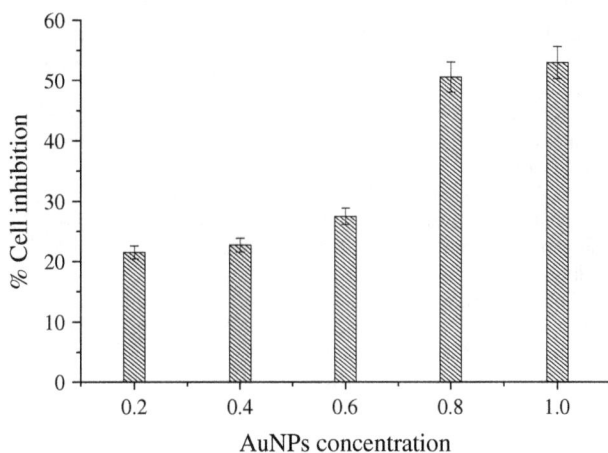

Fig. 11 Cytotoxic effect of AuNPs synthesized using *Dioscorea batatas* rhizome extract on B16/F10 melanoma cells cell line by MTT (dimethyl thiazolyl diphenyl tetrazolium salt) assay

volume and intensity (Fig. 9). The size distribution by volume gives a bell shaped (Fig. 9) pattern which indicates the wide range size distribution of nanoparticles in the sample formulation. The volume % of the samples were found to be in the range of $0.1-1 \times 10^8$(AuNPs) The formed AuNPs are well distributed with respect to volume and intensity, an indication of the formation of well built AuNPs and their mono and poly disparity, respectively.

Antibacterial and cytotoxicity activities of AuNPs

The growth curves of *S. aureus* (41.1, 58.8, 61.3 and 80.77 %) *S. epidermidis* (31.8, 52.88, 62.14 and 86.5 %) and *E. coli* (19.1, 62.43, 69.84 and 82.2 %) in MH broth medium in the presence of AuNPs at 1, 0.5, 0.25 and 0.125 mM concentrations are shown in Fig. 10. Zhang et al. (2008) used an amido-amine coated AuNPs and tested their toxicity on a large suite of gram-negative and positive bacteria. At 2.8 mg/L, AuNPs demonstrated up to a 98 % inhibition of bacterial growth. Gold nanoparticles possess well-developed

surface chemistry, chemical stability and appropriate smaller size, which make them easier to interact with the micro-organisms (Nirmala Grace et al. 2007). Also, the particles interact with the building elements of the outer membrane and might cause structural changes, degradation and finally cell death (Zawrah and Sherein 2011).

The incubation of B16/F10 melanoma cells with synthesized AuNPs significantly reduced the viability of these cells in a dose dependent manner and all concentrations (0.2–1 mM) were found toxic to the cells (Fig. 11). The maximum inhibition of proliferation was 52.98 % at higher concentration (1 mM); 21.51 % minimum cell inhibition observed at lower concentration (0.2 mM). From the results, it is evident that synthesized AuNPs have inhibitory effect on SGT oral cancer cells. It is also important to recognize that a vast majority of gold (I) and gold (III) compounds show varying degrees of cytotoxicity to a variety of cells (Basset et al. 2003). Further size and shape dependent uptake of gold nanoparticles into mammalian cells has been reported and points to the need of in depth study of size and shape dependent antimicrobial and cytotoxic effects of nanoparticles (Devika Chitrani et al. 2006).

Conclusion

The rhizome extract of *Dioscorea batatas* is found to be one of the potential candidates for the green synthesis of AuNPs. The spectroscopic characterization data such as UV–Vis, FT-IR and SEM also support the formation and stability of the bio-synthesized AuNPs. Low bacteria growth observed in 1 mM concentration. 21.51 % cell inhibition observed at lower concentration (0.2 mM) and maximum cell inhibition (52.98 %) was observed at higher concentration (1 mM). This simple, efficient and rapid green synthesis of AuNPs can be used to produce large scale production of gold nanoparticles for their use in various biomedical and biotechnological applications.

References

Arunachalam KD, Annamalai SK, Hari S (2013) One-step green synthesis and characterization of leaf extract-mediated biocompatible silver and gold nanoparticles from *Memecylon umbellatum*. Int J Nanomed 8:1307

Aswathy Aromal S, Philip D (2012) Spectroch. Acta. A: green synthesis of gold nanoparticles using Trigonella foenum-graecum and its size-dependent catalytic activity. Mol Biomol Spectrosc 97:1

Au ALH, Kwok CC, Lee ATC, Kwan YW (2004) Acute simvastatin inhibits K_{ATP} channels of porcine coronary artery myocytes. Eur J Pharmacol 502:123

Bar H, Bhui DK, Sahoo GP, Sarkar P, Pyne S, Misra A (2009) Synthesis of silver nanoparticles using latex of *Jatropha curcas*. Colloid Surf A: Physicochem Eng Aspects 348:212

Basset C, Vadrot J, Denis J, Poupon J, Zafrani ES (2003) Prolonged cholestasis and ductopenia following gold salt therapy. Liver Int 23:89

Choi EM, Hwang JK (2002) Enhancement of oxidative response and cytokine production by yam mucopolysaccharide in murine peritoneal macrophage. Fitoterapia 73:629

Choi EM, Koo SJ, Hwang JK (2003) Synergistic induction of iNOS by IFN-γ and glycoprotein isolated from *Dioscorea batatas*. J Ethnopharmacol 91:1

Devika Chitrani B, Ghazani AA, Chan WCW (2006) Determining the Size and shape dependence of gold nanoparticle uptake into mammalian cells. Nanoletters 6:662

Li-Na M, Dian-Jun L, Zhen-Xin W (2011) Nanomaterials for medical application. Chin J Anal Chem 38:1

Mayer AMS, Rodriguez AD, Berlinck RGC, Fusetani N (2011) Marine compounds with antibacterial, antidiabetic, antifungal, anti-inflammatory, antiprotozoal, antituberculosis, and antiviral activities; affecting the immune and nervous systems, and other miscellaneous mechanisms of action. Comp Biochem Physiol C: Toxicol Pharmacol 153:191–222

Morrison EY, Ragoobirsingh D, Peter SA (2006) The unitarian hypothesis for the aetiology of diabetes mellitus. Med Hypotheses 67:1115

Mulvaney P (1996) Surface plasmon spectroscopy of nanosized metal particles. Langmuir 12:788

Nagajyothi PC, Sreekanth TVM, Prasad TNVKV, Lee KD (2012a) Green synthesis of silver and gold nanoparticles using *Lonicera Japonica* flower extract. Adv Sci Lett 5:124

Nagajyothi PC, Lee SE, An Minh, Lee KD (2012b) Green Synthesis of silver and gold nanoparticles using *Lonicera Japonica* flower extract. Bull Korean Chem Soc 33:2609

Nagajyothi PC, Lee KD, Sreekanth TVM (2013) Plants as green source towards synthesis of nanoparticles. J Optoelectron Adv Mater 15:269

Newman JDS, Blanchard GJ (2006) Formation of gold nanoparticles using amine reducing agents. Langmuir 22:5882

Nirmala Grace A, Pandian K, Collo. Surfa A (2007) Antibacterial efficacy of aminoglycosidic antibiotics protected gold nanoparticles—a brief study. Physicochem Eng Aspects 297:63

Noginov MA, Zhu G, Bahoura M, Adegoke J, Small C, Ritzo BA, Drachev VP, Shalaev VM (2007) The effect of gain and absorption on surface plasmons in metal nanoparticles. Appl Phys B 86:455

Phillip D (2010) Spectroch. Acta. A: *Gnidia glauca* flower extract mediated synthesis of gold nanoparticles and evaluation of its chemocatalytic potential. Mol Biomol Spectrosc 77:807

Prasad TNVKV, Giridhara Krishna T (2012) Soil nanoscience: plenty of room at the bottom. World J Appl Environ Chem 1:72–75

Rajasekharreddy P, Usha Rani P, Sreedhar B (2011) Green synthesis of silver-protein (core–shell) nanoparticles using *Piper betle* L. leaf extract and its ecotoxicological studies on *Daphnia magna*. J Colloids Surf A: Physiochem Eng Aspects 12:1711

Ramgopal M, Saisushma C, Attitalla IH, Alhasin AM (2011) A facile green synthesis of silver nanoparticles using soap nuts. Res J Microbiol 6:432

Sharma VK, Yngard RA, Lin Y (2009) Silver nanoparticles: green synthesis and their antimicrobial activities. Adv Colloid Interface Sci 145:83

Singaravelu G, Arockiamary JS, Ganesh Kumar V, Govindaraju K (2007) A novel extracellular synthesis of monodisperse gold nanoparticles using marine alga, *Sargassum wightii* Greville. Colloid Surf B-Biointerface 57:97

Sunil P, Goldie O, Mewada A, Sharon M (2012) Green synthesis of highly stable gold nanoparticles using Momordica charantia as nano fabricato. Arch Appl Sci Res 4:1135

Tetty CO, Nagajyothi PC, Lee SE, Ocloo A, Minh An TN, Sreekanth TVM, Lee KD (2012) Anti-melanoma, tyrosinase inhibitory and anti-microbial activities of gold nanoparticles synthesized from aqueous leaf extracts of *Teraxacum officinale*. Int J Cosmet Sci 34:150

Thirumurugan A, Jiflin GJ, Rajagomathi G, Neethu Anns T, Ramachandran S, Jaiganesh R (2010) Synthesis of gold nanoparticles of *Azadirachta indica* leaf extract. Int J Biol Technol 1:75

Tikariha S, Singh S, Banerjee S, Vidyarthi AS (2012) Anthelmintic efficacy of gold nanoparticles derived from a phytopathogenic fungus Nigrospora oryzae. Int J Pharma Sci Res 3:1603

Zawrah MF, Sherein I (2011) Antimicrobial activities of gold nanoparticles against major foodborne pathogens. Life Sci J 8:37

Zhanh Y, Peng H, Huang W, Zhou Y, Yan DJ (2008) Facile preparation and characterization of highly antimicrobial colloid Ag or Au nanoparticles. Colloid Interface Sci 325:371

Zhu M (2000) Biocompactibility synthesis of silver and gold nanoparticles. Apparatus analyses. Higher education press, Beijing

25

Molecular scale analysis of dry sliding copper asperities

Bhavin N. Vadgama · Robert L. Jackson ·
Daniel K. Harris

Abstract A fundamental characterization of friction requires an accurate understanding of how the surfaces in contact interact at the nano or atomic scales. In this work, molecular dynamics simulations are used to study friction and deformation in the dry sliding interaction of two hemispherical asperities. The material simulated is copper and the atomic interactions are defined by the embedded atom method potential. The effect of interference, δ, relative sliding velocity, v, asperity size, R, lattice orientation, θ, and temperature control, on the friction characteristics are investigated. Extensive plastic deformation and material transfer between the asperities were observed. The sliding process was dominated by adhesion and resulted in high effective friction coefficient values. The friction force and the effective friction coefficient increased with the interference and asperity size but showed no significant change with an increase in the sliding velocity or with temperature control. The friction characteristics varied strongly with the lattice orientation and an average effective friction coefficient was calculated that compared quantitatively with existing measurements.

Keywords Adhesion · Friction mechanisms ·
Nanotribology · Copper · Contact mechanics

Background

Understanding the physics of friction is of fundamental importance for a wide range of applications and more so for small-scale applications. With the rapid development of surface examining technologies like AFM, FFM and others as well as development of MEMS and NEMS devices, a better understanding of the atomistic mechanisms of sliding friction is essential. When surfaces in contact slide across each other only a small number of micro or nano sized peaks or asperities truly come into contact. The manner in which these contacting asperities interact has a significant influence on the frictional characteristic of the sliding surfaces. The phenomena occurring at the nano-scale are therefore complex and difficult to predict since the tribological properties of sliding contacts are greatly affected by the adhesion and contact deformation. Furthermore, the adhesion is directly proportional to the number of atomic or molecular bonds that are broken and formed at the interface of contacting surfaces during sliding (Landman et al. 2004). Below a certain scale the dependence of the friction and dissipation of frictional heat on factors such as inter-atomic forces, surface topography and composition of materials increases (Achanta et al. 2009). Since the asperities or peaks on rough surfaces occur at multiple scales and may have contact areas with values scaling over many orders of magnitudes, their properties can vary significantly due to scale dependent mechanisms (Jackson 2006). Therefore, the behavior of asperities at the smaller scales may benefit from techniques such as molecular dynamics (MD) to characterize them. Blau (1991) has shown that steady-state friction is highly scale dependent and therefore the friction mechanisms should be modeled considering the entire tribo-system rather than a discrete asperity system. Such an exercise to model the scale effects would be extremely productive if carried out using MD, provided the computational resources allow.

Computer simulations, specifically using MD, have become quite popular among researchers with much work

B. N. Vadgama · R. L. Jackson (✉) · D. K. Harris
Department of Mechanical Engineering, Auburn University,
Auburn, AL 36849, USA
e-mail: jacksr7@auburn.edu

being done in micro and nano tribology. In one of the early works, Harrison et al. (1992) investigated the atomic-scale friction of diamond surfaces using MD. Landman et al. (1996) studied the formation and properties of interfacial junctions in the normal contact between a hemispherical asperity and a flat surface. They also study the shearing of lubricated junctions in asperity–asperity sliding geometry. In a later work (Landman et al. 2004), they reviewed the Amontons' Law with the help of MD and suggested that for adhering surfaces an additional contribution, which is proportional to the real molecular contact area, is present in the basic equation of friction. Song and Srolovitz (2007) performed MD simulations of single asperity contact and deformation on a flat surface over single- and multi-cycle loading and unloading. In an extensive study, Sorensen et al. (1996) conducted MD simulations on atomic-scale friction of sliding copper surfaces. The dependence of the normal load, contact area, sliding velocity, temperature, and lattice mismatch was investigated with an emphasis on observing slip-stick for different tip-surface and surface–surface contacts. They observed atomic scale stick slip for matching as well as non-matching surfaces; however, it decreased with an increase in the contact size. Zhang and Tanaka (1997) performed MD simulations of a diamond tip sliding over a flat copper surface. They found that four distinct deformation regimes existed characterized by no-wear, adhering, ploughing, and cutting of the surface. These were governed by indentation depth, sliding speed, asperity geometry, and surface conditions. Investigation of the simple case of a single asperity sliding over a flat surface using MD has also been carried out by several other research groups (Cho et al. 2005; Jeng et al. 2007; Yang and Komvopoulos 2005; Zhu et al. 2011; Zhang et al. 2001; Ivashchenko and Turchi 2006). However, an important aspect of the frictional sliding process between rough surfaces is the interaction between two separate asperities rather than an asperity and a flat surface. The current work studies two separate hemispherical asperities, which has surprisingly only been studied a few times using molecular dynamics.

It would be useful to study asperity–asperity interaction to gain useful insights on the deformation mechanisms and frictional characteristics in addition to the work done on the asperity-surface case as mentioned above. In the most similar work, Zhong et al. (2003) performed MD simulations of two asperities sliding into each other to study such asperity–asperity interaction. In their work, they modeled a sliding "hard" upper asperity deforming a fixed lower aluminum asperity, which is still very different from the two deformable copper asperities considered in the current work. They studied the effects of a wide range of conditions including sliding velocities, temperatures, and crystal

orientations on the wear process. At the micro-scale, several continuum based semi-analytical as well as finite element based models have also been developed for friction between sliding asperities and cover both elastic as well as plastic deformations (Jackson et al. 2007; Faulkner and Arnell 2000; Boucly et al. 2007). Therefore, in addition, it would be interesting and insightful to see if these models can be compared to MD simulations at the nano-scale. There have also been recent works that blend molecular dynamics and the finite element method, but the current work uses only the MD method (Eid et al. 2011).

The origin of friction is an extremely complex phenomenon and is an open and growing research field. The existing theories developed for the bulk are not always consistent with material behavior at the molecular scale. The relationship between experimentally measured values of friction to the material properties is still not clear which makes the prediction of friction challenging. The objective of this work is to numerically study the atomistic mechanism of friction and contact deformation using molecular dynamics simulations on the dry sliding contact of nano-scale asperities. The material selected for this work is copper, which has been increasingly used in nano-engineering where metallic properties are of importance. The higher electrical and thermal conductivity along with low friction makes copper suitable for use in applications such as MEMS switches (Barriga et al. 2007) or as an additive to enhance the "wear mending" property of a lubricant (Liu et al. 2004). Besides its application in MEMS, copper is also the nano-material of choice to be incorporated in the friction material used in automotive brake pads for its ductility and high thermal conductivity (Osterle et al. 2010). Copper nanoparticles have been used as an additive to enhance the tribological properties of lubricants (Tarasov et al. 2002; Choi et al. 2009). The effects of interference δ, relative sliding velocity v, asperity size R, lattice orientation θ, and temperature control, on the friction characteristics are investigated quantitatively and qualitatively. All the simulations were carried out using the MD code LAMMPS (Plimpton 1995).

Methodology

The geometric 3D model used here to represent asperity–asperity sliding contact is shown in Fig. 1. The model consists of upper and lower sections each having a hemispherical asperity and a rigid base, composed of copper atoms in a FCC lattice structure. The x, y, and z axes are oriented in the [100], [010] and [001] lattice directions, respectively. Periodic boundary conditions are imposed along the x- and z-axis so that the atoms can exit one side

Fig. 1 Asperity–asperity sliding contact simulation model

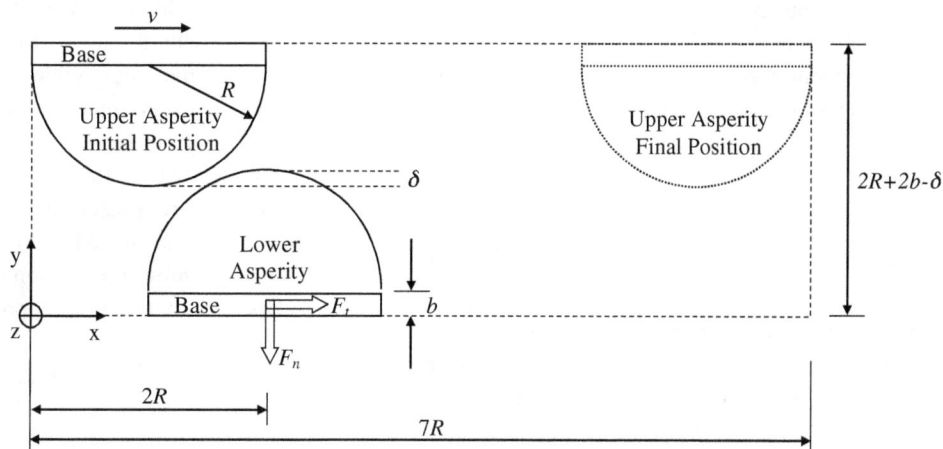

Table 1 Computation parameters for simulations

	$R = 5$ nm	$R = 7.5$ nm	$R = 10$ nm
b	0.5 nm	0.5 nm	0.5 nm
δ/R	0.10 to −0.04	0.10 to −0.04	0.10 to −0.04
Base atoms	~4,600	~10,500	~18,400
Asperity atoms	~22,400	~75,000	~177,000
Total atoms	~54,000	~172,000	~390,800

of the boundary and re-enter on the other side. Non-periodic and shrink-wrapped (the extents of the boundary are set so as to include the atoms in that dimension no matter how far they move (Plimpton 1995) boundary conditions are imposed along the y-axis. The size of the model and the number of atoms in the simulation are chosen to obtain a reasonable balance between nano-scale representation and the computational time and power required. Copper asperities of radii 5, 7.5 and 10 nm were used in this work and the number of corresponding atoms and system dimensions are given in Table 1.

The atomic interactions were described using the embedded atom method (EAM) given by Daw and Baskes (1984) provided in the LAMMPS library. The EAM treats each atom as embedded in a host lattice consisting of all other atoms where the total energy of a system with N atoms is given by,

$$E_{\text{tot}} = \sum_{i=1}^{N} G_i \left[\sum_{j \neq i}^{N} \rho_j (r_{ij}) \right] + \frac{1}{2} \sum_{i,j \neq i}^{N} U_{ij} (r_{ij}). \qquad (1)$$

Here, G_i is the energy required to embed atom i into electron density ρ_j created by surrounding atoms and $U_{ij}(r_{ij})$ is the electrostatic pair-potential between atoms i and j separated by distance r_{ij}. In metals, since the valence electrons may be shared between atoms, local electron densities need to be taken into account and the bonds between atoms are not independent of each other.

Therefore, density-independent pair-potentials like the Lennard–Jones (LJ) potential are not always suitable to capture the physics of metallic bonding (Foiles et al. 1986). The EAM effectively includes the many-body contributions and allows the electron density to vary between surface and the bulk, and as a result accurately describes bulk and defect properties (energy, structural, mechanical, and thermal properties) of metals (specially, fcc metals; Foiles et al. 1986) and metal alloys (Plimpton and Hendrickson 1993, Plimpton 1995) when compared to other interatomic potentials. It should also be noted that EAM only accounts for the contribution of bound electrons to atomic potentials and not that of the valence electrons. An important difference of the current work compared to that of Zhong et al. (2003) is that in latter's work a generic LJ potential was used for the hard (non-deformable) upper asperity and the EAM potential was used for the lower Al asperity.

Newton's equations of motion were numerically integrated using the velocity-Verlet algorithm with a time-step size of 0.002 ps. The time-step was empirically determined to capture the system dynamics as best as possible while keeping the computational time reasonable. The MD simulations were performed in two stages on a high-performance computing cluster using between 40 and 80, 2.8 GHz Intel Xeon processors in parallel. First, for each simulation run, the system was initialized and equilibrated at 300 K temperature for 10 ps (5,000 time-steps). For the remaining time of the simulation, the temperature control was not enforced anywhere on the system. However, it was found that holding the temperature constant on the base only affected the effective friction coefficient in a minor way (as shown in a later section). After the equilibrating cycle, the top asperity was set in motion towards the bottom asperity by imposing an average velocity on the group of atoms in the base region along the x-direction. During the asperity interaction, normal and tangential reaction forces were monitored on the base regions. These forces were averaged over the sliding distance and an effective friction coefficient was calculated for the distance through which the asperities were in contact, given by

$$F = \frac{1}{N} \sum_{i=n_i}^{n_o} f_i \tag{2}$$

$$\mu_{\text{eff}} = \frac{F_t}{F_n}. \tag{3}$$

Here, n_i and n_o are the force values corresponding to the time when the asperities come in contact and get out of contact, respectively and N is the number of force values between n_i and n_o. Note that μ_{eff} is not the same as the friction coefficient measured at the macro scale. However, they may correlate if an average is taken for different asperity sizes at different interferences.

Results and discussion

Asperity–asperity interaction

In order to obtain the sliding motion of the asperities across each other, the atoms in the base region of the upper asperity were imposed with a velocity in the x-direction ([100] direction). The lower asperity was held fixed to its position while the position of the upper asperity was set to get interference with the lower asperity for a range of $0.1R$ to $-0.04R$. The δ/R values of less than zero were included for two reasons: (1) at theoretical zero interference, a positive interference was generated when the system was equilibrated at 300 K as the atoms relaxed and (2) the adhesion was strong enough to pull the atoms to make contact if the two asperity tips were within the attractive range. The simulations were run for a sufficient number of time-steps to ensure that the asperities were completely out of contact. At every time-step, the forces on the atoms in the base regions were summed and recorded to obtain the reaction force components. These values of forces were averaged for the time the asperities remained in contact. A visualization tool was used to note the start and end of the contact and this was also verified by the values of forces, which averaged to zero before and after the contact.

Figure 2 shows the asperities at the start and end of sliding contact for $\delta/R = 0.10$ and $v = 10$ m/s for the asperity of radius 5 nm. In the first part of the sliding process, the upper asperity makes contact with the lower one and a junction is formed. This junction grows through the sliding as more atoms bond to each other and the free surface decreases until the centers of the asperities align. In the second part, this adhered junction starts stretching along the sliding direction as the asperities move apart and

finally breaks after necking. Such formation of necking is consistent with the work by Zhong et al. (2003). They report that the neck glides on the Al surface along a favorable slip system and material is transferred from lower Al tip to the upper LJ tip by adhesion. It can be seen (in Fig. 2) that in the current work both the asperities undergo severe plastic deformation along with several atoms being transferred from one asperity to the other.

Figure 3 shows the friction force which starts from zero, reaches a maximum at approximately $x/R = 0.75$ soon after the asperities align and decreases back to zero as the asperities come out of contact. The first half of the sliding process is characterized mostly by the ploughing of atoms while the second half is dominated by adhesion. This is further supported by the fact that the normal force, as shown in Fig. 4, starts increasing as the asperities come into contact and peaks approximately when the upper and the lower asperities align at $x/R = 0.0$. It then starts decreasing and eventually changes direction. Even before making a contact, as the asperities get closer the adhesion force pulls them into contact. This is also seen at x/R values between -0.75 and -0.50. At the micro-scale, the inverse trend has been observed for the friction and the normal forces (Jackson et al. 2007; Faulkner and Arnell 2000). In the elasto-plastic sliding of micro scale hemispherical asperities without adhesion, it is the friction force that changes direction as the asperities push each other apart at the end of the contact. The normal force does not show this behavior due to the absence of a strong adhesion force. However, here the adhesion was strong enough to pull the atoms of the asperities to make contact even when the interference was negative, i.e. the asperities were separated by a distance of $0.04R$. Similar behavior was observed for copper asperity sliding on a copper work piece where strong adhesion resulted in larger contact area (Zhang et al. 2001; Cha et al. 2004; Li et al. 2003).

Effect of interference, δ

Figure 5 shows the effect of normalized interference (δ/R) on the averaged friction force for all three asperity sizes and relative sliding velocities. Here, the interference δ, is normalized by the asperity radius R and the interference values used in the current work is in the range $0.1\delta/R$ to $-0.04\delta/R$. This range of interferences is similar to that encountered for the single asperities in rough surface contacts. For example, large deformations are usually not as statistically significant in a rough surface contact due to there being very few of them (Jackson and Green 2006).

As the interference increases, the number of atoms, which interact with each other, also increases. This causes increased atom displacement as well as adhesion and results in an increased friction force, which is needed to

Fig. 2 Copper asperities before (*left*) and after (*right*) sliding process for $R = 7.5$ nm, $\delta/R = 0.10$ and $v = 10$ m/s

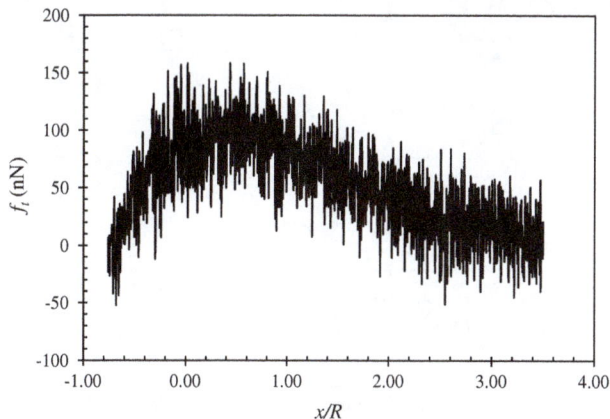

Fig. 3 The friction force during sliding in normalized sliding direction for $R = 7.5$ nm, $\delta/R = 0.10$ and $v = 10$ m/s

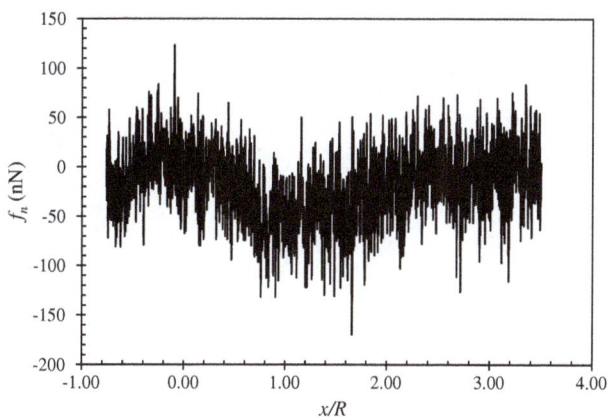

Fig. 4 The normal force during sliding in normalized sliding direction for $R = 7.5$ nm, $\delta/R = 0.10$ and $v = 10$ m/s

plough through the atoms and also overcome the adhesion. Thus, the friction force is directly proportional to the interference. This is not the case with the normal force (see Fig. 6). The normal force, while not showing a definite trend, seems to randomly fluctuate for certain interference values, which are different for each asperity size. This also results in a fluctuation of the effective coefficient of friction with interference (δ/R). The visualizations of the sliding process were analyzed which revealed that this fluctuation was caused due to the vertical alignment of the lattice and the resulting discretization of the smooth surface geometry

for a given value of interference. However, the asperities have the same average continuous geometry for all the values of interference. The number of atoms would either increase or decrease slightly depending on the discretization and result in a biased value of the normal force.

Similar to the current work, the results in a study of friction in nanometric scratching of a rigid diamond tool on a copper work-piece (Zhu et al. 2011) show a steeper increase in the friction force as compared to the normal force with increasing scratching depth. That work reports that at smaller scratching depths friction was dominated by adhesion. In this work since the friction force is rapidly increasing compared to the normal force, the effective friction coefficient also increases with the interference (see Fig. 7).

Effect of asperity radius, R

To study the effect of asperity size on the forces and effective friction coefficient, asperities with radii of 5, 7.5, and 10 nm were considered. As mentioned earlier, these asperity sizes are chosen to obtain a reasonable balance between nano-scale representation and the computational time and power required. The typical surface roughness of MEMS/NEMS surfaces (Bora et al. 2005; Rezvanian et al. 2007; Ansari and Ashurst 2011) was used as a reference for choosing these sizes.

Again, as the asperity size increased, the number of atoms in both of the asperities taking part in the interaction also increased. Note that as the size increases the smoothness of the curvature of the asperity also increases. This further adds to the number of interacting atoms. Figure 8 shows that the friction force and the normal force increase almost linearly with size. The effective coefficient of friction (Fig. 9) also increases with size since the friction force increases more rapidly than the normal force. This was expected and was in agreement to other similar works (Zhang et al. 2001; Stone et al. 2008).

Effect of sliding velocity, v

Three sliding velocities of 10, 50, and 100 m/s were used in this work to quantify its effect on the friction characteristics along with the change in the interference and size.

Fig. 5 The effect of
interference, asperity size, and
relative sliding velocity on the
average friction force

◇v = 100 m/s, R = 5 nm   □v = 100 m/s, R = 7.5 nm   △v = 100 m/s, R = 10 nm
+v = 50 m/s, R = 5 nm   ▬v = 50 m/s, R = 7.5 nm   ▬v = 50 m/s, R = 10 nm
✕v = 10 m/s, R = 5 nm   ✱v = 10 m/s, R = 7.5 nm   ○v = 10 m/s, R = 10 nm

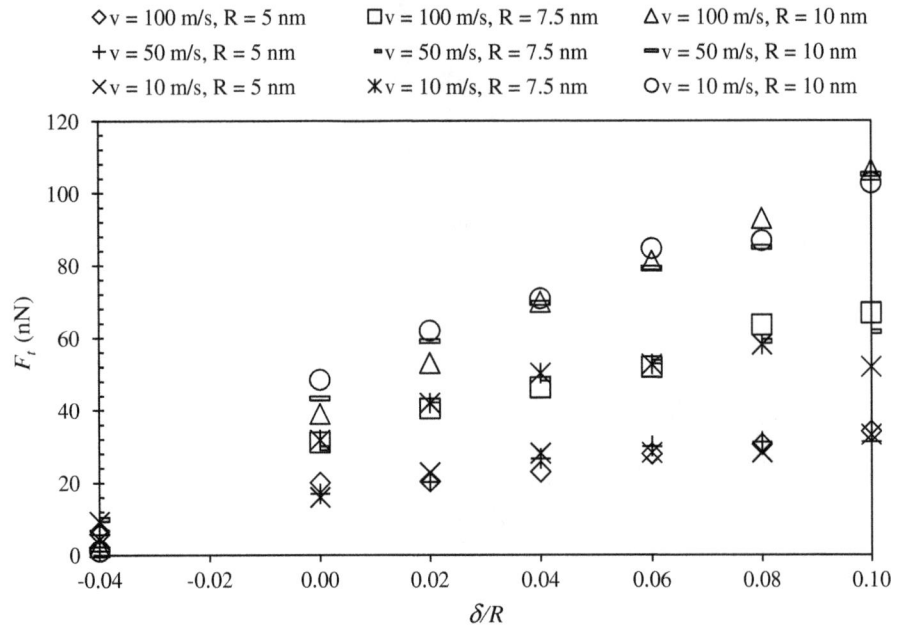

Fig. 6 The effect of
interference, asperity size, and
relative sliding velocity on the
average normal force

◇v = 100 m/s, R = 5 nm   □v = 100 m/s, R = 7.5 nm   △v = 100 m/s, R = 10 nm
+v = 50 m/s, R = 5 nm   ▬v = 50 m/s, R = 7.5 nm   ▬v = 50 m/s, R = 10 nm
✕v = 10 m/s, R = 5 nm   ✱v = 10 m/s, R = 7.5 nm   ○v = 10 m/s, R = 10 nm

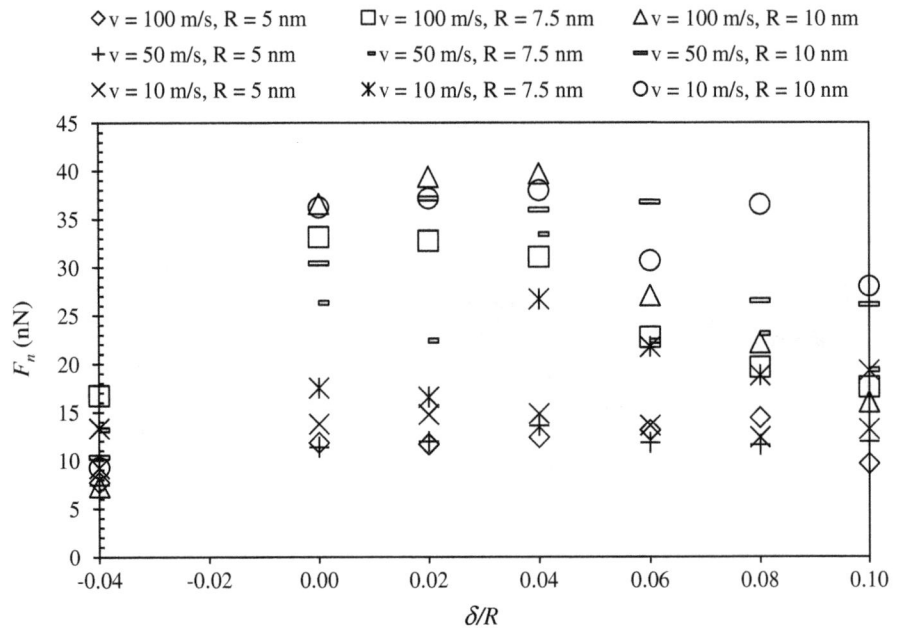

These values of velocity represent the velocities commonly encountered in MEMS/NEMS devices (Ping and NingBo 2007; Karthikeyan et al. 2009; Pei et al. 2007). For example, a high temperature micro gas turbine has a rotational speed that would translate into a sliding velocity of over 500 m/s (Bhushan 2007).

An important observation of this work was that for the sliding velocity of 50 and 100 m/s, due to the sudden acceleration of the asperity atoms in the x-direction at the start of the simulation, the upper asperity oscillated about its center as it travelled towards and across the lower asperity. This caused the normal force and as a result the

friction coefficient to fluctuate severely. To reduce this effect, instead of applying an instantaneous increase in velocity, the atoms were gradually accelerated. A velocity ramping function $v(t) = \tanh(t/C)$, where t is the time, and C is the constant, was used to accelerate the asperity to 50 and 100 m/s before the start of the contact and then was maintained at this velocity for the remaining sliding process. Although not evident during the visualization and in the force curves, some of this oscillation might still be present even after the gradual ramping of the velocity as well as for a sliding velocity of 10 m/s for which such a ramping function was not in place. This could also explain

Fig. 7 The effect of
interference, asperity size, and
relative sliding velocity on the
effective friction coefficient

◇v = 100 m/s, R = 5 nm   □v = 100 m/s, R = 7.5 nm   △v = 100 m/s, R = 10 nm
+v = 50 m/s, R = 5 nm   ●v = 50 m/s, R = 7.5 nm   ●v = 50 m/s, R = 10 nm
✕v = 10 m/s, R = 5 nm   ✳v = 10 m/s, R = 7.5 nm   Ov = 10 m/s, R = 10 nm

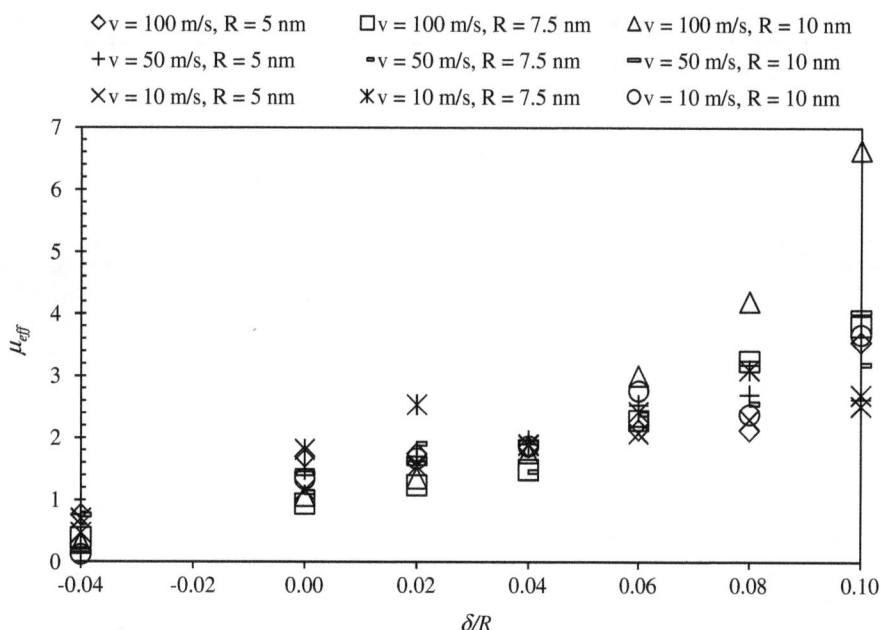

Fig. 8 The effect of asperity
size on the average friction
force and the average normal
force, averaged over δ/R

◇Ft @ v = 100 m/s   □Fn @ v = 100 m/s
✳Ft @ v = 50 m/s   OFn @ v = 50 m/s
△Ft @ v = 10 m/s   ✕Fn @ v = 10 m/s

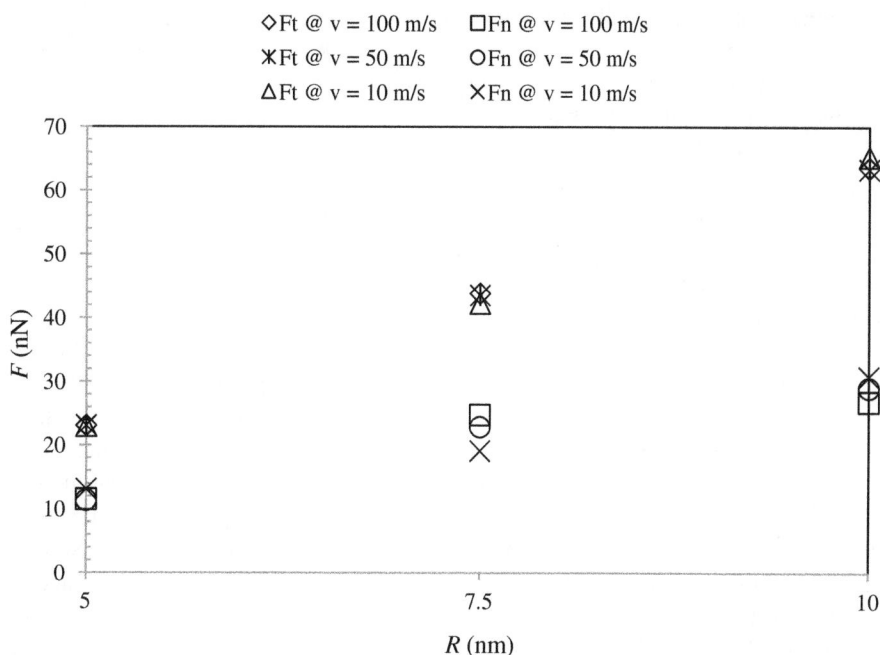

the random fluctuation of the normal force as observed in
Fig. 6, but then one would also expect the tangential force
to fluctuate and it does not. Therefore, it appears more
likely that the previously mentioned lattice alignment
effect causes these fluctuations.

As seen in Figs. 5, 6, 7, 8, 9, the change in the sliding
velocity of the asperity did not produce any consistent and
significant change in the forces and the effective friction
coefficient. This is in contradiction to several reported
results on asperity sliding friction (Yang and Komvopoulos
2005; Karthikeyan et al. 2009; Lin et al. 2007).

Karthikeyan et al. (2009) have reported a fivefold increase
in the friction coefficient when the sliding velocity of
copper block on an iron block was increased from 300 to
1,000 m/s. It should be noted that this sliding velocity is in
a different range than that considered in the current work.
In the sliding of a square prismatic diamond tip on a copper
surface, the friction force increased as the sliding velocity
increased from 10 to 100 m/s while the normal force
remained unchanged (Yang and Komvopoulos 2005).
However, the main difference between the above-mentioned works and the current work is that the sliding

Fig. 9 The effect of asperity size on the effective friction coefficient averaged over δ/R

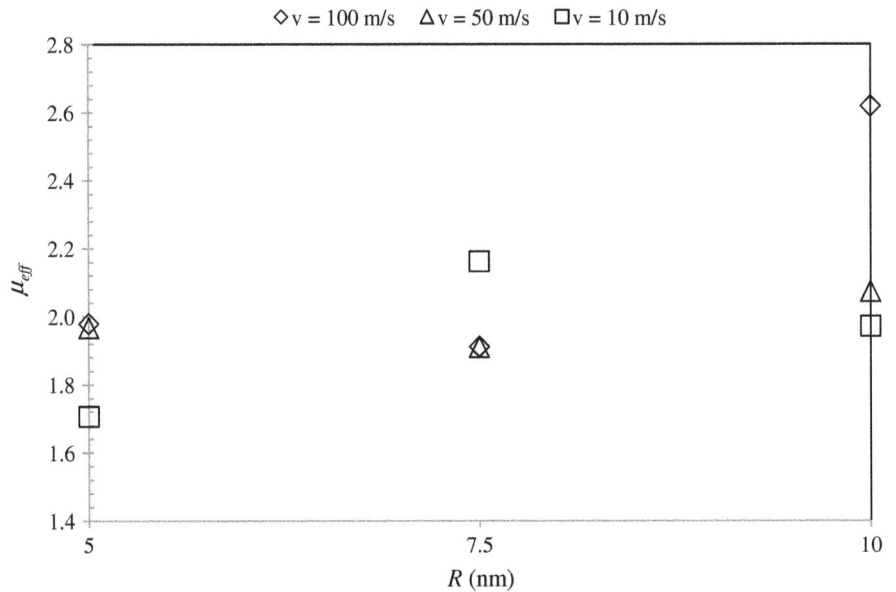

surfaces remain in contact for the entire duration and the contact area is also essentially constant throughout the sliding process in those works. In the current work, asperities remain in contact for only a small duration with the contact area continuously changing. Intuitively, this seems logical since there would be less time for the temperature to rise. In better agreement with the current work, Zhong et al. (2003) performed the simulations with sliding velocities of 50, 100, and 400 m/s and found that at 300 K the wear of the aluminum asperity changed very little with respect to the sliding velocity.

Effect of lattice orientation, θ

All of the simulations so far were performed with the x-, y- and z-axis oriented in the [100], [010] and [001] lattice directions, respectively, such that the (001) planes were parallel to the sliding direction. The normal force fluctuation perhaps caused due the vertical alignment of the lattices and the resulting discretization of the asperity surfaces for certain interference values was observed. If the orientation of the lattice is changed then this phenomena should disappear. This is demonstrated by performing simulations for various lattice orientations, θ, about the z-axis as listed in Table 2. This is probably a more realistic condition since, in a real surface, the lattices are oriented randomly within the asperities and thus, would average out the orientation bias. The slip planes of both the asperities remained parallel to each other as the lattices were reoriented the same for both asperities.

As θ increased, the sliding direction appeared more favorable for slip and the friction force decreased while the normal force increased. The deformation or the material

Table 2 Lattice orientations simulated

xyz direction	θ
[110], [$\bar{1}$10], [001]	45°
[320], [$\bar{1}$30], [001]	34°
[210], [$\bar{1}$20], [001]	27°
[310], [$\bar{1}$30], [001]	18°
[410], [$\bar{1}$40], [001]	14°

transfer did not decrease visibly for the range $\theta = 0°$ to $\theta = 34°$. But at $\theta = 45°$, which corresponds to the (101) planes parallel to the sliding direction, the deformation decreased dramatically with much less material transfer as observed in Fig. 10. Zhong et al. (2003) also reported that the wear decreased significantly when the sliding surfaces where parallel to the (111) planes compared to the (100) planes. Further, Sorensen et al. (1996) have shown that between a Cu tip and Cu surface, with non-matching surfaces parallel to the (111) plane the wear is minimum. Although, the most favorable case of (111) parallel to the sliding direction is not considered in this work, the results are comparable. However, one should also note that the contact plane actually changes in the current work due to the curved nature of both surfaces (i.e. at the initial contact the plane between the two surfaces will be sloped, but as the asperities progress further, the slope eventually becomes zero). In most of the other works that are of an asperity against a flat, the contact plane is always parallel to the sliding plane.

Similar to the case of $\theta = 0°$, Fig. 11 shows the friction force for the case of $\theta = 45°$ which starts from zero, reaches a maximum as the asperities align and decreases

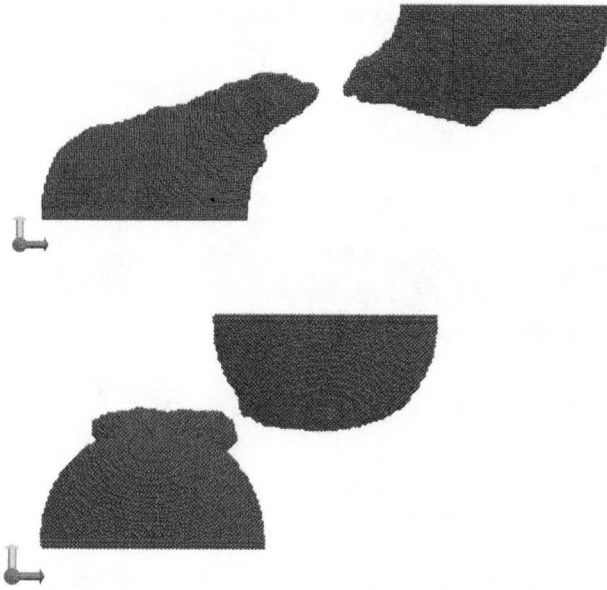

Fig. 10 Deformation at the end of sliding for $\theta = 0°$ (top), $\theta = 45°$ (bottom) for $R = 7.5$ nm, $\delta/R = 0.10$ and $v = 10$ m/s

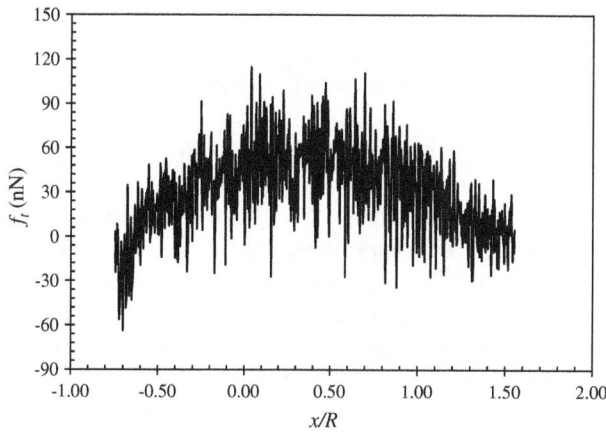

Fig. 11 The friction force during sliding in normalized sliding direction for $R = 7.5$ nm, $\delta/R = 0.10$ and $v = 10$ m/s with $\theta = 45°$

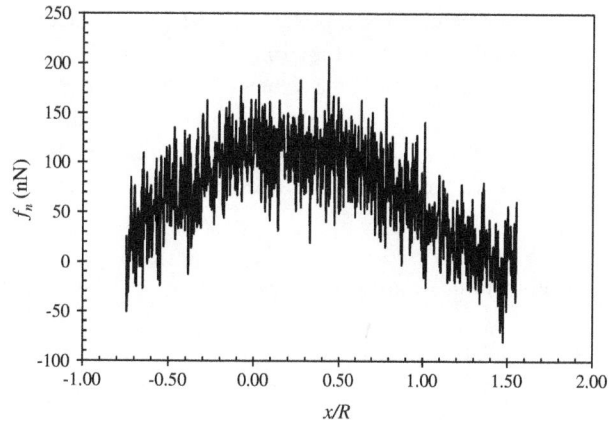

Fig. 12 The normal force during sliding in normalized sliding direction for $R = 7.5$ nm, $\delta/R = 0.10$ and $v = 10$ m/s with $\theta = 45°$

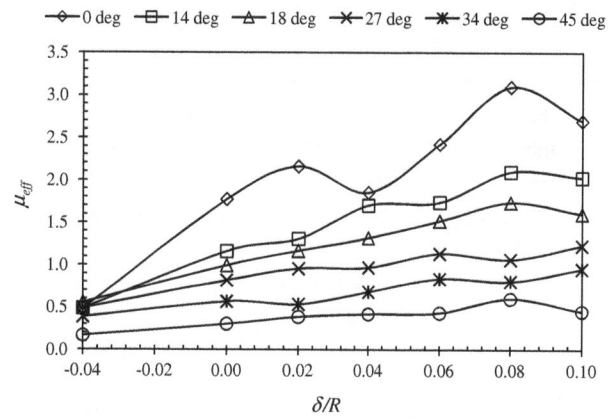

Fig. 13 The effect of lattice orientation on the effective friction coefficient for $R = 7.5$ nm, $v = 10$ m/s

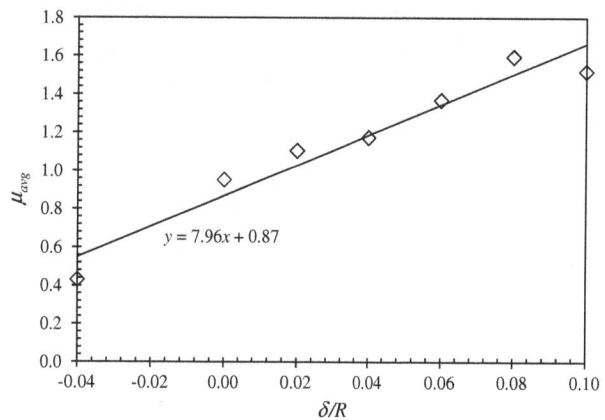

Fig. 14 The average effective friction coefficient as a function of normalized interference for $R = 7.5$ nm, $v = 10$ m/s, *solid line* is a linear fit

back to zero as the asperities come out of contact. However, the normal force does not show the same nature in this case (see Fig. 12). First, the asperities do not get pulled into contact as they come closer and the corresponding jump in the normal force is also not observed. Second, the normal force increases to a maximum as the asperities align and decreases back to zero as the asperities come out of contact without changing the direction. This suggests that the adhesion is comparatively much lower when $\theta = 45°$ to significantly resist the separation of asperities than when $\theta = 0°$.

Figure 13 shows the effect of lattice orientation on the effective friction coefficient for the asperity of radius 7.5 nm and a sliding velocity of 10 m/s. As the lattice is rotated about the z-axis from $0°$ to $45°$ the friction force

decreases while the normal force increases and as a result the effective friction coefficient also decreases on average by a factor of about 6. An average effective friction coefficient was calculated and plotted for the interference values considered as shown in Fig. 14. In some of the early experiments (Buckley 1967), the values of friction coefficient of pure polycrystalline copper in high vacuum were reported to be in the range 0.9–1.2. These are in reasonable agreement with values of 0.4–1.6 obtained in the current work.

Effect of temperature control

All of the simulations performed up to this point in this work did not have any temperature control or thermostat in place during the sliding interaction, which would drain the energy out of the system. However, most of the works reported in the background section of this report had some sort of thermostat present in their system. Therefore, further simulations were carried out with a temperature control in order to observe its effect on the frictional characteristics. An asperity of radius 7.5 nm with a sliding velocity of 10 m/s was simulated for two different lattice orientations: $\theta = 0°$ and $\theta = 45°$. After the entire system was equilibrated at 300 K, the base regions of both the lower and the upper asperities were held at 300 K by scaling the velocities of the atoms.

Figure 15 shows the effective friction coefficient as a function of interference with and without a thermostat. It can be seen that there was little overall change observed by putting a temperature control for both the lattice orientations. Further, it confirms the previous finding of a lower effective friction coefficient at $\theta = 45°$. The effect of temperature control on the asperity temperature for lattice orientation of $\theta = 0°$ and $\theta = 45°$ is shown in Fig. 16. When the thermostat was not in place, the asperity temperature increased by approximately 80 K for the case of $\theta = 45°$ as compared to 230 K for $\theta = 0°$ since the work required to slide the asperities across each other is more in

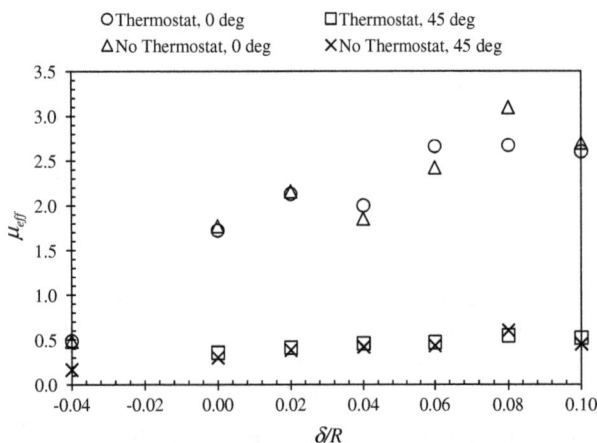

Fig. 16 The effect of temperature control and lattice orientation on the asperity temperature for $R = 7.5$ nm, $v = 10$ m/s

the later case. Thus, the temperature rise is a function of adhesion and this finding is in agreement with the work by Ray and Roy Chowdhury (2010). Also, the duration for which the asperities remained in contact is also much smaller for $\theta = 45°$ for the same reason. This was also the case when the thermostat was in place, however, there was no significant increase in the asperity temperature for either of the orientations.

Electrons and phonons carry out thermal transport in metals, and for pure bulk metals the phononic contribution can be negligible. At the nano scale, the thermal transport decreases due to boundary scattering of electrons and phonons, and size effects (Yuan and Jiang 2006). Since the mean free path for electrons is larger than that of phonons, electrons suffer more scattering than phonons and therefore their contribution to thermal transport decreases (Feng et al. 2009). Even then the contribution of conducting electrons is crucial in thermal transport. As mentioned earlier, although EAM describes metallic interactions effectively it neglects the contribution of conducting electrons and therefore under-predicts the thermal transport. Consequently, the temperature rise observed as mentioned above is probably over predicted, and yet its effect is still relatively small on the friction coefficient.

Conclusions

Molecular dynamics simulations of the dry sliding of copper asperities were performed using the embedded atom potential with LAMMPS. The aim of this work was to study asperity–asperity interaction to gain useful insights on the deformation mechanisms and frictional characteristics in a dry sliding process with a geometry that could be compared and perhaps coupled to continuum models. The

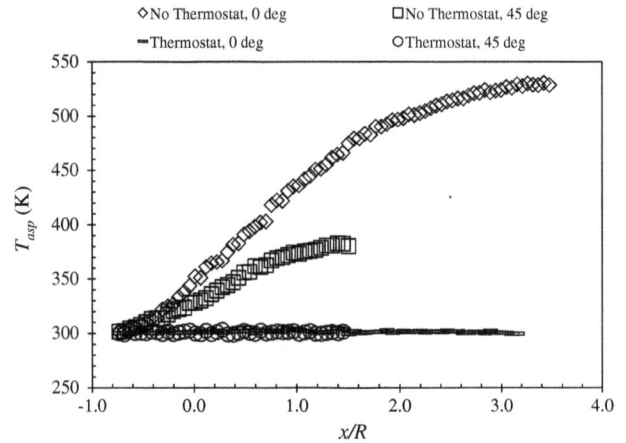

Fig. 15 The effect of temperature control on the effective friction coefficient for $R = 7.5$ nm, $v = 10$ m/s

effect of interference, asperity size, sliding velocity, lattice orientation and temperature control on sliding friction was investigated. The important findings and conclusions are summarized as follows:

1. As the asperities come in contact a junction was formed which grew through the sliding as more atoms bonded to each other. This adhered junction stretched along the sliding direction as the asperities moved apart and finally broke after necking. Extensive deformation and material transfer was observed for most of the cases that were studied.
2. Adhesion dictated the sliding process as opposed to ploughing resulting in higher values of effective friction coefficient, which is consistent with literature.
3. Friction force and effective friction coefficient increased with the interference whereas the friction force, normal force, and effective friction coefficient increased with the asperity size. This is attributed to the increase in the number of interacting atoms. Fluctuations in the normal force were observed due to the vertical lattice alignment and the resulting discretization of the smooth surface geometry for a given value of interference.
4. For the range of 10–100 m/s, the velocity presented no significant change in the friction characteristics as the duration of contact was not long enough to realize its effect.
5. Lattice orientation presented a significant influence on the frictional characteristics with a reduction in the effective friction coefficient by a factor of about 6 for the range of orientation considered. For the case when the sliding direction was parallel to the (101) plane, the least material transfer and deformation were observed.
6. Using a temperature control during the sliding with the base of the asperities at 300 K produced almost no change in the friction characteristics in comparison to no temperature control.

Acknowledgments We would like to thank Steve Plimpton of Sandia National Laboratories and Robert S. Hoy of the Department of Mechanical Engineering and Material Science at Yale University for their valuable suggestions on some details of the simulations.

References

Achanta S, Liskiewicz T, Drees D, Celis J (2009) Friction mechanisms at the micro-scale. Tribol Int 42:1792–1799

Ansari N, Ashurst WR (2011) An easy-to-fabricate improved microinstrument for systematically investigating adhesion between MEMS sidewalls. Appl Surf Sci 257:10917–10925

Barriga J, Fernandez-Diaz B, Juarros A, Ahmed SI-U, Arana JL (2007) Microtribological analysis of gold and copper contacts. Tribol Int 40:1526–1530

Bhushan B (2007) Nanotribology and nanomechanics of MEMS/NEMS and BioMEMS/BioNEMS materials and devices. Microelectron Eng 84:387–412

Blau PJ (1991) Scale effects in steady-state friction. Tribol Trans 34(3):335–342

Bora CK, Flater EE, Street MD, Redmond JM, Starr MJ, Carpick RW et al (2005) Multiscale roughness and modeling of MEMS interfaces. Tribol Lett 19(1)

Boucly V, Nelias D, Green I (2007) Modeling of the rolling and sliding contact between two asperities. J Tribol 129:235–245

Buckley DH; Lewis Research Center, Ohio (1967) Effect of recrystallization on friction properties of some metals in single-crystal and polycrystalline form. Technical Note, National Aeronautics and Space Administration

Cha P, Srolovitz DJ, Vanderlick TK (2004) Molecular dynamics simulation of single asperity contact. Acta Mater 52:3983–3996

Cho MH, Kim SJ, Lim D, Jang H (2005) Atomic scale stick-slip caused by dislocation nucleation and propagation during scratching of a Cu substrate with a nanoindenter: a molecular dynamics simulation. Wear 259:1392–1399

Choi Y, Lee C, Hwang Y, Park M, Lee J, Choi C et al (2009) Tribological behavior of copper nanoparticles as additives in oil. Curr Appl Phys 9:e124–e127

Daw MS, Baskes MI (1984) Embedded-atom method: derivation and application to impurities, surfaces and other defects in metals. Phys Rev B 29(12):6443–6453

Eid H, Adams GG, McGruer NE, Fortini A, Buldyrev S, Srolovitz D (2011) A combined molecular dynamics and finite element analysis of contact and adhesion of a rough sphere and a flat surface. Tribol Trans 54:920–928

Faulkner A, Arnell RD (2000) The development of a finite element model to simulate the sliding interaction between two, three-dimensional, elastoplastic, hemispherical asperities. Wear 242:114–122

Feng B, Li Z, Zhang X (2009) Role of phonon in the thermal and electrical transport in metallic nanofilms. J Appl Phys 105:1–7

Foiles SM, Baskes MI, Daw MS (1986) Embedded-atom-method function for fcc metals Cu, Ag, Au, Ni, Pd, Pt, and their alloys. Phys Rev B 33(12):7983–7991

Harrison JA, White CT, Colton RJ, Brenner DW (1992) Molecular-dynamics simulations of atomic-scale friction of diamond surfaces. Phys Rev B 46(15):9700–9708

Ivashchenko VI, Turchi PE (2006) Atomic-scale sliding friction of amorphous and nanostructured SiC and diamond surfaces. Tribol Trans 49(1):61–65

Jackson RL (2006) The effect of scale-dependent hardness on elasto-plastic asperity contact between rough surfaces. Tribol Trans 49(2):135–150

Jackson RL, Green I (2006) A statistical model of elasto-plastic asperity contact between rough surfaces. Tribol Int 39:906–914

Jackson RL, Duvvuru RS, Meghani H, Mahajan M (2007) An analysis of elasto-plastic sliding spherical asperity interaction. Wear 262:210–219

Jeng Y, Su C, Lay Y (2007) An investigation of nanoscale tribological characteristics under different interaction forces. Appl Surf Sci 253:6754–6761

Karthikeyan S, Agrawal A, Rigney DA (2009) Molecular dynamics simulations of sliding in an Fe–Cu tribopair system. Wear 267:1166–1176

Landman U, Luedtke WD, Gao J (1996) Atomic-scale issues in tribology: interfacial junctions and nano-elastohydrodynamics. Langmuir 12(19):4514–4528

Landman U, Gao J, Luedtke WD, Gourdon D, Ruths M, Israelachvili JN (2004) Frictional forces and Amontons' law: from the molecular to the macroscopic scale. J Phys Chem B 108(11):3410–3425

Li B, Clapp PC, Rifkin JA, Zhang XM (2003) Molecular dynamics calculation of heat dissipation during sliding friction. Int J Heat Mass Transf 46:37–43

Lin J, Fang T, Wu C, Houng K (2007) Contact and frictional behavior of rough surfaces using molecular dynamics combined with fractal theory. Comput Mater Sci 40:480–484

Liu G, Li X, Qin B, Xing D, Guo Y, Fan R (2004) Investigation of the mending effect and mechanism of copper nano-particles on a tribologically stressed surface. Tribol Lett 17(4):961–966

Osterle W, Prietzel C, Kloss H, Dmitriev AI (2010) On the role of copper in brake friction materials. Tribol Int 43:2317–2326

Pei QX, Lu C, Lee HP (2007) Large scale molecular dynamics study of nanometric machining of copper. Comput Mater Sci 41:177–185

Ping Y, NingBo L (2007) Surface sliding simulation in micro-gear train for adhesion problem and tribology design by using molecular dynamics model. Comput Mater Sci 38:678–684

Plimpton S (1995) Fast parallel algorithms for short range molecular dynamics. J Comput Phys 117:1–19

Plimpton SJ, Hendrickson BA (1993) Parallel molecular dynamics with the embedded atom method. Materials theory and modelling. MRS Proceedings 291, Pittsburgh, p 37

Ray S, Roy Chowdhury SK (2010) An analysis of surface temperature rise at the contact between sliding bodies with small-scale surface roughness. Tribol Trans 53:491–501

Rezvanian O, Zikry MA, Brown C, Krim J (2007) Surface roughness, asperity contact and gold RF MEMS switch behavior. J Micromech Microeng 17

Song J, Srolovitz DJ (2007) Atomistic simulation of multicycle asperity contact. Acta Mater 55:4759–4768

Sorensen MR, Jacobsen KW, Stoltze P (1996) Simulations of atomic-scale sliding friction. Phys Rev B 53(4):2101–2113

Stone TW, Horstemeyer MF, Hammi Y, Gullett PM (2008) Contact and friction of single crystal nickel nanoparticles using molecular dynamics. Acta Mater 56:3577–3584

Tarasov S, Kolubaev A, Belyaev S, Lerner M, Tepper F (2002) Study of friction reduction by nanocopper additives to motor oil. Wear 252:63–69

Yang J, Komvopoulos K (2005) A molecular dynamics analysis of surface interference and tip shape and size effects on atomic-scale friction. J Tribol 127:513–521

Yuan SP, Jiang PX (2006) Thermal conductivity of small nickel particles. Int J Thermophys 27(2):581–595

Zhang L, Tanaka H (1997) Towards a deeper understanding of wear and friction on the atomic scale—a molecular dynamics analysis. Wear 211:44–53

Zhang LC, Johnson KL, Cheong WC (2001) A molecular dynamics study of scale effects on the friction of single-asperity contacts. Tribol Lett 10(1–2):23–28

Zhong J, Adams JB, Hector LG (2003) Molecular dynamics simulations of asperity shear in aluminum. J Appl Phys 94(7):4306–4314

Zhu P, Hu Y, Ma T, Wang H (2011) Molecular dynamics study on friction due to ploughing and adhesion in nanometric scratching process. Tribol Lett 41:41–46

Permissions

The contributors of this book come from diverse backgrounds, making this book a truly international effort. This book will bring forth new frontiers with its revolutionizing research information and detailed analysis of the nascent developments around the world.

We would like to thank all the contributing authors for lending their expertise to make the book truly unique. They have played a crucial role in the development of this book. Without their invaluable contributions this book wouldn't have been possible. They have made vital efforts to compile up to date information on the varied aspects of this subject to make this book a valuable addition to the collection of many professionals and students.

This book was conceptualized with the vision of imparting up-to-date information and advanced data in this field. To ensure the same, a matchless editorial board was set up. Every individual on the board went through rigorous rounds of assessment to prove their worth. After which they invested a large part of their time researching and compiling the most relevant data for our readers.

The editorial board has been involved in producing this book since its inception. They have spent rigorous hours researching and exploring the diverse topics which have resulted in the successful publishing of this book. They have passed on their knowledge of decades through this book. To expedite this challenging task, the publisher supported the team at every step. A small team of assistant editors was also appointed to further simplify the editing procedure and attain best results for the readers.

Apart from the editorial board, the designing team has also invested a significant amount of their time in understanding the subject and creating the most relevant covers. They scrutinized every image to scout for the most suitable representation of the subject and create an appropriate cover for the book.

The publishing team has been an ardent support to the editorial, designing and production team. Their endless efforts to recruit the best for this project, has resulted in the accomplishment of this book. They are a veteran in the field of academics and their pool of knowledge is as vast as their experience in printing. Their expertise and guidance has proved useful at every step. Their uncompromising quality standards have made this book an exceptional effort. Their encouragement from time to time has been an inspiration for everyone.

The publisher and the editorial board hope that this book will prove to be a valuable piece of knowledge for researchers, students, practitioners and scholars across the globe.

List of Contributors

Rajeev Kumar
Physics and Engineering of Carbon, Division of Materials Physics and Engineering, CSIR-National Physical Laboratory, Dr. K. S. Krishnan Marg, New Delhi 110012, India

Saroj Kumari
Physics and Engineering of Carbon, Division of Materials Physics and Engineering, CSIR-National Physical Laboratory, Dr. K. S. Krishnan Marg, New Delhi 110012, India

Rakesh B. Mathur
Physics and Engineering of Carbon, Division of Materials Physics and Engineering, CSIR-National Physical Laboratory, Dr. K. S. Krishnan Marg, New Delhi 110012, India

Sanjay R. Dhakate
Physics and Engineering of Carbon, Division of Materials Physics and Engineering, CSIR-National Physical Laboratory, Dr. K. S. Krishnan Marg, New Delhi 110012, India

P. Saravana Kumar
Division of Microbiology, Entomology Research Institute, Loyola College, Nungambakkam, Chennai 600 034, Tamil Nadu, India

C. Balachandran
Division of Microbiology, Entomology Research Institute, Loyola College, Nungambakkam, Chennai 600 034, Tamil Nadu, India

V. Duraipandiyan
Department of Botany and Microbiology, Addriyah Chair for Environmental Studies, College of Science, King Saud University, Po. Box. No. 2455, Riyadh 11451, Saudi Arabia

D. Ramasamy
Regional Research Institute of Unani Medicine, Royapuram, Chennai 600 013, India

S. Ignacimuthu
Division of Microbiology, Entomology Research Institute, Loyola College, Nungambakkam, Chennai 600 034, Tamil Nadu, India
Department of Botany and Microbiology, Addriyah Chair for Environmental Studies, College of Science, King Saud University, Po. Box. No. 2455, Riyadh 11451, Saudi Arabia

Naif Abdullah Al-Dhabi
Department of Botany and Microbiology, Addriyah Chair for Environmental Studies, College of Science, King Saud University, Po. Box. No. 2455, Riyadh 11451, Saudi Arabia

S. Muruganandam
Department of Physics, Presidency College, Chennai 600 005, India

G. Anbalagan
Department of Physics, Presidency College, Chennai 600 005, India

G. Murugadoss
Centre for Nanoscience and Technology, Anna University, Chennai 600 025, India

N. A. Shah
Department of Electronics, Saurashtra University, Rajkot 360005, India

P. S. Solanki
Department of Physics, Saurashtra University, Rajkot 360005, India

Ashish Ravalia
Department of Physics, Saurashtra University, Rajkot 360005, India

D. G. Kuberkar
Department of Physics, Saurashtra University, Rajkot 360005, India

Shib Shankar Dash
Department of Chemistry and Chemical Technology, Vidyasagar University, Midnapore 721 102, West Bengal, India

Braja Gopal Bag
Department of Chemistry and Chemical Technology, Vidyasagar University, Midnapore 721 102, West Bengal, India

Poulami Hota
Department of Chemistry and Chemical Technology, Vidyasagar University, Midnapore 721 102, West Bengal, India

Amrutham Santoshi kumari
Department of Chemistry, University College of Science, Osmania University, Hyderabad 500007, India

Maragoni Venkatesham
Department of Chemistry, University College of Science, Osmania University, Hyderabad 500007, India

Dasari Ayodhya
Department of Chemistry, University College of Science, Osmania University, Hyderabad 500007, India

Guttena Veerabhadram
Department of Chemistry, University College of Science, Osmania University, Hyderabad 500007, India

Satyendra Singh
Department of Physics, Shri Ram College of Engineering and Management, Banmore, Morena 476444, Madhya Pradesh, India

Pankaj Srivastava
Nanomaterials Research Group, Indian Institute of Information Technology and Management (ABV-IIITM), Gwalior 474010, Madhya Pradesh, India

F. Zabihi
Department of Physics, Sari Branch, Islamic Azad University, Sari, Iran

F. Taleshi
Department of Applied Science, Qaemshahr Branch, Islamic Azad University, PO Box 163, Qaemshahr, Iran

A. Salmani
Department of Physics, Sari Branch, Islamic Azad University, Sari, Iran

A. Pahlavan
Department of Physics, Sari Branch, Islamic Azad University, Sari, Iran

N. Dehghan-niarostami
Department of Physics, Sari Branch, Islamic Azad University, Sari, Iran

M. M. Vadadi
Department of Chemistry, Payam Noor University, PO Box 19395-697, Tehran, Iran

Nurapati Pantha
Central Department of Physics, Tribhuvan University, Kirtipur, Kathmandu, Nepal

Kamal Belbase
Central Department of Physics, Tribhuvan University, Kirtipur, Kathmandu, Nepal

Narayan Prasad Adhikari
Central Department of Physics, Tribhuvan University, Kirtipur, Kathmandu, Nepal

S. U. Nandanwar
Department of Chemical Engineering, S. V. National Institute of Technology, Surat 395 007, Gujarat, India

J. Barad
Department of Chemical Engineering, S. V. National Institute of Technology, Surat 395 007, Gujarat, India

S. Nandwani
Department of Chemical Engineering, S. V. National Institute of Technology, Surat 395 007, Gujarat, India

M. Chakraborty
Department of Chemical Engineering, S. V. National Institute of Technology, Surat 395 007, Gujarat, India

R. Lakshmipathy
Environmental and Analytical Chemistry Division, School of Advanced Sciences, VIT University, Vellore 632014, Tamilnadu, India

B. Palakshi Reddy
Department of GEBH, Sree Vidyanikethan Engineering College, Tirupati, AndhraPradesh, India

N. C. Sarada
Environmental and Analytical Chemistry Division, School of Advanced Sciences, VIT University, Vellore 632014, Tamilnadu, India

K. Chidambaram
Centre for Excellence in Nanomaterials, School of Advanced Sciences, VIT University, Vellore 632014, Tamilnadu, India

Sk. Khadeer Pasha
Centre for Excellence in Nanomaterials, School of Advanced Sciences, VIT University, Vellore 632014, Tamilnadu, India

S. Mugundan
Department of Physics, Annamalai University, Annamalainagar, Chidambaram, Tamilnadu 608002, India

B. Rajamannan
Department of Engineering Physics, (FEAT), Annamalai University, Annamalainagar, Chidambaram, Tamilnadu 608002, India

G. Viruthagiri
Department of Physics, Annamalai University, Annamalainagar, Chidambaram, Tamilnadu 608002, India

N. Shanmugam
Department of Physics, Annamalai University, Annamalainagar, Chidambaram, Tamilnadu 608002, India

R. Gobi
Department of Physics, Annamalai University, Annamalainagar, Chidambaram, Tamilnadu 608002, India

P. Praveen
Department of Physics, Annamalai University, Annamalainagar, Chidambaram, Tamilnadu 608002, India

K. R. Nemade
Department of Applied Physics, J D College of Engineering and Management, Nagpur 441 501, India

S. A. Waghuley
Department of Physics, Sant Gadge Baba Amravati University, Amravati 444 602, India

Nagoth Joseph Amruthraj
Department of Plant Biology and Biotechnology, Loyola College, Chennai 600 034, Tamil Nadu, India

John Poonga Preetam Raj
Department of Plant Biology and Biotechnology, Loyola College, Chennai 600 034, Tamil Nadu, India

Antoine Lebel
Department of Plant Biology and Biotechnology, Loyola College, Chennai 600 034, Tamil Nadu, India

Tehmina Naz
Department of Chemistry, Quaid-i-Azam University, Islamabad 45320, Pakistan

Adeel Afzal
Department of Chemistry, Quaid-i-Azam University, Islamabad 45320, Pakistan

Humaira M. Siddiqi
Department of Chemistry, Quaid-i-Azam University, Islamabad 45320, Pakistan

Javeed Akhtar
Department of Physics, COMSATS Institute of Information Technology, Islamabad Campus, Chak Shahzad, Islamabad 44000, Pakistan

Amir Habib
School of Chemical and Materials Engineering, National University of Science and Technology, H-12, Islamabad 44000, Pakistan

Mateusz Banski
Institute of Physics, Wroclaw University of Technology, Wybrzeze Wyspianskiego 27, 50-370 Wroclaw, Poland

Artur Podhorodecki
Institute of Physics, Wroclaw University of Technology, Wybrzeze Wyspianskiego 27, 50-370 Wroclaw, Poland

Bhumi Gaddala
Department of Botany, Sri Venkateswara University, Tirupati, Andhra Pradesh, India

Savithramma Nataru
Department of Botany, Sri Venkateswara University, Tirupati, Andhra Pradesh, India

Andrew R. Markelonis
Department of Chemistry, University of Idaho, Renfrew Hall, Moscow, ID 83844, USA

Joanna S. Wang
Materials and Manufacturing Directorate, Air Force Research Laboratory, 3005 Hobson Way, Wright-Patterson Air Force Base, OH 45433, USA

Bruno Ullrich
Materials and Manufacturing Directorate, Air Force Research Laboratory, 3005 Hobson Way, Wright-Patterson Air Force Base, OH 45433, USA

Chien M. Wai
Department of Chemistry, University of Idaho, Renfrew Hall, Moscow, ID 83844, USA

Gail J. Brown
Materials and Manufacturing Directorate, Air Force Research Laboratory, 3005 Hobson Way, Wright-Patterson Air Force Base, OH 45433, USA

Durga Prameela Gaddam
Department of Botany, S.V. University, Tirupati, A.P. 517 502, India

Nagalakshmi Devamma
Department of Botany, S.V. University, Tirupati, A.P. 517 502, India

Tollamadugu Naga Venkata Krishna Vara Prasad
Nanotechnology Laboratory, Institute of Frontier Technology, Regional Agricultural Research Station, Acharya N G Ranga Agricultural University, Tirupati, A.P. 517 502, India

Adnan Mujahid
Institute of Chemistry, University of Punjab, Quaid-e-Azam Campus, Lahore 54590, Pakistan

Aimen Idrees Khan
Institute of Chemistry, University of Punjab, Quaid-e-Azam Campus, Lahore 54590, Pakistan

Adeel Afzal
Affiliated Colleges at Hafr Al-Batin, King Fahd University of Petroleum and Minerals, P.O. Box 1803, Hafr Al-Batin 31991, Saudi Arabia

Tajamal Hussain
Institute of Chemistry, University of Punjab, Quaid-e-Azam Campus, Lahore 54590, Pakistan

Muhammad Hamid Raza
Institute of Chemistry, University of Punjab, Quaid-e-Azam Campus, Lahore 54590, Pakistan

Asma Tufail Shah
Interdisciplinary Research Centre for Biomedical Materials, COMSATS Institute of Information Technology, Defence Road, Off. Raiwind Road, Lahore 45600, Pakistan

Waheed uz Zaman
Institute of Chemistry, University of Punjab, Quaid-e-Azam Campus, Lahore 54590, Pakistan

T. Kathiraven
Centre of Advance Study, Marine Biology, Faculty of Marine Sciences, Annamalai University, Parangipettai 608 502, Tamilnadu, India

A. Sundaramanickam
Centre of Advance Study, Marine Biology, Faculty of Marine Sciences, Annamalai University, Parangipettai 608 502, Tamilnadu, India

N. Shanmugam
Department of Physics, Annamalai University, Annamalai Nagar 608 002, Tamilnadu, India

T. Balasubramanian
Centre of Advance Study, Marine Biology, Faculty of Marine Sciences, Annamalai University, Parangipettai 608 502, Tamilnadu, India

M. Mohamed Rafi
Department of Physics, C. Abdul Hakeem College, Melvisharam 632509, Tamilnadu, India

K. Syed Zameer Ahmed
Department of Biochemistry, C. Abdul Hakeem College, Melvisharam 632509, Tamilnadu, India

K. Prem Nazeer
PG and Research Department of Physics, Islamiah College, Vaniyambadi 635752, Tamilnadu, India
Department of Physics, C. Abdul Hakeem College, Melvisharam 632509, Tamilnadu, India

D. Siva Kumar
Department of Physics, TPGIT, Vellore 632002, Tamilnadu, India

M. Thamilselvan
Department of Physics, C. Abdul Hakeem College, Melvisharam 632509, Tamilnadu, India

Puneet Rana
Department of Mathematics, Jaypee Institute of Information Technology, Noida, Uttar Pradesh, India

O. Anwar Bég
Gort Engovation-Propulsion, Nanomechanics and Biophysics, Southmere Avenue, Bradford BD73NU, UK

Rajeev Kumar
Physics and Engineering of Carbon, Division of Material Physics and Engineering, CSIR-National Physical Laboratory, Dr. K. S. Krishnan Marg, 110012 New Delhi, India

Saroj Kumari
Physics and Engineering of Carbon, Division of Material Physics and Engineering, CSIR-National Physical Laboratory, Dr. K. S. Krishnan Marg, 110012 New Delhi, India

Sanjay R. Dhakate
Physics and Engineering of Carbon, Division of Material Physics and Engineering, CSIR-National Physical Laboratory, Dr. K. S. Krishnan Marg, 110012 New Delhi, India

T. V. M. Sreekanth
Department of Life Chemistry, Catholic University of Daegu, Hayang-eup, Gyeongsan-si, Geyongbuk 712-702, Republic of Korea

P. C. Nagajyothi
Department of Nanomaterials Chemistry, Dongguk University, Seokjang-dong 707, Gyeongju 780 714, Republic of Korea

N. Supraja
Regional Agricultural Research Station, Nanotechnology Laboratory, Institute of Frontier Technology, Acharya N G Ranga Agricultural University, Tirupati, AP 517 502, India

T. N. V. K. V. Prasad
Regional Agricultural Research Station, Nanotechnology Laboratory, Institute of Frontier Technology, Acharya N G Ranga Agricultural University, Tirupati, AP 517 502, India

Bhavin N. Vadgama
Department of Mechanical Engineering, Auburn University, Auburn, AL 36849, USA

Robert L. Jackson
Department of Mechanical Engineering, Auburn University, Auburn, AL 36849, USA

Daniel K. Harris
Department of Mechanical Engineering, Auburn University, Auburn, AL 36849, USA